VOLUME ONE HUNDRED AND FIFTY TWO

INTERNATIONAL REVIEW OF NEUROBIOLOGY

Stress and Brain Health: In Clinical Conditions

INTERNATIONAL REVIEW OF NEUROBIOLOGY

VOLUME 152

SERIES EDITOR

PATRICIA JANAK

Janak Lab, Dunning Hall
Baltimore, MD, USA

PETER JENNER

Division of Pharmacology and Therapeutics
GKT School of Biomedical Sciences
King's College, London, UK

VOLUME ONE HUNDRED AND FIFTY TWO

INTERNATIONAL REVIEW OF NEUROBIOLOGY

Stress and Brain Health: In Clinical Conditions

Edited by

ANGELA CLOW
Psychology, School of Social Sciences, University of Westminster, London, England

NINA SMYTH
Psychology, School of Social Sciences, University of Westminster, London, England

ELSEVIER

ACADEMIC PRESS
An imprint of Elsevier

Academic Press is an imprint of Elsevier
50 Hampshire Street, 5th Floor, Cambridge, MA 02139, United States
525 B Street, Suite 1650, San Diego, CA 92101, United States
The Boulevard, Langford Lane, Kidlington, Oxford OX5 1GB, United Kingdom
125 London Wall, London, EC2Y 5AS, United Kingdom

First edition 2020

ISBN: 978-0-12-821116-8
ISSN: 0074-7742

For information on all Academic Press publications
visit our website at https://www.elsevier.com/books-and-journals

Publisher: Zoe Kruze
Acquisitions Editor: Sam Mahfoudh
Editorial Project Manager: Peter Llewellyn
Production Project Manager: Abdulla Sait
Cover Designer: Greg Harris

Typeset by SPi Global, India

Contents

Contributors

Mustafa al'Absi
Department of Family Medicine and Biobehavioral Health, University of Minnesota Medical School, Duluth, MN, United States

Qadeer Arshad
ENT Department, Leicester Royal Infirmary; inAmind Laboratory, Department of Psychology, Neuroscience and Behaviour, University of Leicester, Leicester; Academic Department of Neuro-otology, Charing Cross Hospital Campus, Imperial College London, London, United Kingdom

Trina Basu
Department of Neuroscience, Tufts University School of Medicine, Boston, MA, United States

J. Douglas Bremner
Emory University School of Medicine, Atlanta; Atlanta Veterans Administration Medical Center, Decatur, GA, United States

Rachel E. Conley
Northwestern University, Evanston, IL, United States

Juan Manuel Cortell-Tormo
University of Alicante, Alicante, Spain

Mayank Dutia
Biomedical Sciences, Edinburgh Medical School, University of Edinburgh, Edinburgh, United Kingdom

Alison Fixsen
University of Westminster, London, United Kingdom

Nicola Gartland
School of Psychology, University of Leeds, Leeds, United Kingdom

Peter Jenner
King's College London, Neurodegenerative Diseases Research Group, Institute of Pharmaceutical Sciences, Faculty of Health Sciences and Medicine, London, United Kingdom

Jamie Maguire
Department of Neuroscience, Tufts University School of Medicine, Boston, MA, United States

Carmen Manchado-Lopez
University of Alicante, Alicante, Spain

Vijay A. Mittal
Northwestern University, Evanston, IL, United States

Maria Antonietta Nettis
King's College London, Institute of Psychiatry, Psychology and Neuroscience, Department of Psychological Medicine; National Institute for Health Research, Biomedical Research Centre, South London and Maudsley NHS Foundation Trust and King's College London, London, United Kingdom

Daryl B. O'Connor
School of Psychology, University of Leeds, Leeds, United Kingdom

Rory C. O'Connor
Suicidal Behavior Research Laboratory, Institute of Health & Wellbeing, University of Glasgow, Glasgow, United Kingdom

Carmine Maria Pariante
King's College London, Institute of Psychiatry, Psychology and Neuroscience, Department of Psychological Medicine; National Institute for Health Research, Biomedical Research Centre, South London and Maudsley NHS Foundation Trust and King's College London, London, United Kingdom

Marie Polley
University of Westminster, London, United Kingdom

K. Ray Chaudhuri
King's College London, Department of Neurosciences, Institute of Psychiatry, Psychology & Neuroscience; Parkinson Foundation Centre of Excellence, King's College Hospital, London, United Kingdom

Peter Rea
ENT Department, Leicester Royal Infirmary, Leicester, United Kingdom

Jay A. Salpekar
Kennedy Krieger Institute, Johns Hopkins University Medical School, Baltimore, MD, United States

Yougan Saman
ENT Department, Leicester Royal Infirmary; inAmind Laboratory, Department of Psychology, Neuroscience and Behaviour, University of Leicester, Leicester, United Kingdom

Gema Sanchis-Soler
University of Alicante, Alicante, Spain

Swami Thangaraj
Kennedy Krieger Institute, Johns Hopkins University Medical School, Baltimore, MD, United States

Juan Tortosa-Martínez
University of Alicante, Alicante, Spain

Teresa Vargas
Northwestern University, Evanston, IL, United States

Daniel J. van Wamelen
King's College London, Department of Neurosciences, Institute of Psychiatry, Psychology & Neuroscience; Parkinson Foundation Centre of Excellence, King's College Hospital, London, United Kingdom

Yi-Min Wan
King's College London, Department of Neurosciences, Institute of Psychiatry, Psychology & Neuroscience; Parkinson Foundation Centre of Excellence, King's College Hospital, London, United Kingdom; Department of Psychiatry, Ng Teng Fong General Hospital, Singapore, Singapore

Matthew T. Wittbrodt
Emory University School of Medicine, Atlanta, GA, United States

CHAPTER ONE

Stress, the brain, and trauma spectrum disorders

J. Douglas Bremner[a,b,*], Matthew T. Wittbrodt[a]
[a]Emory University School of Medicine, Atlanta, GA, United States
[b]Atlanta Veterans Administration Medical Center, Decatur, GA, United States
*Corresponding author: e-mail address: jdbremn@emory.edu

Contents

Abstract

This chapter reviews the relationship between stress and brain function in patients with neuropsychiatric disorders, with an emphasis on disorders that have most clearly been linked to traumatic stress exposure. These disorders, which have been described as trauma spectrum disorders, include posttraumatic stress disorder (PTSD), a subgroup of major depression, borderline personality disorder (BPD) and dissociative disorders; they share in common a neurobiological footprint, including smaller hippocampal volume, and are distinguished from other disorders that may share symptom similarities, like some of the anxiety disorders, but are not as clearly linked to stress. The relationship between environmental events such as stressors, especially in early childhood, and their effects on brain and neurobiology is important to understand in approaching these disorders as well as the development of therapeutic interventions. Addressing patients with stress-related disorders from multiple developmental (age at onset of trauma) as well as levels of analysis (cognitive, cultural, neurobiological) approaches will provide the most complete picture and result in the most successful treatment outcomes.

International Review of Neurobiology, Volume 152
ISSN 0074-7742
https://doi.org/10.1016/bs.irn.2020.01.004

1. Lasting effects of traumatic stress

Traumatic stress can have lasting effects on the individual. Brain and physiological systems that helped us survive when we were wandering in small bands on the savannah are no longer as necessary in modern times. A sympathetic nervous system that allowed us to respond a split second faster to a predator even at the expense of long-term exaggerated responsiveness was favored in terms of survival and passing on genes to the next generation. Today, a hyperactive sympathetic nervous system may impair our ability to perform mundane office-based tasks with no real added value in a world essentially devoid of predators. The task of those in the mental health field is to optimize the health of our patients, which includes adjustment for the potentially crippling effects of neurobiological systems that previously were essential for survival but now may have detrimental effects.

Physiological response systems involved in survival underlie the pathophysiology of stress-related psychiatric disorders. Neural and physiological systems make us react quickly, fight back or run away. These stress response systems are preserved even when they are no longer as valuable. Reactions that once may have prevented death from an attack of a predator now simply make it impossible to function in daily life. Repeated episodes of stress lead to increased responsivity of these systems. This may have been useful in previous times, for example, the body shifting to an increased chronic presence of a band of lions in the region. In the modern age, stress can lead to psychiatric disorders with resultant pathology. We review here the relationship between stress, brain and physiology, and neuropsychiatric disorders. We focus on disorders with a clear link to stress, that have been described elsewhere as Trauma Spectrum Disorders (Bremner, 2002a). Those disorders include posttraumatic stress disorder (PTSD), a subgroup of major depression, Borderline Personality Disorder (BPD), and Dissociative Identity Disorder (DID). With the Diagnostic and Statistical Manual (DSM)-5, PTSD was moved from the category of Anxiety Disorders to Trauma- and Stressor-related Disorders (American Psychiatric Association, 2013). Included in this category is Acute Stress Disorder (ASD), which is seen as an early form of PTSD as opposed to a unique disorder (Bremner, 1999). Other disorders in this category include Adjustment Disorders, which have not been the subject of extensive research, and childhood attachment disorders that are beyond the scope of the current review. Neurobiological studies have shown that the other anxiety disorders, including Panic Disorder, Generalized

Anxiety Disorder, and Obsessive Compulsive Disorder, in addition to lacking the clear link to stress, do not have the same neurobiological profile as the Trauma Spectrum Disorders (Bremner, 2002a; Heim, Bremner, & Nemeroff, 2006). These disorders are therefore not considered in detail here.

2. Trauma spectrum disorders

The model of Trauma Spectrum Disorders posits that a group of psychiatric disorders related to psychological trauma, usually in childhood, share an underlying neurobiological footprint (Bremner, 1999, 2002a, 2016). Although there is often symptom overlap, a common cause, such as childhood trauma, can lead to multiple outcomes, illustrating the principle of multifinality (Fig. 1) (Cicchetti & Rogosch, 1996). The differences in outcomes, which may be related to genetic, cultural, or concomitant environmental factors, can often be quite striking. For instance, patients with trauma

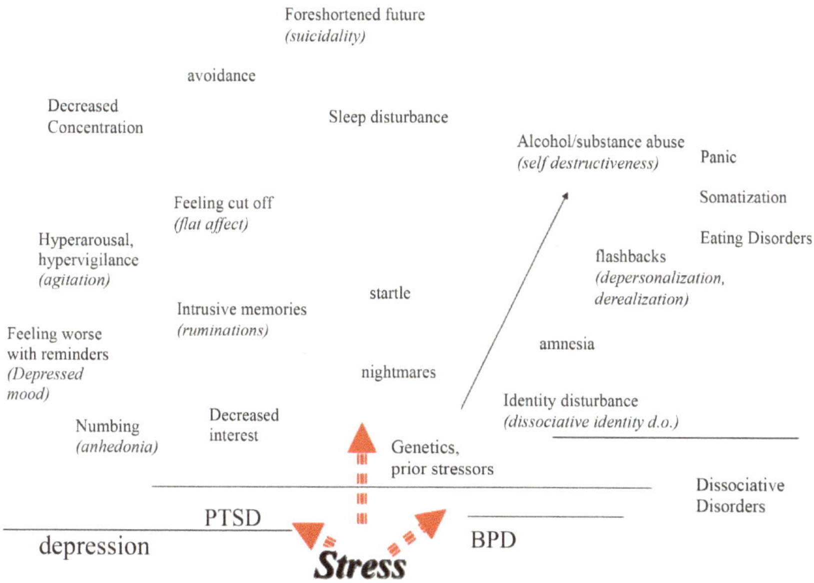

Fig. 1 Stress can lead to multiple psychiatric outcomes. However many of the symptoms of these disorders overlap. The symptoms above the line over posttraumatic stress disorder (PTSD) include the symptoms of that disorder. However, it can be seen that they overlap with symptoms of depression, represented by the symptoms listed above the line over that disorder. Other outcomes include dissociative disorders, which overlap with Borderline Personality Disorder (BPD). Genetics and prior experience modulate how stress manifests itself in psychiatric symptoms.

spectrum disorders share a common neurobiological footprint in smaller volume of a brain area involved in memory called the hippocampus, although they will show differing physiological responses to different stimuli depending on the nature of their specific psychiatric disorder.

In the trauma spectrum disorders model, patients with PTSD, BPD, DID and a subgroup of depression share common changes in brain and neurobiology that underlie the symptoms of these disorders. Although these are separate disorders, they also share a high degree of symptom overlap, and are often co-morbid with one another (Fig. 1), a fact that is not surprising given their overlapping neurobiological correlates (Bremner, 2002a, 2016). The DSM, the modern Bible of Psychiatry, was based on the medical disease model, and treats mental disorders as discrete entities (Bremner, 2006), forcing the use of terms like "high co-morbidity" to account for the overlapping nature of these disorders. The use of this term implies that these are discreet disorders, and their overlapping nature is an anomaly. The DSM attempts to side-step these issues by describing the same phenomena using differing terminology. For instance, psychomotor retardation and loss of interest in things you used to enjoy in major depression and feeling cut off from others and loss of pleasure in PTSD are all aspects of anhedonia that likely share altered function in mesocortical and mesolimbic dopamine pathways. Other symptoms are identical, like sleep disturbance and irritability. The utility of a model lies in its ability to predict and describe phenomena, therefore the trauma spectrum model is superior to the medical model in describing stress-related mental disorders as it more accurately describes these disorders as they actually exist in clinical practice and correctly predicts future behaviors. This likely accounts for the sparse use of DSM by clinicians engaged in actual clinical practice, outside of compliance with bureaucratic requirements.

The trauma spectrum model also has relevance on a neurobiological level. The model predicts that brain circuits and systems involved in the stress response and fear memory will be affected in these disorders, and that they will share a similar neurobiological footprint. A priori we identified a network of brain areas based on animal studies, a knowledge of functional neuroanatomy and findings from patients with specific neurological deficits, that were likely involved in the trauma spectrum disorders (Bremner, Krystal, Southwick, & Charney, 1995). These included the amygdala, hippocampus, thalamus, and medial prefrontal cortex (anterior cingulate). Close to a decade of research later, brain imaging studies in patients with

PTSD and other stress-related psychiatric disorders corroborated the initial hypotheses (Bremner, 2003), and another decade of research replicated those initial studies (Campanella & Bremner, 2016). Neurohormonal systems like cortisol and norepinephrine also show similar alterations stress-related psychiatric disorders (Heim et al., 2006).

Hippocampal volume has been a particularly useful marker of the trauma spectrum disorders (Bremner & Vermetten, 2012). The hippocampus is sensitive to stress and plays a key role in declarative memory (Bremner & Vermetten, 2012). In a series of studies we and others showed smaller hippocampal volume as measured with magnetic resonance imaging (MRI) in patients with a history of childhood abuse and the diagnoses of PTSD (Bremner, Randall, et al., 1995), women with BPD and early life trauma (Driessen et al., 2000; Irle, Lange, & Sachsse, 2005; Schmahl, Vermetten, Elzinga, & Bremner, 2003), and women with abuse and DID (Vermetten, Schmahl, Lindner, Loewenstein, & Bremner, 2006). Women with early childhood sexual abuse and depression had smaller hippocampal volume compared to women with depression without early abuse, and non-abused non-depressed women (Vythilingam et al., 2002). Smaller hippocampal volume was not seen in patients with anxiety disorders, including panic disorder (Vythilingam et al., 2000), obsessive-compulsive disorder (OCD) or other anxiety disorders (Bremner, 2005a; Cannistraro & Rauch, 2003). These studies show a common correlate of stress-related mental disorders in a brain area known to be sensitive to stress that is specific to these disorders.

The amygdala is involved in the encoding of fear memories and also plays a role in some symptoms of stress-related psychiatric disorders. Women with abuse and BPD (Driessen et al., 2000; Schmahl, Vermetten, et al., 2003) and women with abuse and DID (Vermetten et al., 2006) showed smaller amygdala volume on MRI, which is seen in some studies of PTSD but not others (Bremner & Vermetten, 2012; Campanella & Bremner, 2016). Other neurobiological overlaps between PTSD and BPD associated with childhood abuse include alterations in function of the hypothalamic-pituitary adrenal (HPA) axis and dysregulation of the prefrontal-limbic axis (Donegan et al., 2003; Driessen et al., 2004; Juengling et al., 2003; Lange, Kracht, Herholz, Sachsse, & Irle, 2005; Schmahl & Bremner, 2006; Schmahl, Elzinga, et al., 2003; Schmahl, McGlashan, & Bremner, 2002; Schmahl, Vermetten, Elzinga, & Bremner, 2004). In summary, findings are consistent with a common neurobiology and neurocircuitry underlying the trauma spectrum disorders, including PTSD, DID, BPD, and depression related to early abuse.

3. Psychologic trauma at various levels of development

A proper understanding of the effects of psychological trauma on neurobiology and the individual requires a knowledge of the developmental epoch at which the trauma occurred (Bremner, 2006, 2016). As the individual progresses through development from infancy to young adulthood and possibly moves from one physical and cultural environment to the next, changes are taking place on a neurological, social, individual, psychological, and cultural level (Feiring & Lewis, 1996). Some might start at different points with pathways that lead them to the same place (equifinality), or travel the same path with different ultimate outcomes (multipotentiality) (Cicchetti & Dawson, 2002; Cicchetti & Rogosch, 1996). The same childhood trauma may lead to two different outcomes, with one individual showing resilience and overcoming and perhaps thriving in spite of their environment while others succumb to psychopathology and dysfunction (Vythilingam et al., 2009). Childhood trauma is associated with an increase in risk for injected drug use (Anda et al., 2006; Dube et al., 2003), although not all patients with Opioid Use Disorders (OUDs) have a history of childhood trauma. Amongst women with depression, those with the "childhood trauma footprint" appear to respond better to combinations of psychotherapy and medication, illustrating the utility of the multi-level assessment in clinical practice (Richters & Cicchetti, 1993; Richters & Hinshaw, 1999).

4. A psycho-evolutionary view of stress, the brain, and neuropsychiatric disorders

Another factor to consider in reviewing stress, the brain, and neuropsychiatric disorders is the interaction between evolution and mental disorders. Fundamental to understanding stress-related psychiatric disorders is the role of neurophysiological systems involved in survival and their role in ancient and modern times. In the pre-agricultural age, humans roved the Savannah in small bands of hunter-gatherers, and evolution up to that point involved changes in the human brain that allowed individuals to run faster and fight back harder. Other advantages conveyed on humans included developing the capacity for abstract thought, language, and advanced social relationships. These processes required the development of a much larger frontal cortex than other animals, and along with that came side effects of anxiety and depression. Whether these are accidental by-products of evolution or served their own roles in specific situations is unclear. Taking an example from another area of medical science, the heterozygous gene for sickle cell anemia confers resistance to malaria, which is advantageous in African countries where this disease is common. Homozygous expression of the gene, however, confers sickle cell anemia. Thus a biological alteration that is advantageous in some situations is not in others. Similarly, depressive behaviors may be advantageous during sickness (in order to preserve energy) or during times of extreme threat (removing the individual from threat). In other situations, however, depressive behavior may be maladaptive. Similarly, avoidance and hyper responsiveness seen in PTSD may have allowed members of primitive bands to survive threats when groups of predators were in the vicinity. Whatever behaviors were helpful in these prehistoric periods, however, are probably lost in the transition to modern culture, where life in high rise buildings and days spent puzzling over financial accounts rather than chasing animals through the bush came to predominate.

In earlier times fear-related behaviors may have meant the difference between life and death. If there was a pack of tigers in an area where a band of hunter-gatherers was camped, hyper-vigilance could promote survival. These behaviors would be expected to persist, even after the threat had been removed. Children who exist in an environment of real or perceived threat respond in similar ways would sense the type of environment they were born into, and their developmental evolution would be adjusted accordingly. Children in a harsh, deprived, or non-supportive environment would grow up being mistrustful and socially withdrawn. This might have survival value in protecting the child from potentially harmful persons in the absence of a functional parent to protect them.

Symptoms of depression also probably have had a behavioral function at some point in time. They are a by-product of the development of the capacity in humans to form close social attachments. When those attachments are broken, through death or separation, there is a natural process of grieving where individuals recover from the disruption of the social connection and form the capacity to form new connections. If they do not go through this process, they may not go on to reproduce with others in the future. In addition, depression causes a social withdrawal and decrease in activity. Death of a partner or parent may put the survivor at greater risk without the added protection, so withdrawal from potentially dangerous situations would also have survival value. Similar behavioral effects occur with illnesses such as infections, which probably have similar survival value in allowing the conservation of energy and resources.

5. Neurohormonal responses to trauma

Trauma can have lasting effects on neurotransmitter and neurohormonal systems involved in the stress response (Bremner & Pearce, 2016). Cortisol and norepinephrine systems are altered in PTSD (Bremner, 2002b; Bremner & Vermetten, 2001; Yehuda, 2006) and underlie symptoms of PTSD and other stress-related mental disorders (Baker et al., 2005; Bremner, 2002a; Lindqvist et al., 2014; Neylan et al., 2005).

The hypothalamic-pituitary-adrenal (HPA) axis is activated by stress. Corticotropin-releasing factor (CRF) is released from the hypothalamus, with stimulation of adrenocorticotropic hormone (ACTH) release from the pituitary resulting in turn in cortisol release from the adrenal. CRF also acts centrally to create fear-related behaviors and triggers other neurochemical responses to stress, including the noradrenergic neurotransmitter system (Melia & Duman, 1991). Early life stress leads to long-term sensitization of the HPA axis and CRF to further stressors (Coplan et al., 1996; Levine, Weiner, & Coe, 1993; Makino, Smith, & Gold, 1995; Plotsky & Meaney, 1993; Stanton, Gutierrez, & Levine, 1988). Stress also leads to over activation of the norepinephrine system, which plays a critical role in the fight or flight response (Bremner, Krystal, Southwick, & Charney, 1996a, 1996b) (Fig. 2).

As noted above, a network of brain areas involved in memory and fear, including the hippocampus, medial prefrontal cortex, and amygdala, interact with these neurochemical systems to mediate the brain's response to stress (Bremner, 2005b, 2011; Rauch, Shin, & Phelps, 2006). The medial prefrontal cortex includes the anterior cingulate, orbitofrontal cortex, and adjacent areas. Inhibition of the amygdala by the medial prefrontal cortex represents the brain circuitry underlying extinction of fear responses (Milad & Quirk, 2002; Quirk, Garcia, & Gonzalez-Lima, 2006). This brain area, together with the hypothalamus, also modulates peripheral neurohormonal responses to stress via relay stations in the spinal cord (Diorio, Viau, & Meaney, 1993; Feldman, Conforti, & Weidenfeld, 1995; Frysztak & Neafsey, 1994).

Altered HPA axis function is associated with PTSD (Yehuda, 2002). Women with childhood sexual abuse-related PTSD had decreased baseline cortisol based on 24-h diurnal assessments of plasma with 10 min sampling periods, a flattening of the normal diurnal cortisol curve, and increased pulsatility of cortisol reflecting dysregulation of CRF release, compared to women with abuse without PTSD, and women without abuse or PTSD (Bremner, Vermetten, & Kelley, 2007). Lower baseline cortisol was correlated

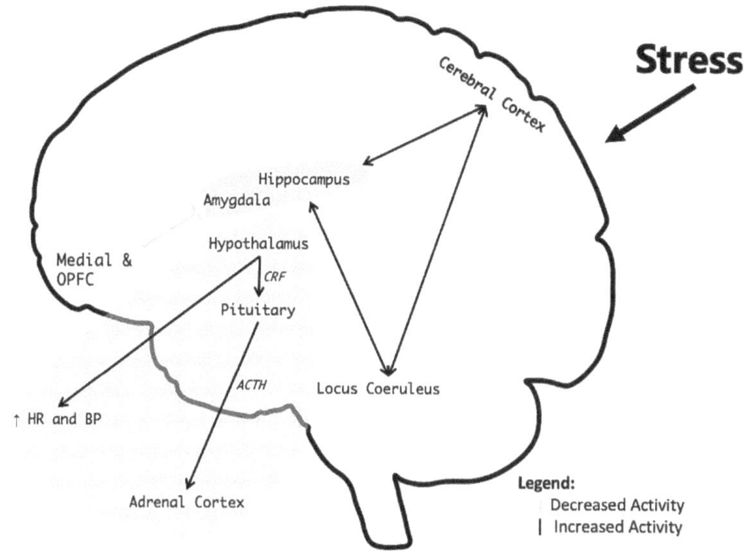

Fig. 2 Model of how stress affects brain function. Brain areas involved in stress and memory including the hippocampus, amygdala, and medial and orbitofrontal cortex (OPFC) (including anterior cingulate) mediate symptoms of stress-related psychiatric disorders. Neurochemical systems also mediate the stress response and underlie psychiatric disorders. Most of the norepinephrine cell bodies are located in the locus coeruleus (pons) with neurons that distribute throughout the brain and activate the fight or flight response. Stress information is processed by the cerebral cortex with pathways through the hypothalamus that activate peripheral sympathetic and cardiovascular responses, including increased Heart Rate (HR) and Blood Pressure (BP), as well as the hypothalamic-pituitary-adrenal (HPA) axis, with activation of Corticotropin Releasing Factor (CRF) from the hypothalamus, which releases Adrenocorticotropic hormone (ACTH) from the pituitary, which in turn releases cortisol from the Adrenal Cortex. Norepinephrine and cortisol play critical roles in the stress response, and dysfunction of these systems underlie symptoms of stress-related psychiatric disorders.

with increased symptoms of PTSD in the PTSD group (Bremner et al., 2007). In a similar study of comprehensive sampling over 24 h, combat veterans with PTSD showed lower plasma cortisol levels and a flattening of the normal diurnal curve (Yehuda, Teicher, Levengood, Trestman, & Siever, 1994), while other studies showed either normal or low cortisol measured in 24 h urine collections (Yehuda, Boisoneau, Lowy, & Giller, 1995; Yehuda, Kahana, et al., 1995; Yehuda, Teicher, Trestman, Levengood, & Siever, 1996) and a blunting of the normal rise in cortisol after awakening in PTSD (de Kloet, Vermetten, Heijnen, et al., 2007; Neylan et al., 2005; Wessa, Rohleder, Kirschbaum, & Flor, 2006). Other findings included increased suppression of cortisol to low doses of dexamethasone, which together with other studies reflects altered sensitivity of glucocorticoid receptors in the pituitary and/or hippocampus (Bremner, Vythilingam, Vermetten, Newcomer, & Charney, 2004; de Kloet, Vermetten, Bikker, et al., 2007; de Kloet, Vermetten, Heijnen, et al., 2007; Goenjian et al., 1996; Stein, Yehuda, Koverola, & Hanna, 1997; Vermetten, 2008; Yehuda, Halligan, Grossman, Golier, & Wong, 2002; Yehuda, Levengood, et al., 1996; Yehuda, Lowy, Southwick, Shaffer, & Giller, 1991; Yehuda et al., 1993), PTSD patients showed increased cortisol in response to hearing personalized scripts of their childhood abuse (Elzinga, Schmahl, Vermetten, van Dyck, & Bremner, 2003) as well as with "neutral" mental stress tasks involving cognitive stressors like mental arithmetic (Bremner, Vythilingam, Vermetten, Adil, et al., 2003). Women with depression and a history of childhood abuse had increased cortisol response to a stressful cognitive challenge relative to controls (Heim et al., 2000) and a blunted ACTH response to CRF challenge (Heim, Newport, Bonsall, Miller, & Nemeroff, 2001). PTSD was associated with increases in CRF concentrations in the cerebrospinal fluid (CSF) (Baker et al., 2005, 1999; Bremner, Licinio, et al., 1997; Sautter et al., 2003) and plasma (de Kloet et al., 2008), and a blunted ACTH response to CRF challenge, suggesting decreased pituitary sensitivity to CRF (Smith et al., 1989).

Increased noradrenergic and peripheral sympathetic function is associated with PTSD (Bremner et al., 1996a, 1996b). Symptoms irritability, increased startle, hyperarousal and sleep disturbance, all indicate excessive sympathetic and noradrenergic function (Bremner et al., 1996b). Consistent with this, baseline concentrations of norepinephrine and its metabolites in urine and plasma in PTSD were found to be increased (De Bellis et al., 1999; De Bellis, Lefter, Trickett, & Putnam, 1994; Lemieux & Coe, 1995; Mason, Giller, & Kosten, 1988; Yehuda, Siever, & Teicher, 1998) or unchanged (Blanchard, Kolb, Prins, Gates, & McCoy, 1991; McFall, Veith, & Murburg, 1992; Mellman, Kumar, Kulick-Bell, Kumar, & Nolan, 1995; Pitman & Orr, 1990; Southwick et al., 1993) and increased in CSF (Geracioti et al., 2001). Increased sympathetic reactivity based on heart rate, blood pressure, and skin conductance (Blanchard, Kolb, Gerardi, Ryan, & Pallmeyer, 1986; Blanchard, Kolb, Pallmeyer, & Gerardi, 1982; Bremner et al., 1996b; Malloy, Fairbank, & Keane, 1983; McFall, Murburg, Ko, & Veith, 1990; Orr et al., 1998; Orr, Lasko, Shalev, & Pitman, 1995; Orr, Pitman, Lasko, & Herz, 1993; Orr & Roth, 2000), as well as increased norepinephrine and epinephrine in plasma (Blanchard et al., 1991; McFall et al., 1992), was seen in PTSD patients exposed to traumatic reminders in the form of combat slides and sounds or personalized trauma scripts, while administration of the $alpha_2$ adrenergic receptor antagonist, yohimbine, which increases firing of noradrenergic neurons in the locus coeruleus, increased PTSD symptoms and anxiety, as well plasma concentrations of the norepinephrine metabolite, 3-methoxy-4-hydroxyphenylglycol (MHPG) (Southwick et al., 1997, 1993). PTSD patients had a decrease in brain function in the medial prefrontal cortex with yohimbine not seen in healthy controls (decreased frontal lobe function occurs with excessive release of norepinephrine in that brain area) (Bremner, Innis, et al., 1997). Increased heart rate in the emergency room after trauma predicted subsequent development of chronic PTSD (Shalev et al., 1998). These findings all support increased noradrenergic function in PTSD.

6. The neural circuitry of PTSD

Brain imaging studies have been consistent in supporting the model of a specific circuitry of PTSD (Bremner, 2011). Exposure to memories of childhood sexual abuse with personalized traumatic scripts in women with childhood sexual abuse-related PTSD resulted in decreased blood flow in the medial prefrontal cortex/anterior cingulate, including Brodmann's area 25, or subcallosal gyrus, and area 32, as well as decreased blood flow in parietal cortex and inferior frontal gyrus (Bremner, Narayan, et al., 1999; Shin et al., 1999), hippocampus and visual association cortex (Bremner, Narayan, et al., 1999). Emotional memory tasks, such as retrieval of word pairs like "blood-stench" resulted in similar decreases in blood flow in this medial prefrontal cortex area, as well as hippocampus, and fusiform gyrus/inferior temporal gyrus, with increased activation in posterior cingulate, left inferior parietal cortex, left middle frontal gyrus, and visual association and motor cortex (Bremner et al., 2003b). Decreased medial prefrontal cortical/anterior cingulate activation was seen in women with abuse-related PTSD during performance of the emotional Stroop task (i.e., naming the color of a word such as "rape") (Bremner, Vermetten, et al., 2004). Other studies replicated the finding of decreased medial prefrontal cortex/anterior function during exposure to traumatic reminders in the form of traumatic slides and/or sounds or traumatic scripts using PET, SPECT or fMRI (Bremner, Narayan, et al., 1999; Bremner, Staib, et al., 1999; Britton, Phan, Taylor, Fig, & Liberzon, 2005; Lanius et al., 2001, 2003; Liberzon et al., 1999; Lindauer et al., 2004; Phan, Britton, Taylor, Fig, & Liberzon, 2006; Semple et al., 2000; Shin et al., 1997, 1999, , 2004, 2001, 2005; Yang, Wu, Hsu, & Ker, 2004). Studies have also shown increased amygdala function with classical fear conditioning or other fear-related tasks (Armony, Corbo, Clement, & Brunet, 2005; Bremner et al., 2005; Liberzon et al., 1999; Pissiota et al., 2002; Protopopescu et al., 2005; Rauch et al., 1996, 2000; Shin et al., 2004, 2005; Vermetten, Schmahl, Southwick, & Bremner, 2007). Other findings in studies of traumatic reminder exposure include decreased function with exposure to traumatic reminders in hippocampus (Bremner, Narayan, et al., 1999), thalamus (Lanius et al., 2001, 2003), visual association cortex (Bremner, Narayan, et al., 1999; Lanius et al., 2003; Shin et al., 1997, 2004), parietal cortex (Bremner, Narayan, et al., 1999; Rauch et al., 1996; Sakamoto et al., 2005; Shin et al., 1997, 1999), inferior frontal gyrus (Bremner, Narayan, et al., 1999; Lanius et al., 2003; Rauch et al., 1996; Sakamoto et al., 2005; Shin et al., 1997, 1999, 2001), and increased function in insula (Bruce et al., 2013; Nicholson et al., 2016), posterior cingulate (Bremner, Narayan, et al., 1999; Bremner, Staib, et al., 1999; Lanius et al., 2001; Shin et al., 1997), insula (Vermetten et al., 2007), and parahippocampal gyrus (Bremner, Narayan, et al., 1999; Bremner, Staib, et al., 1999; Liberzon et al., 1999).

Brain imaging and physiological assessments were used to test the hypothesis that a common network of brain areas mediated symptoms of trauma spectrum disorders. We studied women with a history of childhood abuse and the diagnosis of BPD, with half of them having a co-morbid diagnosis of PTSD, although BPD was felt to be the primary diagnosis. Exposure to scripts of a hypothetical situation where they were abandoned in a mall led to marked increases in sympathetic responses as measured by heart rate and electrodermal responses, whereas exposure to personalized scripts of traumatic events from their childhood led to a much less exaggerated response (Schmahl, Elzinga, & Bremner, 2002; Schmahl, Elzinga, et al., 2004). Women with PTSD related to childhood abuse without the diagnosis of BPD, on the other hand, had no response to abandonment scripts, but had pronounced physiological responses to the personalized traumatic scripts (Schmahl, Elzinga, et al., 2004). This was in spite of the fact that the severity of traumas between the two groups of women were comparable. Brain imaging showed that the women with PTSD and BPD had a similar footprint in neural responses (failure of response of the medial prefrontal cortex/anterior cingulate) to the particular script (abandonment/trauma) for which they were most reactive (Bremner, Narayan, et al., 1999; Schmahl, Elzinga, et al., 2003; Schmahl, Vermetten, et al., 2004). These studies illustrate how similar events lead to different outcomes, however brain circuits mediating the effects overlap, suggesting that other factors like prior experience or genetics overlap with neurobiology to determine outcome (Fig. 2).

Studies of structural MRI showed smaller hippocampal volume across the trauma spectrum disorders (Bremner, 2016), including abuse-related PTSD (Bremner, Randall et al., 1997, Bremner et al., 2003a), DID with early abuse (Vermetten et al., 2006), BPD with early abuse (Driessen et al., 2000; Schmahl, Vermetten, et al., 2003), and depression with early abuse (Vythilingam et al., 2002). Patients with another anxiety disorder, panic disorder, did not show a reduction in hippocampal volume (Vythilingam et al., 2000). Patients with trauma spectrum disorders exhibit other alterations in stress responsive neurohormonal systems. HPA axis dysfunction, and a functional dysregulation of the prefrontal-limbic axis, are seen in BPD that are similar to PTSD (Donegan et al., 2003; Driessen et al., 2004; Juengling et al., 2003; Lange et al., 2005; Schmahl & Bremner, 2006; Schmahl, Elzinga, et al., 2003; Schmahl, McGlashan, et al., 2002; Schmahl, Vermetten, et al., 2004). These findings are consistent with a common neurobiology and neurocircuitry underlying the trauma spectrum disorders, including PTSD, DID, BPD, and depression related to early abuse, as well as some differences. The partial overlap of neurobiology then parallels the partial overlap in symptomatology in patients with trauma-related mental disorders.

7. Conclusions

Stress can have effects on the brain that lead to the symptoms of many stress-related psychiatric disorders. Treatments that target the underlying neurobiology can be helpful in their treatment. For instance, serotonin is one of the neurochemical systems associated with symptoms of PTSD and depression, and treatment with selective serotonin reuptake inhibitors (SSRIs) can be helpful for many patients. Other non-medication treatments can be helpful as well, including psychotherapy and behavioral interventions like meditation training. Meditation adds a sense of control (Astin, 1997; Weissbecker et al., 2002), allowing trauma victims to gain control of their thoughts and memories in a non-emotional way, disrupting patterns of ruminative thinking (Teasdale, Segal, & Williams, 1995) which can be associated with pathological arousal (Teasdale et al., 2000). Interventions like these may modify the way traumatic memories are encoded and consolidated and therefore reduce risk of chronic symptoms of PTSD. Other interventions that block sympathetic arousal with reminders, like neuromodulation, may also be useful (Bremner & Rapaport, 2017). Research on neurobiological correlates of stress-related psychiatric disorders will be helpful in the development of new treatments for these disabling disorder.

Acknowledgment

The work presented in this review was supported by Grants from NIH R01 MH056120, R01 HL088726, K24 MH076955, and P01 HL101398.

References

American Psychiatric Association. (2013). *The diagnostic and statistical manual of mental disorders, fifth edition (DSM-5)* (5th ed.). Washington, DC: American Psychiatric Association.

Anda, R. F., Felitti, V. J., Walker, J., Whitfield, C., Bremner, J. D., Perry, B. D., et al. (2006). The enduring effects of childhood abuse and related experiences in childhood: A convergence of evidence from neurobiology and epidemiology. *European Archives of Psychiatry and Clinical Neuroscience, 256*(3), 174–186.

Armony, J. L., Corbo, V., Clement, M. H., & Brunet, A. (2005). Amygdala response in patients with acute PTSD to masked and unmasked emotional facial expressions. *The American Journal of Psychiatry, 162*(10), 1961–1963.

Astin, J. (1997). Stress reduction through mindfulness meditation: Effects on psychological symptomatology, sense of control, and spiritual experiences. *Psychotherapy and Psychosomatics, 66*, 97–106.

Baker, D. G., Ekhator, N. N., Kasckow, J. W., Dashevsky, B., Horn, P. S., Bednarik, L., et al. (2005). Higher levels of basal serial CSF cortisol in combat veterans with post-traumatic stress disorder. *The American Journal of Psychiatry, 162*, 992–994.

Baker, D. G., West, S. A., Nicholson, W. E., Ekhator, N. N., Kasckow, J. W., Hill, K. K., et al. (1999). Serial CSF corticotropin-releasing hormone levels and adrenocortical activity in combat veterans with posttraumatic stress disorder. *The American Journal of Psychiatry, 156*, 585–588.

Blanchard, E. B., Kolb, L. C., Gerardi, R. J., Ryan, P., & Pallmeyer, T. P. (1986). Cardiac response to relevant stimuli as an adjunctive tool for diagnosing post-traumatic stress disorder in Vietnam veterans. *Behavior Therapy, 17*(5), 592–606.

Blanchard, E. B., Kolb, L. C., Pallmeyer, T. P., & Gerardi, R. J. (1982). A psychophysiological study of post-traumatic stress disorder in Vietnam veterans. *Psychiatric Quarterly, 54*(4), 220–229.

Blanchard, E. B., Kolb, L. C., Prins, A., Gates, S., & McCoy, G. C. (1991). Changes in plasma norepinephrine to combat-related stimuli among Vietnam veterans with post-traumatic stress disorder. *Journal of Nervous and Mental Disease, 179*, 371–373.

Bremner, J. D. (1999). Acute and chronic responses to stress: Where do we go from here? [Editorial]. *The American Journal of Psychiatry, 156*(3), 349–351.

Bremner, J. D. (2002a). *Does stress damage the brain? Understanding trauma-related disorders from a mind-body perspective*. New York: W.W. Norton.

Bremner, J. D. (2002b). Neuroimaging of childhood trauma. *Seminars in Clinical Neuropsychiatry, 7*, 104–112.

Bremner, J. D. (2003). Functional neuroanatomical correlates of traumatic stress revisited 7 years later, this time with data. *Psychopharmacology Bulletin, 37*(2), 6–25.

Bremner, J. D. (2005a). *Brain imaging handbook*. New York: W.W. Norton.

Bremner, J. D. (2005b). The neurobiology of childhood sexual abuse in women with post-traumatic stress disorder. In K. A. Kendall-Tackett (Ed.), *Handbook of women, stress and trauma* (pp. 181–206). New York: Brunner-Routledge.

Bremner, J. D. (2006). Traumatic stress from a multiple-levels-of-analysis perspective. In D. Cicchetti & D. J. Cohen (Eds.), *Vol. 2 Developmental psychopathology* (pp. 656–676). Hoboken, NJ: John Wiley & Sons.

Bremner, J. D. (2011). Stress and human neuroimaging studies. In C. D. Conrad (Ed.), *The handbook of stress: Neuropsychological effects on the brain* (pp. 446–462). Chicester, West Sussex, UK: Wiley-Blackwell Press.

Bremner, J. D. (2016). Traumatic stress from a multi-level developmental psychopathology perspective. In D. Cicchetti (Ed.), *Developmental psychopathology, Vol. 3: Maladaptation and psychopathology* (3rd ed., pp. 386–424). Hoboken, N.J.: Wiley.

Bremner, J. D., Innis, R. B., Ng, C. K., Staib, L., Duncan, J., Bronen, R., et al. (1997). PET measurement of cerebral metabolic correlates of yohimbine administration in posttraumatic stress disorder. *Archives of General Psychiatry, 54*, 246–256.

Bremner, J. D., Krystal, J. H., Southwick, S. M., & Charney, D. S. (1995). Functional neuroanatomical correlates of the effects of stress on memory. *Journal of Traumatic Stress, 8*, 527–554.

Bremner, J. D., Krystal, J. H., Southwick, S. M., & Charney, D. S. (1996a). Noradrenergic mechanisms in stress and anxiety: I. Preclinical studies. *Synapse, 23*, 28–38.

Bremner, J. D., Krystal, J. H., Southwick, S. M., & Charney, D. S. (1996b). Noradrenergic mechanisms in stress and anxiety: II. Clinical studies. *Synapse, 23*, 39–51.

Bremner, J. D., Licinio, J., Darnell, A., Krystal, J. H., Owens, M., Southwick, S. M., et al. (1997). Elevated CSF corticotropin-releasing factor concentrations in posttraumatic stress disorder. *The American Journal of Psychiatry, 154*, 624–629.

Bremner, J. D., Narayan, M., Staib, L. H., Southwick, S. M., McGlashan, T., & Charney, D. S. (1999). Neural correlates of memories of childhood sexual abuse in women with and without posttraumatic stress disorder. *The American Journal of Psychiatry, 156*, 1787–1795.

Bremner, J. D., & Pearce, B. (2016). Neurotransmitter, neurohormonal, and neuropeptidal function in stress and PTSD. In J. D. Bremner (Ed.), *Posttraumatic stress disorder: From neurobiology to treatment* (pp. 181–232). Hoboken, New Jersey: Wiley-Blackwell.

Bremner, J. D., Randall, P. R., Scott, T. M., Bronen, R. A., Delaney, R. C., Seibyl, J. P., et al. (1995). MRI-based measurement of hippocampal volume in patients with combat-related posttraumatic stress disorder. *American Journal of Psychiatry, 152*, 973–981.

Bremner, J. D., Randall, P. R., Vermetten, E., Staib, L., Bronen, R. A., Mazure, C. M., et al. (1997). Magnetic resonance imaging-based measurement of hippocampal volume in posttraumatic stress disorder related to childhood physical and sexual abuse: A preliminary report. *Biological Psychiatry, 41*, 23–32.

Bremner, J. D., & Rapaport, M. H. (2017). Vagus nerve stimulation: Back to the future. *The American Journal of Psychiatry, 174*(7), 609–610. https://doi.org/10.1176/appi.ajp.2017.17040422.

Bremner, J. D., Staib, L., Kaloupek, D., Southwick, S. M., Soufer, R., & Charney, D. S. (1999). Neural correlates of exposure to traumatic pictures and sound in Vietnam combat veterans with and without posttraumatic stress disorder: A positron emission tomography study. *Biological Psychiatry, 45*, 806–816.

Bremner, J. D., & Vermetten, E. (2001). Stress and development: Behavioral and biological consequences. *Development and Psychopathology, 13*, 473–489.

Bremner, J. D., & Vermetten, E. (2012). The hippocampus and post-traumatic stress disorders. In T. Bartsch (Ed.), *The clinical neurobiology of the hippocampus: An integrative view* (pp. 262–272): Oxford University Press.

Bremner, D., Vermetten, E., & Kelley, M. E. (2007). Cortisol, dehydroepiandrosterone, and estradiol measured over 24 hours in women with childhood sexual abuse-related posttraumatic stress disorder. *Journal of Nervous and Mental Disease, 195*(11), 919–927.

Bremner, J. D., Vermetten, E., Schmahl, C., Vaccarino, V., Vythilingam, M., Afzal, N., et al. (2005). Positron emission tomographic imaging of neural correlates of a fear acquisition and extinction paradigm in women with childhood sexual abuse-related posttraumatic stress disorder. *Psychological Medicine, 35*(6), 791–806.

Bremner, J. D., Vermetten, E., Vythilingam, M., Afzal, N., Schmahl, C., Elzinga, B. E., et al. (2004). Neural correlates of the classical color and emotional stroop in women with abuse-related posttraumatic stress disorder. *Biological Psychiatry, 55*(6), 612–620.

Bremner, J. D., Vythilingam, M., Vermetten, E., Adil, J., Khan, S., Nazeer, A., et al. (2003). Cortisol response to a cognitive stress challenge in posttraumatic stress disorder (PTSD) related to childhood abuse. *Psychoneuroendocrinology, 28*, 733–750.

Bremner, J. D., Vythilingam, M., Vermetten, E., Newcomer, J. W., & Charney, D. S. (2004). Effects of dexamethasone on declarative memory function in posttraumatic stress disorder (PTSD). *Psychiatry Research, 129*(1), 1–10.

Bremner, J. D., Vythilingam, M., Vermetten, E., Southwick, S. M., McGlashan, T., Nazeer, A., et al. (2003a). MRI and PET study of deficits in hippocampal structure and function in women with childhood sexual abuse and posttraumatic stress disorder (PTSD). *American Journal of Psychiatry, 160*, 924–932.

Bremner, J. D., Vythilingam, M., Vermetten, E., Southwick, S. M., McGlashan, T., Staib, L., et al. (2003b). Neural correlates of declarative memory for emotionally valenced words in women with posttraumatic stress disorder (PTSD) related to early childhood sexual abuse. *Biological Psychiatry, 53*, 289–299.

Britton, J. C., Phan, K. L., Taylor, S. F., Fig, L. M., & Liberzon, I. (2005). Corticolimbic blood flow in posttraumatic stress disorder during script-driven imagery. *Biological Psychiatry, 57*(8), 832–840.

Bruce, S. E., Buchholz, K. R., Brown, W. J., Yan, L., Durbin, A., & Sheline, Y. L. (2013). Altered emotional interference processing in the amygdala and insula in women with post-traumatic stress disorder. *Neuroimage. Clinical, 2*, 43–49.

Campanella, C., & Bremner, J. D. (2016). Neuroimaging of PTSD. In J. D. Bremner (Ed.), *Posttraumatic stress disorder: From neurobiology to treatment* (pp. 291–320). Hoboken, New Jersey: Wiley-Blackwell.

Cannistraro, P. A., & Rauch, S. L. (2003). Neural circuitry of anxiety: Evidence from structural and functional neuroimaging studies. *Psychopharmacology Bulletin, 37*(4), 8–25.

Cicchetti, D., & Dawson, G. (2002). Editorial: Multiple levels of analysis. *Development and Psychopathology, 14*, 417–420.

Cicchetti, D., & Rogosch, F. A. (1996). Equifinality and multifinality in developmental psychopathology. *Development and Psychopathology, 8*, 597–600.

Coplan, J. D., Andrews, M. W., Rosenblum, L. A., Owens, M. J., Friedman, S., Gorman, J. M., et al. (1996). Persistent elevations of cerebrospinal fluid concentrations of corticotropin-releasing factor in adult nonhuman primates exposed to early-life stressors: Implications for the pathophysiology of mood and anxiety disorders. *Proceedings of the National Academy of Sciences of the United States of America, 93*, 1619–1623.

De Bellis, M. D., Baum, A. S., Keshavan, M. S., Eccard, C. H., Boring, A. M., Jenkins, F. J., et al. (1999). A.E. Bennett Research Award: Developmental traumatology: Part I: Biological stress systems. *Biological Psychiatry, 45*, 1259–1270.

De Bellis, M. D., Lefter, L., Trickett, P. K., & Putnam, F. W. (1994). Urinary catecholamine excretion in sexually abused girls. *Journal of the American Academy of Child and Adolescent Psychiatry, 33*, 320–327.

de Kloet, C. S., Vermetten, E., Bikker, A., Meulman, E., Geuze, E., Kavelaars, A., et al. (2007). Leukocyte glucocorticoid receptor expression and immunoregulation in veterans with and without post-traumatic stress disorder. *Molecular Psychiatry, 12*(5), 443–453. https://doi.org/10.1038/sj.mp.4001934.

de Kloet, C. S., Vermetten, E., Geuze, E., Lentjes, E. G., Heijnen, C. J., Stalla, G. K., et al. (2008). Elevated plasma corticotrophin-releasing hormone levels in veterans with post-traumatic stress disorder. *Progress in Brain Research, 167*, 287–291. https://doi.org/10.1016/s0079-6123(07)67025-3.

de Kloet, C. S., Vermetten, E., Heijnen, C. J., Geuze, E., Lentjes, E. G., & Westenberg, H. G. (2007). Enhanced cortisol suppression in response to dexamethasone administration in traumatized veterans with and without posttraumatic stress disorder. *Psychoneuroendocrinology, 32*(3), 215–226. https://doi.org/10.1016/j.psyneuen.2006.12.009.

Diorio, D., Viau, V., & Meaney, M. J. (1993). The role of the medial prefrontal cortex (cingulate gyrus) in the regulation of hypothalamic-pituitary-adrenal responses to stress. *Journal of Neuroscience, 13*(9), 3839–3847.

Donegan, N. H., Sanislow, C. A., Blumberg, H. P., Fulbright, R. K., Lacadie, C., Skudlarski, P., et al. (2003). Amygdala hyperreactivity in borderline personality disorder: Implications for emotional dysregulation. *Biological Psychiatry, 54*, 1284–1293.

Driessen, M., Beblo, T., Mertens, M., Piefke, M., Rullkoetter, N., Silva-Saavedra, A., et al. (2004). Posttraumatic stress disorder and fMRI activation patterns of traumatic memory in patients with borderline personality disorder. *Biological Psychiatry, 55*(6), 603–611.

Driessen, M., Herrmann, J., Stahl, K., Zwaan, M., Meier, S., Hill, A., et al. (2000). Magnetic resonance imaging volumes of the hippocampus and the amygdala in women with borderline personality disorder and early traumatization. *Archives of General Psychiatry, 57*, 1115–1122.

Dube, S. R., Felitti, V. J., Dong, M., Chapman, D. P., Giles, W. H., & Anda, R. F. (2003). Childhood abuse, neglect, and household dysfunction and the risk of illicit drug use: The adverse childhood experiences study. *Pediatrics, 111*(3), 564–572.

Elzinga, B. M., Schmahl, C. S., Vermetten, E., van Dyck, R., & Bremner, J. D. (2003). Higher cortisol levels following exposure to traumatic reminders in abuse-related PTSD. *Neuropsychopharmacology, 28*(9), 1656–1665.

Feiring, C., & Lewis, M. (1996). Finality in the eye of the beholder: Multiple sources, multiple time points, multiple paths. *Development and Psychopathology, 8*, 721–733.

Feldman, S., Conforti, N., & Weidenfeld, J. (1995). Limbic pathways and hypothalamic neurotransmitters mediating adrenocortical responses to neural stimuli. *Neuroscience and Biobehavioral Reviews, 19*(2), 235–240.

Frysztak, R. J., & Neafsey, E. J. (1994). The effect of medial frontal cortex lesions on cardiovascular conditioned emotional responses in the rat. *Brain Research, 643*, 181–193.

Geracioti, T. D. J., Baker, D. G., Ekhator, N. N., West, S. A., Hill, K. K., Bruce, A. B., et al. (2001). CSF norepinephrine concentrations in posttraumatic stress disorder. *American Journal of Psychiatry, 158*, 1227–1230.

Goenjian, A. K., Yehuda, R., Pynoos, R. S., Steinberg, A. M., Tashjian, M., Yang, R. K., et al. (1996). Basal cortisol, dexamethasone suppression of cortisol, and MHPG in adolescents after the 1988 earthquake in Armenia. *The American Journal of Psychiatry, 153*, 929–934.

Heim, C., Bremner, J. D., & Nemeroff, C. B. (2006). Trauma spectrum disorders. In M. S. Runge & C. Patterson (Eds.), *Principles of molecular medicine* (pp. 1203–1210). Totowa, N.J: Humana Press.

Heim, C., Newport, D. J., Bonsall, R., Miller, A. H., & Nemeroff, C. B. (2001). Altered pituitary-adrenal axis responses to provocative challenge tests in adult survivors of childhood abuse. *American Journal of Psychiatry, 158*, 575–581.

Heim, C., Newport, D. J., Heit, S., Graham, Y. P., Wilcox, M., Bonsall, R., et al. (2000). Pituitary-adrenal and autonomic responses to stress in women after sexual and physical abuse in childhood. *Journal of the American Medical Association, 284*, 592–597.

Irle, E., Lange, C., & Sachsse, U. (2005). Reduced size and abnormal asymmetry of parietal cortex in women with borderline personality disorder. *Biological Psychiatry, 57*(2), 173–182.

Juengling, F., Schmahl, C. G., Hesslinger, B., Eberg, D., Bremner, J. D., Gostomzyk, M., et al. (2003). Positron emission tomography in female patients with borderline personality disorder. *Journal of Psychiatric Research, 37,* 109–115.

Lange, C., Kracht, L., Herholz, K., Sachsse, U., & Irle, E. (2005). Reduced glucose metabolism in temporo-parietal cortices of women with borderline personality disorder. *Psychiatry Research: Neuroimaging, 139,* 115–126.

Lanius, R. A., Williamson, P. C., Densmore, M., Boksman, K., Gupta, M. A., Neufeld, R. W., et al. (2001). Neural correlates of traumatic memories in posttraumatic stress disorder: A functional MRI investigation. *The American Journal of Psychiatry, 158,* 1920–1922.

Lanius, R. A., Williamson, P. C., Hopper, J., Densmore, M., Boksman, K., Gupta, M. A., et al. (2003). Recall of emotional states in posttraumatic stress disorder: An fMRI investigation. *Biological Psychiatry, 53*(3), 204–210.

Lemieux, A. M., & Coe, C. L. (1995). Abuse-related posttraumatic stress disorder: Evidence for chronic neuroendocrine activation in women. *Psychosomatic Medicine, 57,* 105–115.

Levine, S., Weiner, S. G., & Coe, C. L. (1993). Temporal and social factors influencing behavioral and hormonal responses to separation in mother and infant squirrel monkeys. *Psychoneuroendocrinology, 4,* 297–306.

Liberzon, I., Taylor, S. F., Amdur, R., Jung, T. D., Chamberlain, K. R., Minoshima, S., et al. (1999). Brain activation in PTSD in response to trauma-related stimuli. *Biological Psychiatry, 45,* 817–826.

Lindauer, R. J., Booij, J., Habraken, J. B., Uylings, H. B., Olff, M., Carlier, I. V., et al. (2004). Cerebral blood flow changes during script-driven imagery in police officers with posttraumatic stress disorder. *Biological Psychiatry, 56*(11), 853–861.

Lindqvist, D., Wolkowitz, O. M., Mellon, S., Yehuda, R., Flory, J. D., Henn-Haase, C., et al. (2014). Proinflammatory milieu in combat-related PTSD is independent of depression and early life stress. *Brain, Behavior, and Immunity, 42,* 81–88. https://doi.org/10.1016/j.bbi.2014.06.003.

Makino, S., Smith, M. A., & Gold, P. W. (1995). Increased expression of corticotropin-releasing hormone and vasopressin messenger-ribonucleic acid (messenger RNA) in the hypothalamic paraventricular nucleus during repeated stress-association with reduction in glucocorticoid messenger-RNA levels. *Endocrinology, 136,* 3299–3309.

Malloy, P. F., Fairbank, J. A., & Keane, T. M. (1983). Validation of a multimethod assessment of posttraumatic stress disorders in Vietnam veterans. *Journal of Consulting and Clinical Psychology, 51*(4), 488–494.

Mason, J. W., Giller, E. L., & Kosten, T. R. (1988). Elevation of urinary norepinephrine/cortisol ratio in posttraumatic stress disorder. *Journal of Nervous and Mental Disease, 176,* 498–502.

McFall, M. E., Murburg, M. M., Ko, G. N., & Veith, R. C. (1990). Autonomic responses to stress in Vietnam combat veterans with posttraumatic stress disorder. *Biological Psychiatry, 27,* 1165–1175.

McFall, M. E., Veith, R. C., & Murburg, M. M. (1992). Basal sympathoadrenal function in posttraumatic stress disorder. *Biological Psychiatry, 31,* 1050–1056.

Melia, K. R., & Duman, R. S. (1991). Involvement of corticotropin-releasing factor in chronic stress regulation of the brain noradrenergic system. *Proceedings of the National Academy of Sciences of the United States of America, 88,* 8382–8386.

Mellman, T. A., Kumar, A., Kulick-Bell, R., Kumar, M., & Nolan, B. (1995). Nocturnal/daytime urine norepinephrine measures and sleep in combat-related PTSD. *Biological Psychiatry, 38,* 174–179.

Milad, M. R., & Quirk, G. J. (2002). Neurons in medial prefrontal cortex signal memory for fear extinction. *Nature, 420,* 70–73.

Neylan, T. C., Brunet, A., Pole, N., Best, S. R., Metzler, T. J., Yehuda, R., et al. (2005). PTSD symptoms predict waking salivary cortisol levels in police officers. *Psychoneuroendocrinology, 30*(4), 373–381.

Nicholson, A. A., Sapru, I., Densmore, M., Frewen, P. A., Neufeld, R. W. J., Théberge, J., et al. (2016). Unique insula subregion resting-state functional connectivity with amygdala complexes in posttraumatic stress disorder and its dissociative subtype. *Psychiatry Research: Neuroimaging, 250*, 61–72.

Orr, S. P., Lasko, N. B., Metzger, L. J., Ahern, C. E., Berry, N. J., & Pitman, R. K. (1998). Psychophysiologic assessment of women with posttraumatic stress disorder resulting from childhood sexual abuse. *Journal of Consulting and Clinical Psychology, 66*(6), 906–913.

Orr, S. P., Lasko, N. B., Shalev, A. Y., & Pitman, R. K. (1995). Physiological responses to loud tones in Vietnam veterans with posttraumatic stress disorder. *Journal of Abnormal Psychology, 104*, 75–82.

Orr, S. P., Pitman, R. K., Lasko, N. B., & Herz, L. R. (1993). Psychophysiological assessment of posttraumatic stress disorder imagery in World War II and Korean combat veterans. *Journal of Abnormal Psychology, 102*, 152–159.

Orr, S. P., & Roth, W. T. (2000). Psychophysiological assessment: Clinical applications for PTSD. *Journal of Affective Disorders, 61*, 225–240.

Phan, K. L., Britton, J. C., Taylor, S. F., Fig, L. M., & Liberzon, I. (2006). Corticolimbic blood flow during nontraumatic emotional processing in posttraumatic stress disorder. *Archives of General Psychiatry, 63*(2), 184–192.

Pissiota, A., Frans, O., Fernandez, M., Von Knorring, L., Fischer, H., & Fredrikson, M. (2002). Neurofunctional correlates of posttraumatic stress disorder: A PET symptom provocation study. *European Archives of Psychiatry and Clinical Neuroscience, 252*, 68–75.

Pitman, R. K., & Orr, S. P. (1990). Twenty-four hour urinary cortisol and catecholamine excretion in combat-related posttraumatic stress disorder. *Biological Psychiatry, 27*, 245–247.

Plotsky, P. M., & Meaney, M. J. (1993). Early, postnatal experience alters hypothalamic corticotropin-releasing factor (CRF) mRNA, median eminence CRF content and stress-induced release in adult rats. *Molecular Brain Research, 18*(3), 195–200.

Protopopescu, X., Pan, H., Tuescher, O., Cloitre, M., Goldstein, M., Engelien, W., et al. (2005). Differential time courses and specificity of amygdala activity in posttraumatic stress disorder subjects and normal control subjects. *Biological Psychiatry, 57*(5), 464–473.

Quirk, G. J., Garcia, R., & Gonzalez-Lima, F. (2006). Prefrontal mechanisms in extinction of conditioned fear. *Biological Psychiatry, 60*(4), 337–343.

Rauch, S. L., Shin, L. M., & Phelps, E. A. (2006). Neurocircuitry models of posttraumatic stress disorder and extinction: Human neuroimaging research—Past, present, and future. *Biological Psychiatry, 60*(4), 376–382.

Rauch, S. L., van der Kolk, B. A., Fisler, R. E., Alpert, N. M., Orr, S. P., Savage, C. R., et al. (1996). A symptom provocation study of posttraumatic stress disorder using positron emission tomography and script driven imagery. *Archives of General Psychiatry, 53*, 380–387.

Rauch, S. L., Whalen, P. J., Shin, L. M., McInerney, S. C., Macklin, M. L., Lasko, N. B., et al. (2000). Exaggerated amygdala response to masked facial stimuli in posttraumatic stress disorder: A functional MRI study. *Biological Psychiatry, 47*(9), 769–776.

Richters, J. E., & Cicchetti, D. (1993). Mark Twain meets DSM-III-R: Conduct disorder, development, and the concept of harmful dysfunction. *Development and Psychopathology, 5*, 5–29.

Richters, J. E., & Hinshaw, S. P. (1999). The abduction of disorder in psychiatry. *Journal of Abnormal Psychology, 108*(3), 438–445.

Sakamoto, H., Fukuda, R., Okuaki, T., Rogers, M., Kasai, K., Machida, T., et al. (2005). Parahippocampal activation evoked by masked traumatic images in posttraumatic stress disorder: A functional MRI study. *NeuroImage*, *26*(3), 813–821.

Sautter, F. J., Bissette, G., Wiley, J., Manguno-Mire, G., Schoenbachler, B., Myers, L., et al. (2003). Corticotropin-releasing factor in posttraumatic stress disorder (PTSD) with secondary psychotic symptoms, nonpsychotic PTSD, and healthy control subjects. *Biological Psychiatry*, *54*(12), 1382–1388.

Schmahl, C. G., & Bremner, J. D. (2006). Neuroimaging in borderline personality disorder. *Journal of Psychiatric Research*, *40*(5), 419–427.

Schmahl, C. G., Elzinga, B. M., & Bremner, J. D. (2002). Individual differences in psychophysiological reactivity in adults with childhood abuse. *Clinical Psychology and Psychotherapy*, *9*, 271–276.

Schmahl, C. G., Elzinga, B., Ebner, U., Haaf, B., Sanislow, C., McGlashan, T. H., et al. (2004). Psychophysiological reactivity to traumatic and abandonment scripts in borderline personality disorder and posttraumatic stress disorder: A preliminary report. *Psychiatry Research*, *126*(1), 33–42.

Schmahl, C. G., Elzinga, B. M., Vermetten, E., Sanislow, C., McGlashan, T. H., & Bremner, J. D. (2003). Neural correlates of memories of abandonment in women with and without borderline personality disorder. *Biological Psychiatry*, *54*, 42–51.

Schmahl, C. G., McGlashan, T., & Bremner, J. D. (2002). Neurobiological correlates of borderline personality disorder. *Psychopharmacology Bulletin*, *36*, 69–87.

Schmahl, C. G., Vermetten, E., Elzinga, B. M., & Bremner, J. D. (2003). Magnetic resonance imaging of hippocampal and amygdala volume in women with childhood abuse and borderline personality disorder. *Psychiatry Research: Neuroimaging*, *122*, 193–198.

Schmahl, C. G., Vermetten, E., Elzinga, B. E., & Bremner, J. D. (2004). A positron emission tomography study of memories of childhood abuse in borderline personality disorder. *Biological Psychiatry*, *55*(7), 759–765.

Semple, W. E., Goyer, P., McCormick, R., Donovan, B., Muzic, R. F., Rugle, L., et al. (2000). Higher brain blood flow at amygdala and lower frontal cortex blood flow in PTSD patients with comorbid cocaine and alcohol abuse compared to controls. *Psychiatry*, *63*, 65–74.

Shalev, A. Y., Sahar, T., Freedman, S., et al. (1998). A prospective study of heart rate responses following trauma and the subsequent development of posttraumatic stress disorder. *Archives of General Psychiatry*, *55*, 553–559.

Shin, L. M., Kosslyn, S. M., McNally, R. J., Alpert, N. M., Thompson, W. L., Rauch, S. L., et al. (1997). Visual imagery and perception in posttraumatic stress disorder: A positron emission tomographic investigation. *Archives of General Psychiatry*, *54*, 233–237.

Shin, L. M., McNally, R. J., Kosslyn, S. M., Thompson, W. L., Rauch, S. L., Alpert, N. M., et al. (1999). Regional cerebral blood flow during script-driven imagery in childhood sexual abuse-related PTSD: A PET investigation. *The American Journal of Psychiatry*, *156*, 575–584.

Shin, L. M., Orr, S. P., Carson, M. A., Rauch, S. L., Macklin, M. L., Lasko, N. B., et al. (2004). Regional cerebral blood flow in the amygdala and medial prefrontal cortex during traumatic imagery in male and female Vietnam veterans with PTSD. *Archives of General Psychiatry*, *61*(2), 168–176.

Shin, L. M., Whalen, P. J., Pitman, R. K., Bush, G., Macklin, M. L., Lasko, N. B., et al. (2001). An fMRI study of anterior cingulate function in posttraumatic stress disorder. *Biological Psychiatry*, *50*, 932–942.

Shin, L. M., Wright, C. I., Cannistraro, P. A., Wedig, M. M., McMullin, K., Martis, B., et al. (2005). A functional magnetic resonance imaging study of amygdala and medial prefrontal cortex responses to overtly presented fearful faces in posttraumatic stress disorder. *Archives of General Psychiatry*, *62*(3), 273–281.

Smith, M. A., Davidson, R., Ritchie, J. C., Kudler, H., Lipper, S., Chappell, P., et al. (1989). The corticotropin-releasing hormone test in patients with posttraumatic stress disorder. *Biological Psychiatry*, *26*, 349–355.
Southwick, S. M., Krystal, J. H., Bremner, J. D., Morgan, C. A., Nicolaou, A., Nagy, L. M., et al. (1997). Noradrenergic and serotonergic function in posttraumatic stress disorder. *Archives of General Psychiatry*, *54*, 749–758.
Southwick, S. M., Krystal, J. H., Morgan, C. A., Johnson, D., Nagy, L. M., Nicolaou, A., et al. (1993). Abnormal noradrenergic function in posttraumatic stress disorder. *Archives of General Psychiatry*, *50*(4), 266–274.
Stanton, M. E., Gutierrez, Y. R., & Levine, S. (1988). Maternal deprivation potentiates pituitary-adrenal stress responses in infant rats. *Behavioral Neuroscience*, *102*, 692–700.
Stein, M. B., Yehuda, R., Koverola, C., & Hanna, C. (1997). Enhanced dexamethasone suppression of plasma cortisol in adult women traumatized by childhood sexual abuse. *Biological Psychiatry*, *42*, 680–686.
Teasdale, J. D., Segal, A., & Williams, J. M. G. (1995). How does cognitive therapy prevent depressive relapse and why should attentional control (mindfulness) training help? *Behavior Research and Therapy*, *33*(1), 25–39.
Teasdale, J. D., Segal, Z. V., Williams, J. M. G., Ridgeway, V. A., Soulsby, J. M., & Lau, M. A. (2000). Prevention of relapse/recurrence in major depression by mindfulness-based cognitive therapy. *Journal of Consulting and Clinical Psychology*, *68*, 615–623.
Vermetten, E. (2008). Epilogue: Neuroendocrinology of PTSD. *Progress in Brain Research*, *167*, 311–313. https://doi.org/10.1016/s0079-6123(07)67030-7.
Vermetten, E., Schmahl, C., Lindner, S., Loewenstein, R. J., & Bremner, J. D. (2006). Hippocampal and amygdalar volumes in dissociative identity disorder. *The American Journal of Psychiatry*, *163*, 1–8.
Vermetten, E., Schmahl, C., Southwick, S. M., & Bremner, J. D. (2007). Positron tomographic emission study of olfactory induced emotional recall in veterans with and without combat-related posttraumatic stress disorder. *Psychopharmacology Bulletin*, *40*(1), 8–30.
Vythilingam, M., Anderson, E., Goddard, A., Woods, S. W., Staib, L. H., Charney, D. S., et al. (2000). Temporal lobe volumes in panic disorder-a quantitative magnetic resonance imaging study. *Psychiatry Research: Neuroimaging*, *99*, 75–82.
Vythilingam, M., Heim, C., Newport, C. D., Miller, A. H., Vermetten, E., Anderson, E., et al. (2002). Childhood trauma associated with smaller hippocampal volume in women with major depression. *The American Journal of Psychiatry*, *159*, 2072–2080.
Vythilingam, M., Nelson, E. E., Scaramozza, M., Waldeck, T., Hazlett, G., Southwick, S. M., et al. (2009). Reward circuitry in resilience to severe trauma: An fMRI investigation of resilient special forces soldiers. *Psychiatry Research*, *172*(1), 75–77. https://doi.org/10.1016/j.pscychresns.2008.06.008.
Weissbecker, I., Salmon, P., Studts, J. L., Floyd, A. R., Dedert, E. A., & Sephton, S. E. (2002). Mindfulness-based stress reduction and sense of coherence among women with fibromyalgia. *Journal of Clinical Psychology in Medical Settings*, *9*, 297–307.
Wessa, M., Rohleder, N., Kirschbaum, C., & Flor, H. (2006). Altered cortisol awakening response in posttraumatic stress disorder. *Psychoneuroendocrinology*, *31*(2), 209–215.
Yang, P., Wu, M. T., Hsu, C. C., & Ker, J. H. (2004). Evidence of early neurobiological alternations in adolescents with posttraumatic stress disorder: A functional MRI study. *Neuroscience Letters*, *370*(1), 13–18.
Yehuda, R. (2002). Post-traumatic stress disorder. *New England Journal of Medicine*, *346*, 108–114.
Yehuda, R. (2006). Advances in understanding neuroendocrine alterations in PTSD and their therapeutic implications. *Annals of the New York Academy of Sciences*, *1071*, 137–166.

Yehuda, R., Boisoneau, D., Lowy, M. T., & Giller, E. L., Jr. (1995). Dose response changes in plasma cortisol and lymphocyte glucocorticoid receptors following dexamethasone administration in combat veterans with and without posttraumatic stress disorder. *Archives of General Psychiatry*, *52*, 583–593.

Yehuda, R., Halligan, S. L., Grossman, R., Golier, J. A., & Wong, C. (2002). The cortisol and glucocorticoid receptor response to low dose dexamethasone administration in aging combat veterans and holocaust survivors with and without posttraumatic stress disorder. *Biological Psychiatry*, *52*(2), 393–403.

Yehuda, R., Kahana, B., Binder-Brynes, K., Southwick, S., Mason, J. W., & Giller, E. L. (1995). Low urinary cortisol excretion in holocaust survivors with posttraumatic stress disorder. *The American Journal of Psychiatry*, *152*, 982–986.

Yehuda, R., Levengood, R. A., Schmeidler, J., Wilson, S., Guo, L. S., & Gerber, D. (1996). Increased pituitary activation following metyrapone administration in posttraumatic stress disorder. *Psychoneuroendocrinology*, *21*, 1–16.

Yehuda, R., Lowy, M. T., Southwick, S. M., Shaffer, D., & Giller, E. L. (1991). Lymphocyte glucocorticoid receptor number in posttraumatic stress disorder. *The American Journal of Psychiatry*, *148*(4), 499–504.

Yehuda, R., Siever, L. J., & Teicher, M. H. (1998). Plasma norepinephrine and 3-methoxy-4-hydroxyphenylglycol concentrations and severity of depression in combat posttraumatic stress disorder and major depressive disorder. *Biological Psychiatry*, *44*, 56–63.

Yehuda, R., Southwick, S. M., Krystal, J. H., Bremner, J. D., Charney, D. S., & Mason, J. (1993). Enhanced suppression of cortisol with low dose dexamethasone in posttraumatic stress disorder. *American Journal of Psychiatry*, *150*, 83–86.

Yehuda, R., Teicher, M. H., Levengood, R. A., Trestman, R. L., & Siever, L. J. (1994). Circadian regulation of basal cortisol levels in posttraumatic stress disorder. *Annals of the New York Academy of Sciences*, *746*(1), 378–380.

Yehuda, R., Teicher, M. H., Trestman, R. L., Levengood, R. A., & Siever, L. J. (1996). Cortisol regulation in posttraumatic stress disorder and major depression: A chronobiological analysis. *Biological Psychiatry*, *40*(2), 79–88.

CHAPTER TWO

Is there neuroinflammation in depression? Understanding the link between the brain and the peripheral immune system in depression

Maria Antonietta Nettis[a,b], **Carmine Maria Pariante**[a,b,*]
[a]King's College London, Institute of Psychiatry, Psychology and Neuroscience, Department of Psychological Medicine, London, United Kingdom
[b]National Institute for Health Research, Biomedical Research Centre, South London and Maudsley NHS Foundation Trust and King's College London, London, United Kingdom
*Corresponding author: e-mail address: carmine.pariante@kcl.ac.uk

Contents

Abstract

Compelling evidence have highlighted the role of inflammation as a possible mechanism linking environmental stress to the development of depression. In particular, the communication between the peripheral and the brain immune system might lead to brain inflammatory processes, in turn causing impaired neurogenesis and neural plasticity. As a consequence, measuring brain inflammation and its possible correlation with peripheral inflammatory processes has become the focus (and a challenge) for a number of recent studies. In this chapter we review the evidence on the link between

International Review of Neurobiology, Volume 152
ISSN 0074-7742
https://doi.org/10.1016/bs.irn.2019.12.004

stress, peripheral and brain inflammation and the way to measure it, through preclinical, post-mortem and clinical models of depression and in healthy humans. We describe the concept of microglial activation as a marker of neuroinflammation and the potential use of anti-inflammatory treatments in depression. The paper concludes by highlighting the unresolved questions and challenges for future studies.

1. The immune pathway to the brain

The role of the immune system in the pathogenesis of depression is one of the most important recent developments in translational mental health, offering a unique opportunity to identify new therapeutic targets (Slavich & Irwin, 2014). There is strong evidence demonstrating increased levels of peripheral inflammatory biomarkers in patients with depression (Baumeister, Russell, Pariante, & Mondelli, 2014). Such evidence includes the presence of increased levels of pro-inflammatory markers (Tumor necrosis factor-α (TNF-α), c-reactive protein (CRP), Interleukin-6 (IL-6)) among depressed individuals (Baumeister et al., 2014); the higher risk of depression in inflammatory and autoimmune diseases (Benros et al., 2013; Dickens, McGowan, Clark-Carter, & Creed, 2002; Korczak, Pereira, Koulajian, Matejcek, & Giacca, 2011); the association between a history of early life trauma—one of the more powerful risk factors for depression—with increased inflammation (Baumeister, Akhtar, Ciufolini, Pariante, & Mondelli, 2016; Danese & Lewis, 2017); and the ability of immune-modulators, such as interferon (IFN)-alpha, to induce depressive symptoms in up 50% of patients receiving it for hepatitis or cancer (Raison & Miller, 2011).

Inflammation is usually defined as the body's primary response to physical injury or infection. However, it is now clear that psychological stress can trigger significant increases in inflammatory activity in the absence of physical injury (Glaser & Kiecolt-Glaser, 2005; Slavich & Irwin, 2014). In particular, a large body of literature has highlighted the association between both early-life stress and adult-life stress with increased levels of inflammatory markers (Slavich & Irwin, 2014). In turn, increases in inflammation can elicit profound changes in behavior, which include the initiation of depressive symptoms such as sad mood, anhedonia, fatigue, psychomotor retardation, and social-behavioral withdrawal. These findings thus support the hypothesis that environmental conditions, may regulate molecular processes that initiate biological changes and behavioral changes that increase risk for depression.

Over the last years, a number of biological pathways have been suggested to explain how inflammation can lead to behavioral changes, culminating in the development of depression. In particular, the ability of peripheral cytokines and immune cells to act on the brain (Dantzer, 2009; Schedlowski, Engler, & Grigoleit, 2014), has been investigated.

Cytokines are small proteins, secreted by various cell types, with several roles during development, homeostasis, and immune regulation. During inflammation, cytokines serve as chemotactic factors that moderate cell-to-cell communication and regulate immune cell differentiation. Animal studies suggest different cytokines pathways to the brain. Two major pathways have been described: the 'humoral pathway' and the 'neural pattern.' The former involves cytokines passage through leaky regions in the blood-brain-barrier (BBB), such as the circumventricular organs, and the binding of cytokines to saturable transport molecules on the BBB; the latter involves the binding of cytokines to peripheral afferent nerve fibers, such as the vagus nerve, that in turn stimulate ascending catecholaminergic fibers in the brain and/or are translated back into central cytokine signals (Miller & Raison, 2016). Also, a cellular pathway may play an important role as well, involving the recruitment of activated immune cells to the brain vasculature and parenchyma (cellular pathway) (Miller & Raison, 2016). In humans, the same pathways have just started to be investigated, as well as the role of the brain immune response in the development of depression. Most importantly, a variety of approaches (neuroimaging, postmortem, cerebrospinal fluid analyses) have been used to investigate the presence of neuroinflammation in the brain of depressed patients, with different approaches at times producing different results.

In this chapter, we will review the (often conflicting) evidence around the presence of brain inflammation (neuroinflammation) in depression, and then discuss the mechanisms of communication between the brain and the peripheral immune system which might underpin the neuroinflammation.

2. Markers of neuroinflammation in depression

So far, few studies have investigated neuroinflammation in depression, but these studies focus on different markers, with resulting inconsistent results. For example, some research focuses on cytokines levels in the cerebrospinal fluid (CSF), as patients with depression tend to have increased levels of interleukin-6 (IL-6), tumor necrosis factor-alpha (TNF-α) and interleukin-8 (IL-8) (Enache, Pariante, & Mondelli, 2019).

Other studies investigate functions of brain immune cells, such as microglia, astrocytes and oligodendroglia, using neuroimaging or postmortem immunocytochemistry.

Microglia are the brain resident macrophages and are responsible for the immune response in the central nervous system (Frost & Schafer, 2016; Miller & Raison, 2016). In the healthy brain, microglia present with a quiescent phenotype where processes extend through the local environment to detect context-specific changes. During this phase, microglia release neurotrophic factors, provide axonal guidance and regulate local cells proliferation. However, in response to inflammatory stimuli, activated microglial cells undergo morphological changes and secrete pro-inflammatory cytokines, including IL-6 and TNF-α. The difference between these 2 phenotypes, has led to the distinction between the M1—pro-inflammatory—and the M2 anti-inflammatory phenotype. However, this view has been recently challenged, as another hypothesis is that microglial phenotype might vary along a continuum (Ransohoff, 2016).

Astrocytes are the most numerous glial cells and can be activated by pro-inflammatory cytokines produced by microglia. Astrocytes have several active roles in the brain, including phagocytic properties and secretion of pro-inflammatory cytokines (IL-6, TNF-α) (Rajkowska and Stockmeier, 2013).

Finally, oligodendroglia are glial cells deriving from oligodendrocytes progenitor cells and playing a key role in myelination (Mechawar & Savitz, 2016). Post-injury, it can produce cytokines (IL-1) and chemokines (monocytes chemoattractant protein 1) (Mechawar & Savitz, 2016).

Among the brain immune cells, microglia plays a key role, as microglial activation has been shown, so far, to be a particularly promising marker of neuroinflammation, and thus we will review here the studies that have examined these cells in depression.

3. Is there microglia activation in depression?

The opportunity to measure neuroinflammation allows to investigate the hypothetical model linking inflammation to the development of depression. According to this model, the communication between peripheral pro-inflammatory cytokines and the brain culminates in microglial activation and in the development of depressive symptoms (Fig. 1). Activated microglial cells not only promote the release of cytokines and chemokines, but also reduce levels of serotonin and generate oxidative stress molecules, altering

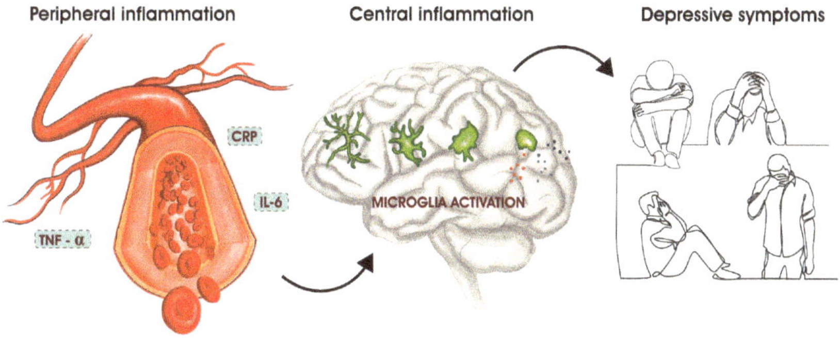

Fig. 1 Model of the link between peripheral inflammation, brain microglial activation and depression.

local metabolic processes. Moreover, microglia are the predominant cells expressing the enzyme kynurenine 3-monooxygenase (KMO) of the kynurenine pathway (KP, explained more in details below) which generates neurotoxic kynurenine metabolites (Madeeh Hashmi, Awais Aftab, Mazhar, Umair, & Butt, 2013).

In rodents, microglia activation is often measured in terms of an increase in the microglia or macrophage cell surface marker, allograft inflammatory factor 1 (Iba1) in different brain regions; post-mortem studies in humans usually focus on immunostaining or gene expression of major histocompatibility complexes (MHC or HLA), class II, instead, because the induction of this marker in microglia is considered a sensitive marker of neuroinflammation (Mondelli, Vernon, Turkheimer, Dazzan, & Pariante, 2017).

More interestingly, microglial activation can be measured in vivo, by using positron emission tomography (PET). Some PET radioligands are able to target the translocator protein (TSPO), a protein located on outer mitochondrial membranes in microglia. TSPO is involved either directly or indirectly in numerous biological functions, including mitochondrial cholesterol transport and steroid hormone biosynthesis, porphyrin transport and heme synthesis, apoptosis, cell proliferation, and anion transport (Papadopoulos, Fan, & Zirkin, 2018). As described below, several TSPO-radioligands have been synthesized so far, with the purpose of detecting microglial activation (and thus, neuroinflammation) with an increasing signal-to-noise ratio. This is because TSPO appears to be overexpressed when microglial cells are activated during inflammatory processes (Albrecht, Granziera, Hooker, & Loggia, 2016). Moreover, increased TSPO expression is associated with increased expression of Iba1, in preclinical and post-mortem models of brain immune activation (Gulyas et al., 2009; Israel et al., 2016). However, TSPO is expressed in several cells in the Central Nervous System, so that a recent study suggested that increased TSPO expression in humans may reflect other types of responses to immune stimuli, rather than activation of microglia, for example, myeloid cells proliferation or an increased monocyte recruitment (Owen et al., 2017). For these and other reasons explained below, the interpretation of TSPO neuroimaging data requests particular caution.

3.1 Preclinical evidence

The initial evidence about microglial activation in models of stress and depression came from pre-clinical studies. Of course, findings from rodent models cannot provide a comprehensive understanding of the biology of human psychiatric disorders, but they allow the access to brain samples or tissues which would be invasive in humans, and at the same time also to investigate both genetic and environmental risk factors linked to psychiatric conditions in a controlled experimental setting. Moreover, they are not affected by potential confounders, such as exposure to medication, that usually complicate clinical studies in humans.

A number of studies have investigated the activation of microglia in animals by using either a model of psychosocial stress or systemic inflammatory stimuli.

In the first model, a recent review on rodent studies (Calcia et al., 2016) reported increases in the expression of Iba1 in the rodent hippocampus, under stress conditions, depending on the type and duration of the exposure to stress. Interestingly, an increased concentration of Iba1 is present in areas relevant to the neurobiology of depression, such as prefrontal cortex, nucleus accumbens, amygdala, and paraventricular nucleus.

In the model case, experiments in animals have demonstrated that microglial activation can be induced by a peripheral immune challenge; for example, many studies have used *Escherichia coli* lipopolysaccharide (LPS) or interferon-alpha (IFN-α) as systemic challenges to elicit microglial activation in rodents (Hoogland, Houbolt, van Westerloo, van Gool, & van de Beek, 2015; Norden, Trojanowski, Villanueva, Navarro, & Godbout, 2016). In particular, studies with IFN-α show an activation of microglia associated with depressive-like symptoms, which could both be prevented by the antibiotic minocycline (Zheng, Kaneko, & Sawamoto, 2015).

Animal studies can also be very useful to understand the mechanisms by which the exposure to stress or immune stimuli in animals corresponds to changes in TSPO in vivo, so that this information can be used for clinical studies. Recently, LPS-induced systemic inflammation in baboons was shown to induce an activation of microglia, as shown with a robust increase in TSPO binding (measured with [^{11}C]PBR28 PET) at 1 and 4h after intravenous injection of LPS (Hannestad et al., 2012).

3.2 Post-mortem studies

A number of studies have investigated brain immune cells alterations in depression using post-mortem brain tissue. However, post-mortem studies have produced inconsistent results. So far, five out of the eight studies performed did not found any changes in microglia related to depression, and three reported an increase in density of activated microglia in patients compared to controls. Interestingly, increased microglia activation has been reported in suicide victims suffering from depression (Torres-Platas, Cruceanu, Chen, Turecki, & Mechawar, 2014). Out of 24 studies on astrocytes in depression, 13 studies found a decrease in the astrocytes specific markers, 3 reported an increase, whereas 11 studies found similar levels between patients with MDD and controls. Finally, evidence on oligodendrocytes seems to point toward a reduction of the oligodendrocytes density in the prefrontal cortex and anterior cingulate of suicide victims with history of childhood abuse, and no difference between non-suicidal depressed patients and controls in other areas such as the white matter of ventromedial and ventral prefrontal cortex, substantia nigra, hippocampus (Enache et al., 2019).

The reason of such inconsistency between studies might be related to the differences in study design, in the analyzed brain areas, in the disorder stage and in the selection of specific patients subgroups. Indeed, it is interesting that some studies found evidence of cellular changes only in suicide victims suffering from depression or also suffering from bipolar disorder and schizophrenia (Schnieder et al., 2014), suggesting an association between suicide and neuroinflammation, independently from the psychiatric diagnosis. This is probably because stress could activate an immune response in the brain that could in turn put an individual at high risk for suicide. Similarly, inconsistency has been found in terms of cytokines expression in post-mortem brains, but the overall evidence pointed at an increased IL-6 mRNA in depressed patients victim of suicide (Hoyo-Becerra et al., 2013; Pandey et al., 2012).

4. Clinical studies

CSF studies report contrasting results on the levels of IL-6, TNF-α, IL-8 and IL1β in patients with depression. A meta-analysis of these studies reported significantly increased IL-6, TNF-α, IL-8 levels in patients compared with controls, but only IL-6 results did survive sensitivity analyses. No significant difference has been found in terms of chemokines between the two clinical groups (Enache et al., 2019).

As mentioned above, in vivo activation of microglia can now be investigated with PET radioligands targeting TSPO. First generation PET radiotracers, such as PK11195, are characterized by a low signal-to-noise ratio. Therefore, higher affinity, second generation ligands have been developed such as PBR28. However, their binding affinity depends on the rs6971 nucleotide polymorphism on the TSPO gene (Owen et al., 2012), and the quantification of the brain signal is confounded by the presence of abundant TSPO in endothelial cells and by the high ligand affinity for plasma proteins (Turkheimer et al., 2015).

As far as studies in patients with depression are concerned, the overall evidence seems to point toward an increased TSPO binding in depressed patients compared with healthy volunteers, (Holmes et al., 2018; Li, Sagar, & Keri, 2018; Richards et al., 2018; Setiawan et al., 2015; Su et al., 2016), despite some negative results (Hannestad et al., 2013). However, previous PET studies in depressed patients failed to find a correlation between central and peripheral inflammatory markers, so that the mechanisms underpinning these co-existing findings are still unclear (Enache et al., 2019). Moreover, the meaning of such TSPO increase is still controversial. As the research by Owen and his colleagues suggested, (Owen et al., 2017), TSPO signal can be driven by other factors, such as recruitment of peripheral monocytes into the parenchyma, adherence of circulating leucocytes to the vascular endothelium and the expression of TSPO in various CNS cells, including vascular endothelial cells and neurons. Indeed, it is likely that TSPO changes reflect not only alterations in the brain immune cells, but also anomalies in cells metabolism, mitochondria function or oxidative stress, which have been associated with depression (Bansal & Kuhad, 2016; Black, Bot, Scheffer, Cuijpers, & Penninx, 2015). Unfortunately, the lack of cellular specificity is often neglected when interpreting PET studies, as well as the assessment of blood-brain barrier permeability and the correction for the endothelial component of TSPO signal.

5. The importance of human models of the link between peripheral and brain inflammation

Studies on clinical populations are unfortunately confounded by the medications effect, the heterogeneity of disease stage and several physical comorbidities that patients with depression may present. Thus, developing models of the link between the body and the brain immune system in healthy humans is increasingly needed.

Such models have started to be tested on the basis of previous TSPO-PET studies in non-human primates (Hannestad et al., 2012; Hillmer et al., 2017) showing a significant correlation between peripheral cytokines levels and the TSPO binding of the tracer [^{11}C] PBR28 in the brain, following LPS administration.

In order to replicate such findings in humans, a PET study was performed following LPS injection in healthy human males (Sandiego et al., 2015). Results showed that the tracer TSPO binding increased by 30–60% at 3 h after LPS, suggesting microglia activation. However, in this experiment, the TSPO binding did not correlate with peripheral inflammation nor with sickness behavior measurements.

In a recent study from our group (under revision), we used the same approach of peripherally inducing a systemic inflammation in healthy humans to study the effects on brain microglia, but we used interferon (IFN)-α as immune challenge. IFN-α is a pro-inflammatory cytokine with antiviral, anticancer, and immunomodulatory effects, approved for treatment of several malignancies and chronic hepatitis C (Machado et al., 2017; Piper, Wen, & Xenakis, 2001). Like LPS, IFN-α induces a systemic immune response associated with sickness behavior, but has the advantage of being a well-established model of immune-related depression (Bonaccorso et al., 2001; Raison, Demetrashvili, Capuron, & Miller, 2005). Preclinical studies have also shown that peripheral IFN-α treatment in mice induces activation of microglia, and that this activation is specifically associated with depressive-like behavior (Wachholz et al., 2016; Zheng et al., 2015). In this study, we also applied a methodology to correct for all limitations affecting TSPO studies, including the endothelial and the plasma protein binding of the radioligand. Surprisingly, however, IFN-α administration was not followed by an increase in TSPO binding and the signal was significantly affected by the radioligand peripheral plasma binding.

Such results question the role of TSPO as a sensitive target for measuring peripherally induced neuroinflammation, and add new information to the ongoing debate on TSPO-imaging data interpretation. Indeed, as recently outlined by Notter, Coughlin, Sawa, and Meyer (2018), altered levels of TSPO might not necessarily reflect selective changes in microglial activity. It is also very likely that the pathological meanings of altered TSPO binding or expression are disease-specific, and therefore, not easily generalizable across different neuropathologies or inflammatory conditions. For example, TSPO might increase and reflect microglial activation after LPS injection or in conditions characterized by genuine neuroinflammatory reactions, such as multiple sclerosis and neurodegenerative disorders. However, its role and changes in psychiatric disorders, typically associated to low-grade inflammation, might be different. This could explain why studies in clinical populations have failed to find a correlation between peripheral and brain immune biomarkers.

5.1 New potential measures of neuroinflammation

The several limitations of TSPO binding radiotracers have led to identifying other molecular biomarkers of activated microglia that would likely be complementary to TSPO PET imaging, for example, the P2X7 receptor (Purinergic ion channel receptor 7) (Tronel et al., 2017). This receptor expression is increased in microglia in animal models of neurodegenerative disorders and has been also proposed as marker of microglia M1 polarization. Studying this new target could help clarifying whether different subtypes of microglia, with their related pro/anti-inflammatory roles, exist or whether they vary along a continuum.

Other promising new targets cannot be imaged with PET, yet, but these molecules can be a useful indicators of microglial activation, for instance, the enzymes of the kynurenine pathway, which are overexpressed by activated microglia (Tronel et al., 2017). The kynurenine pathway (KP) regulates tryptophan metabolism in both periphery and CNS. KP is stimulated by inflammatory molecules, and, in the brain, products of KP can be neuroprotective (picolinic acid, kynurenic acid) or neurotoxic (quinolinic acid, 3-hydroxykinurenine) (Tronel et al., 2017).

Interestingly, Indoleamine 2,3-dioxygenase-1 (IDO-1), the first limiting enzymes of KP, is expressed in immune peripheral and brain cells, in particular in microglia cells, astrocytes, and neurons. IDO-1 is strongly induced in primary microglial cells by LPS and IFN-γ (Wang, Lawson, Kelley, & Dantzer, 2010) and is overexpressed in neurodegenerative diseases, such as Multiple Sclerosis (Mancuso et al., 2015), Alzheimer's disease (Bonda et al., 2010), and Parkinson's disease (Bonda et al., 2010). This all make IDO-1 a serious candidate to follow microglia activation in vivo. Other methods currently under development include: the synthesis of labeled tryptophan analogs to follow its metabolism in animal models; the synthesis of IDO-1 tracers; and the synthesis of novel tracers for the kynurenine-3-monooxygenase (KMO), another enzyme involved in KP and regulating the production of neurotoxic products. Notably, KMO is overexpressed, in vitro, in macrophages and microglia activated by IFN-γ (Tronel et al., 2017). Although they show great potentials, the study of these new targets of microglial activation is still at its embryonal stage.

5.1.1 MRI techniques potential

MRI methods represent a cheaper and less invasive alternative to PET imaging to measure neuroinflammation. They rely on physical properties of water and white matter and can be used to provide indirect measures of inflammation in the brain, such as BBB permeability, cellular infiltration, gliotic reaction and interstitial microenvironment change (Quarantelli, 2015).

For example, T2 relaxation time signal might provide indirect evidence of neuroinflammation. This is because T2-weighted images are sensitive to molecular motion and to interactions between neighboring molecules. As such, they provide high contrast for cerebrospinal fluid which has application in pathologies characterized by water accumulated within tissue. However, the T2 signal increase could be due to other molecular changes that occur in tissue composition and in iron, so the exact interpretation of its contrast remains uncertain.

Another example is Diffusion tensor imaging (DTI) that investigates white matter microstructure. DTI is an unspecific technique as well, as the two main measures provided, fractional anisotropy and mean diffusivity, are not specific, and can reflect alterations in fiber organization, myelination or membrane permeability. More recently, free-water fraction DTI has been proposed to reflect more accurately the presence of neuroinflammation in white matter (Pasternak, Sochen, Gur, Intrator, & Assaf, 2009).

Overall, MRI studies investigating of neuroinflammation in patients with depression are still lacking, but hopefully, the development of new MRI techniques will eventually help to dissect the various components of the neuroinflammatory processes involved in the pathophysiology of depression by examining a much larger number of patients, due to their lack of invasiveness.

6. Clinical implications

Given the role of inflammation in the development of depression (or at least in a subgroup of depressed individuals), anti-inflammatory agents have been suggested as potential adjunctive treatment for depression. Clinical trials on patients with depression suggest that anti-inflammatory medications have stronger antidepressant effects compared with placebo, either as monotherapy or as add-on treatment to antidepressants (Kohler, Krogh, Mors, & Benros, 2016). In particular, the second generation tetracyclic antibiotic minocycline has recently gained interest because of its ability to cross the BBB and to exert neuroprotective activities, including increased neurogenesis and down-regulation of the KP (Pae, Marks, Han, & Patkar, 2008). Moreover, some studies have shown that minocycline reduces microglial activation after stress or immune stimulation. Another study showed that minocycline prevents both microglial activation and depressive behavior due to chronic stress in rats (Mondelli et al., 2017).

To date, four clinical trials have assessed the antidepressant effect of minocycline as an add-on therapy (2 randomized controlled trials (RCTs) and 2 open label studies) and as monotherapy. Overall, a clear antidepressant effect is observed for minocycline compared with placebo, with no difference in side effects (Rosenblat & McIntyre, 2018). However, these studies presented small sample sizes and did not select patients a-priori on the basis of inflammation. Indeed, a similar study conducted by Raison et al. (2013) on the cytokines inhibitor Infliximab indicated that only patients with CRP levels >5 mg/L, indicating some levels of peripheral inflammation, showed improvement in depressive symptoms when compared with placebo. Similarly, it would be important to identify specific subgroups that would particularly benefit from minocycline. That is why a clinical trial using minocycline on depressed patients selected for increased levels of CRP is currently ongoing in our lab (EudraCT Number: 2015-003413-26).

Also P2X7 receptor inhibitors are currently be tested as a promising anti-inflammatory treatment for treatment resistant depression. Ongoing clinical trials are testing the efficacy of these new drugs on depressive symptoms, alongside standard antidepressants, in patients who do not respond to their current treatment. Again, a clinical trial testing a CNS-penetrant P2RX7 antagonist by Johnson and Johnson, JNJ-54175446, is currently ongoing in our laboratory and in other UK centres as part of a large multicenter consortium (EudraCT Number: 2018-001884-21).

Finally, there is now evidence that cognitive- and meditation-based interventions that are known to alleviate stress may also lead to reductions in inflammation. These interventions include mindfulness-based cognitive therapy, tai chi and yoga (Slavich & Irwin, 2014).

Moreover, in case of early-life stress, it is possible to early intervene on inflammation in order to prevent the biological embedding of disease. For example, a study from Pace et al. (2013) has shown that engagement with cognitively-based compassion training (CBCT) may buffer the detrimental effects of early life adversities on inflammation in adolescents placed in foster care (Pace et al., 2013). In this study, reductions in morning salivary CRP concentrations were associated with the number of CBCT practice sessions in the adolescents receiving CBCT. Thus, the degree of engagement with the practice—appeared to be influencing inflammation levels.

7. Conclusions

In conclusion, investigating the relationship between central and peripheral inflammation offers a key opportunity to understand the link between environmental stress and depression, and to create more personalized treatments. However, a correlation between peripheral and central measures of inflammation has not been found yet, and, more importantly, selective biomarkers or neuroinflammation are currently missing. Further research should invest in identifying more specific and possibly less invasive ways to measure inflammation in the brain. This could also help selecting patients who would benefit from specific anti-inflammatory treatment.

Acknowledgment

This research was funded by the National Institute of Health Research (NIHR) Maudsley Biomedical Research Centre (BRC), London, UK.

Conflicts of interest

Dr. Nettis declare no conflicts of interest. Dr. Pariante has received research funding from Janssen Pharmaceutical NV/Janssen Pharmaceutical Companies of Johnson & Johnson and speaker's fees from Lundbeck. Dr. Pariante has also received consultation fees from Consultant to Eleusis Benefit Corporation.

References

Albrecht, D. S., Granziera, C., Hooker, J. M., & Loggia, M. L. (2016). In vivo imaging of human Neuroinflammation. *ACS Chemical Neuroscience*, 7(4), 470–483. https://doi.org/10.1021/acschemneuro.6b00056.

Bansal, Y., & Kuhad, A. (2016). Mitochondrial dysfunction in depression. *Current Neuropharmacology*, *14*(6), 610–618.

Baumeister, D., Akhtar, R., Ciufolini, S., Pariante, C. M., & Mondelli, V. (2016). Childhood trauma and adulthood inflammation: A meta-analysis of peripheral C-reactive protein, interleukin-6 and tumour necrosis factor-alpha. *Molecular Psychiatry*, *21*(5), 642–649. https://doi.org/10.1038/mp.2015.67.

Baumeister, D., Russell, A., Pariante, C. M., & Mondelli, V. (2014). Inflammatory biomarker profiles of mental disorders and their relation to clinical, social and lifestyle factors. *Social Psychiatry and Psychiatric Epidemiology*, *49*(6), 841–849. https://doi.org/10.1007/s00127-014-0887-z.

Benros, M. E., Waltoft, B. L., Nordentoft, M., Ostergaard, S. D., Eaton, W. W., Krogh, J., et al. (2013). Autoimmune diseases and severe infections as risk factors for mood disorders: A nationwide study. *JAMA Psychiatry*, *70*(8), 812–820. https://doi.org/10.1001/jamapsychiatry.2013.1111.

Black, C. N., Bot, M., Scheffer, P. G., Cuijpers, P., & Penninx, B. W. (2015). Is depression associated with increased oxidative stress? A systematic review and meta-analysis. *Psychoneuroendocrinology*, *51*, 164–175. https://doi.org/10.1016/j.psyneuen.2014.09.025.

Bonaccorso, S., Puzella, A., Marino, V., Pasquini, M., Biondi, M., Artini, M., et al.Maes, M., (2001). Immunotherapy with interferon-alpha in patients affected by chronic hepatitis C induces an intercorrelated stimulation of the cytokine network and an increase in depressive and anxiety symptoms. *Psychiatry Research*, *105*(1–2), 45–55.

Bonda, D. J., Mailankot, M., Stone, J. G., Garrett, M. R., Staniszewska, M., Castellani, R. J., et al.Smith, M. A., (2010). Indoleamine 2,3-dioxygenase and 3-hydroxykynurenine modifications are found in the neuropathology of Alzheimer's disease. *Redox Report*, *15*(4), 161–168. https://doi.org/10.1179/174329210X12650506623645.

Calcia, M. A., Bonsall, D. R., Bloomfield, P. S., Selvaraj, S., Barichello, T., & Howes, O. D. (2016). Stress and neuroinflammation: A systematic review of the effects of stress on microglia and the implications for mental illness. *Psychopharmacology*, *233*(9), 1637–1650. https://doi.org/10.1007/s00213-016-4218-9.

Danese, A., & Lewis, S. J. (2017). Psychoneuroimmunology of early-life stress: The hidden wounds of childhood trauma? *Neuropsychopharmacology*, *42*(1), 99–114. https://doi.org/10.1038/npp.2016.198.

Dantzer, R. (2009). Cytokine, sickness behavior, and depression. *Immunology and Allergy Clinics of North America*, *29*(2), 247–264. https://doi.org/10.1016/j.iac.2009.02.002.

Dickens, C., McGowan, L., Clark-Carter, D., & Creed, F. (2002). Depression in rheumatoid arthritis: A systematic review of the literature with meta-analysis. *Psychosomatic Medicine*, *64*(1), 52–60.

Enache, D., Pariante, C., & Mondelli, V. (2019). Markers of central inflammation in major depressive disorder: A systematic review and meta-analysis of studies examining cerebrospinal fluid, positron emission tomography and post-mortem brain tissue. *Brain, Behavior, and Immunity*, *81*, 24–40. https://doi.org/10.1016/j.bbi.2019.06.015.

Frost, J. L., & Schafer, D. P. (2016). Microglia: Architects of the developing nervous system. *Trends in Cell Biology*, *26*(8), 587–597. https://doi.org/10.1016/j.tcb.2016.02.006.

Glaser, R., & Kiecolt-Glaser, J. K. (2005). Stress-induced immune dysfunction: Implications for health. *Nature Reviews Immunology*, *5*(3), 243–251. https://doi.org/10.1038/nri1571.

Gulyas, B., Makkai, B., Kasa, P., Gulya, K., Bakota, L., Varszegi, S., et al. (2009). A comparative autoradiography study in post mortem whole hemisphere human brain slices taken from Alzheimer patients and age-matched controls using two radiolabelled DAA1106 analogues with high affinity to the peripheral benzodiazepine receptor (PBR) system. *Neurochemistry International*, *54*(1), 28–36. https://doi.org/10.1016/j.neuint.2008.10.001.

Hannestad, J., DellaGioia, N., Gallezot, J. D., Lim, K., Nabulsi, N., Esterlis, I., et al. (2013). The neuroinflammation marker translocator protein is not elevated in individuals with mild-to-moderate depression: A [(1)(1)C]PBR28 PET study. *Brain, Behavior, and Immunity*, *33*, 131–138. https://doi.org/10.1016/j.bbi.2013.06.010.

Hannestad, J., Gallezot, J. D., Schafbauer, T., Lim, K., Kloczynski, T., Morris, E. D., et al. (2012). Endotoxin-induced systemic inflammation activates microglia: [(1)(1)C]PBR28 positron emission tomography in nonhuman primates. *NeuroImage*, *63*(1), 232–239. https://doi.org/10.1016/j.neuroimage.2012.06.055.

Hillmer, A. T., Holden, D., Fowles, K., Nabulsi, N., West, B. L., Carson, R. E., et al. (2017). Microglial depletion and activation: A [(11)C]PBR28 PET study in nonhuman primates. *EJNMMI Research*, 7(1), 59. https://doi.org/10.1186/s13550-017-0305-0.

Holmes, S. E., Hinz, R., Conen, S., Gregory, C. J., Matthews, J. C., Anton-Rodriguez, J. M., et al. (2018). Elevated translocator protein in anterior cingulate in major depression and a role for inflammation in suicidal thinking: A positron emission tomography study. *Biological Psychiatry*, *83*(1), 61–69. https://doi.org/10.1016/j.biopsych.2017.08.005.

Hoogland, I. C., Houbolt, C., van Westerloo, D. J., van Gool, W. A., & van de Beek, D. (2015). Systemic inflammation and microglial activation: Systematic review of animal experiments. *Journal of Neuroinflammation*, *12*, 114. https://doi.org/10.1186/s12974-015-0332-6.

Hoyo-Becerra, C., Huebener, A., Trippler, M., Lutterbeck, M., Liu, Z. J., Truebner, K., et al. (2013). Concomitant interferon alpha stimulation and TLR3 activation induces neuronal expression of depression-related genes that are elevated in the brain of suicidal persons. *PLoS One*, *8*(12), e83149. https://doi.org/10.1371/journal.pone.0083149.

Israel, I., Ohsiek, A., Al-Momani, E., Albert-Weissenberger, C., Stetter, C., Mencl, S., et al. (2016). Combined [(18)F]DPA-714 micro-positron emission tomography and autoradiography imaging of microglia activation after closed head injury in mice. *Journal of Neuroinflammation*, *13*(1), 140. https://doi.org/10.1186/s12974-016-0604-9.

Kohler, O., Krogh, J., Mors, O., & Benros, M. E. (2016). Inflammation in depression and the potential for anti-inflammatory treatment. *Current Neuropharmacology*, *14*(7), 732–742.

Korczak, D. J., Pereira, S., Koulajian, K., Matejcek, A., & Giacca, A. (2011). Type 1 diabetes mellitus and major depressive disorder: Evidence for a biological link. *Diabetologia*, *54*(10), 2483–2493.

Li, H., Sagar, A. P., & Keri, S. (2018). Translocator protein (18kDa TSPO) binding, a marker of microglia, is reduced in major depression during cognitive-behavioral therapy. *Progress in Neuro-Psychopharmacology & Biological Psychiatry*, *83*, 1–7. https://doi.org/10.1016/j.pnpbp.2017.12.011.

Machado, M. O., Oriolo, G., Bortolato, B., Kohler, C. A., Maes, M., Solmi, M., et al. (2017). Biological mechanisms of depression following treatment with interferon for chronic hepatitis C: A critical systematic review. *Journal of Affective Disorders*, *209*, 235–245. https://doi.org/10.1016/j.jad.2016.11.039.

Madeeh Hashmi, A., Awais Aftab, M., Mazhar, N., Umair, M., & Butt, Z. (2013). The fiery landscape of depression: A review of the inflammatory hypothesis. *Pakistan Journal of Medical Sciences*, *29*(3), 877–884.

Mancuso, R., Hernis, A., Agostini, S., Rovaris, M., Caputo, D., Fuchs, D., et al. (2015). Indoleamine 2,3 dioxygenase (IDO) expression and activity in relapsing-remitting multiple sclerosis. *PLoS One*, *10*(6), e0130715. https://doi.org/10.1371/journal.pone.0130715.

Mechawar, N., & Savitz, J. (2016). Neuropathology of mood disorders: Do we see the stigmata of inflammation? *Translational Psychiatry*, *6*(11), e946. https://doi.org/10.1038/tp.2016.212.

Miller, A. H., & Raison, C. L. (2016). The role of inflammation in depression: From evolutionary imperative to modern treatment target. *Nature Reviews. Immunology*, *16*(1), 22–34. https://doi.org/10.1038/nri.2015.5.
Mondelli, V., Vernon, A. C., Turkheimer, F., Dazzan, P., & Pariante, C. M. (2017). Brain microglia in psychiatric disorders. *Lancet Psychiatry*, *4*(7), 563–572. https://doi.org/10.1016/S2215-0366(17)30101-3.
Norden, D. M., Trojanowski, P. J., Villanueva, E., Navarro, E., & Godbout, J. P. (2016). Sequential activation of microglia and astrocyte cytokine expression precedes increased Iba-1 or GFAP immunoreactivity following systemic immune challenge. *Glia*, *64*(2), 300–316. https://doi.org/10.1002/glia.22930.
Notter, T., Coughlin, J. M., Sawa, A., & Meyer, U. (2018). Reconceptualization of translocator protein as a biomarker of neuroinflammation in psychiatry. *Molecular Psychiatry*, *23*(1), 36–47. https://doi.org/10.1038/mp.2017.232.
Owen, D. R., Narayan, N., Wells, L., Healy, L., Smyth, E., Rabiner, E. A., et al. (2017). Pro-inflammatory activation of primary microglia and macrophages increases 18 kDa translocator protein expression in rodents but not humans. *Journal of Cerebral Blood Flow and Metabolism*, *37*(8), 2679–2690. https://doi.org/10.1177/0271678X17710182.
Owen, D. R., Yeo, A. J., Gunn, R. N., Song, K., Wadsworth, G., Lewis, A., et al. (2012). An 18-kDa translocator protein (TSPO) polymorphism explains differences in binding affinity of the PET radioligand PBR28. *Journal of Cerebral Blood Flow and Metabolism*, *32*(1), 1–5. https://doi.org/10.1038/jcbfm.2011.147.
Pace, T. W., Negi, L. T., Dodson-Lavelle, B., Ozawa-de Silva, B., Reddy, S. D., Cole, S. P., et al. (2013). Engagement with cognitively-based compassion training is associated with reduced salivary C-reactive protein from before to after training in foster care program adolescents. *Psychoneuroendocrinology*, *38*(2), 294–299. https://doi.org/10.1016/j.psyneuen.2012.05.019.
Pae, C. U., Marks, D. M., Han, C., & Patkar, A. A. (2008). Does minocycline have antidepressant effect? *Biomedicine & Pharmacotherapy*, *62*(5), 308–311. https://doi.org/10.1016/j.biopha.2007.12.005.
Pandey, G. N., Rizavi, H. S., Ren, X., Fareed, J., Hoppensteadt, D. A., Roberts, R. C., et al. (2012). Proinflammatory cytokines in the prefrontal cortex of teenage suicide victims. *Journal of Psychiatric Research*, *46*(1), 57–63. https://doi.org/10.1016/j.jpsychires.2011.08.006.
Papadopoulos, V., Fan, J., & Zirkin, B. (2018). Translocator protein (18 kDa): An update on its function in steroidogenesis. *Journal of Neuroendocrinology*, *30*(2). https://doi.org/10.1111/jne.12500.
Pasternak, O., Sochen, N., Gur, Y., Intrator, N., & Assaf, Y. (2009). Free water elimination and mapping from diffusion MRI. *Magnetic Resonance in Medicine*, *62*(3), 717–730. https://doi.org/10.1002/mrm.22055.
Piper, J. M., Wen, T. T., & Xenakis, E. M. (2001). Interferon therapy in primary care. *Primary Care Update for Ob/Gyns*, *8*(4), 163–169.
Quarantelli, M. (2015). MRI/MRS in neuroinflammation: Methodology and applications. *Clinical and Translational Imaging*, *3*, 475–489. https://doi.org/10.1007/s40336-015-0142-y.
Raison, C. L., Demetrashvili, M., Capuron, L., & Miller, A. H. (2005). Neuropsychiatric adverse effects of interferon-alpha: Recognition and management. *CNS Drugs*, *19*(2), 105–123. https://doi.org/10.2165/00023210-200519020-00002.
Raison, C. L., & Miller, A. H. (2011). Is depression an inflammatory disorder? *Current Psychiatry Reports*, *13*(6), 467–475. https://doi.org/10.1007/s11920-011-0232-0.

Raison, C. L., Rutherford, R. E., Woolwine, B. J., Shuo, C., Schettler, P., Drake, D. F., et al. (2013). A randomized controlled trial of the tumor necrosis factor antagonist infliximab for treatment-resistant depression: The role of baseline inflammatory biomarkers. *JAMA Psychiatry*, *70*(1), 31–41. https://doi.org/10.1001/2013.jamapsychiatry.4.

Rajkowska, G., & Stockmeier, C. A. (2013). Astrocyte pathology in major depressive disorder: Insights from human postmortem brain tissue. *Current Drug Targets*, *14*, 1225–1236.

Ransohoff, R. M. (2016). A polarizing question: Do M1 and M2 microglia exist? *Nature Neuroscience*, *19*(8), 987–991. https://doi.org/10.1038/nn.4338.

Richards, E. M., Zanotti-Fregonara, P., Fujita, M., Newman, L., Farmer, C., Ballard, E. D., et al. (2018). PET radioligand binding to translocator protein (TSPO) is increased in unmedicated depressed subjects. *EJNMMI Research*, *8*(1), 57. https://doi.org/10.1186/s13550-018-0401-9.

Rosenblat, J. D., & McIntyre, R. S. (2018). Efficacy and tolerability of minocycline for depression: A systematic review and meta-analysis of clinical trials. *Journal of Affective Disorders*, *227*, 219–225.

Sandiego, C. M., Gallezot, J. D., Pittman, B., Nabulsi, N., Lim, K., Lin, S. F., et al. (2015). Imaging robust microglial activation after lipopolysaccharide administration in humans with PET. *Proceedings of the National Academy of Sciences of the United States of America*, *112*(40), 12468–12473. https://doi.org/10.1073/pnas.1511003112.

Schedlowski, M., Engler, H., & Grigoleit, J. S. (2014). Endotoxin-induced experimental systemic inflammation in humans: A model to disentangle immune-to-brain communication. *Brain, Behavior, and Immunity*, *35*, 1–8. https://doi.org/10.1016/j.bbi.2013.09.015.

Schnieder, T. P., Trencevska, I., Rosoklija, G., Stankov, A., Mann, J. J., Smiley, J., et al. (2014). Microglia of prefrontal white matter in suicide. *Journal of Neuropathology and Experimental Neurology*, *73*(9), 880–890. https://doi.org/10.1097/NEN.0000000000000107.

Setiawan, E., Wilson, A. A., Mizrahi, R., Rusjan, P. M., Miler, L., Rajkowska, G., et al. (2015). Role of translocator protein density, a marker of neuroinflammation, in the brain during major depressive episodes. *JAMA Psychiatry*, *72*(3), 268–275. https://doi.org/10.1001/jamapsychiatry.2014.2427.

Slavich, G. M., & Irwin, M. R. (2014). From stress to inflammation and major depressive disorder: A social signal transduction theory of depression. *Psychological Bulletin*, *140*(3), 774–815. https://doi.org/10.1037/a0035302.

Su, L., Faluyi, Y. O., Hong, Y. T., Fryer, T. D., Mak, E., Gabel, S., et al. (2016). Neuroinflammatory and morphological changes in late-life depression: The NIMROD study. *The British Journal of Psychiatry*, *209*(6), 525–526. https://doi.org/10.1192/bjp.bp.116.190165.

Torres-Platas, S. G., Cruceanu, C., Chen, G. G., Turecki, G., & Mechawar, N. (2014). Evidence for increased microglial priming and macrophage recruitment in the dorsal anterior cingulate white matter of depressed suicides. *Brain, Behavior, and Immunity*, *42*, 50–59. https://doi.org/10.1016/j.bbi.2014.05.007.

Tronel, C., Largeau, B., Santiago Ribeiro, M. J., Guilloteau, D., Dupont, A. C., & Arlicot, N. (2017). Molecular targets for PET imaging of activated microglia: The current situation and future expectations. *International Journal of Molecular Sciences*, *18*(4). https://doi.org/10.3390/ijms18040802.

Turkheimer, F. E., Rizzo, G., Bloomfield, P. S., Howes, O., Zanotti-Fregonara, P., Bertoldo, A., et al. (2015). The methodology of TSPO imaging with positron emission tomography. *Biochemical Society Transactions*, *43*(4), 586–592. https://doi.org/10.1042/BST20150058.

Wachholz, S., Esslinger, M., Plumper, J., Manitz, M. P., Juckel, G., & Friebe, A. (2016). *Brain, Behavior, and Immunity*, *55*, 105–113. https://doi.org/10.1016/j.bbi.2015.09.016.

Wang, Y., Lawson, M. A., Kelley, K. W., & Dantzer, R. (2010). Primary murine microglia are resistant to nitric oxide inhibition of indoleamine 2,3-dioxygenase. *Brain, Behavior, and Immunity*, *24*(8), 1249–1253. https://doi.org/10.1016/j.bbi.2010.04.015.

Zheng, L. S., Kaneko, N., & Sawamoto, K. (2015). Minocycline treatment ameliorates interferon-alpha- induced neurogenic defects and depression-like behaviors in mice. *Frontiers in Cellular Neuroscience*, *9*, 5. https://doi.org/10.3389/fncel.2015.00005.

CHAPTER THREE

Chronic stress, structural exposures and neurobiological mechanisms: A stimulation, discrepancy and deprivation model of psychosis

Teresa Vargas, Rachel E. Conley, Vijay A. Mittal*
Northwestern University, Evanston, IL, United States
*Corresponding author: e-mail address: vijay.mittal@northwestern.edu

Contents

International Review of Neurobiology, Volume 152
ISSN 0074-7742
https://doi.org/10.1016/bs.irn.2019.11.004

Abstract

Chronic stress exposure has been established as a key vulnerability factor for developing psychotic disorders, including schizophrenia. A structural, or systems level perspective, has often been lacking in conceptualizations of chronic stress for psychotic disorders. The current review thus identified three subtypes of structural exposures. *Stimulation* exposures included urban environments, population density and crime exposure, with intermediary mechanisms of lack of safety and high attentional demands. Underlying neural mechanisms included threat neural circuits. *Discrepancy* exposures included environmental ethnic density, income inequality, and social fragmentation, with intermediary mechanisms of lack of belonging and social exclusion, and neural mechanisms including the oxytocin system. *Deprivation* exposures included environments lacking socioeconomic, educational, or material resources, with intermediary mechanisms of lack of needed environmental enrichment, and underlying neural mechanisms of over-pruning and protracted PFC development. Delineating stressor etiology at the systems level is a necessary step in reducing barriers to effective interventions and health policy.

1. Introduction

Psychotic disorders, including schizophrenia, are chronic and highly debilitating, estimated as one of the top 15 leading causes of disability worldwide (Vos et al., 2017). Psychotic disorders are characterized by marked decreases in cognitive, social and role functioning, accompanied by symptoms such as delusions, hallucinations, and disorganized communication. Risk factors for developing a psychotic disorder include genetic and biological pre-existing vulnerabilities, as well as chronic stress exposure, which can over time exacerbate pre-existing vulnerabilities and drive illness onset. Thus, chronic stress exposure has long been implicated as a central etiological factor in the development of psychotic disorders. Through animal and human models, decades of insightful research have yielded a strong understanding of the underlying neurobiological systems implicated in the effects of chronic stress exposure. Simultaneously, the literature is rich with regards to measuring specific individual level stressors, including exposure to bullying, school performance stress, parental instability, among other life events. As individuals we exist within a larger environmental and social context (i.e., structural level factors) which, if harmful, may confer chronic stress at the systems level (Bronfenbrenner, 1994; Glass & McAtee, 2006). Certain local or even country level structural characteristics (such as neighborhood socioeconomic status and crime exposure) may systemically affect the health and well-being of the individuals inhabiting these environments, instituting barriers to healthy living and development. Since these structural impacts are not captured when exclusively considering individual level stressors, it then becomes necessary to more thoroughly incorporate structural level stressors when considering models for risk of developing psychopathology.

The psychosis literature has widely begun to explore structural characteristics, and found compelling evidence of associations of certain structural characteristics with psychotic disorder incidence. Given the literature, one can deduce that exposure to adverse structural level factors engages converging mechanisms that underlie the harmful effects of stress on physical and mental health. Nonetheless, it is critical to consider that stressors with phenomenologically differing components may also impact neural function and neurodevelopment in distinct manners. The research on neural effects of different stressors is well developed; however, to our knowledge there have been no attempts to synthesize the literature on structural stressors and psychosis risk with the research on distinct neurobiological effects of certain stressors. Further understanding the mechanistic neural effects of different structural exposures is essential from an epidemiological and etiological standpoint. Increasing our understanding of structural barriers to mental health stands to inform health policy, as well as prevention and intervention efforts at the societal level. Thus, in the current review, we first aim to discuss converging biological mechanisms of chronic stress exposure, which are theoretically impacted by structural stressors broadly defined. We then will propose three kinds of structural environmental exposure. Relations between the three exposures and risk for developing psychosis will be discussed in the context of a *stimulation, discrepancy* and *deprivation* model of chronic stress in psychosis, along with possible intermediary mechanisms and neurobiological underpinnings of these associations. Before we can discuss converging biological mechanisms of chronic stress exposure, along with the three areas and subsequent conceptual model, it is first necessary to discuss certain caveats that are central to any discussion of these subjects.

1.1 Co-occurrence and convergence of structural exposures

The review aims to differentiate three types of structural stressors: stimulation, *deprivation*, and *discrepancy*. *Stimulation* includes environments that activate threat neural systems and confer feelings of lack of safety and an excess of sensory inputs/attentional demands. *Deprivation* describes environments that have a lack of necessary stimuli for brain health and normative maturation. *Discrepancy* describes environments that foster a sense of social exclusion, isolation, or lack of belonging. Before describing these in more detail, it is crucial to acknowledge caveats related to overlap in exposures. Far from being isolated exposures, there is reason to believe the structural stressors that will be discussed co-occur rather frequently. For example, it is intuitive to imagine *stimulation* exposures such as high crime environments very often co-occurring with *deprivation* exposures such as low neighborhood socioeconomic status. Therefore, we do not seek to hypothesize that these exposures occur independently of each other. Rather, the review seeks to highlight the evidence that these exposures each individually effect risk for developing a psychotic disorder. As will be discussed in more detail below, the literature has often controlled for exposure to co-occurring structural factors, and found this not to alter associations with increased psychosis risk for each of the three structural exposures discussed below. Similarly, with regards to neural and intermediary mechanisms underlying each exposure, while the review seeks to differentiate between neural systems that could be affected by each specific exposure, it is also critical to acknowledge neural systems that could be affected through chronic stress exposure by all structural stressor types. The review poses that while converging neural effects exist, the impact of specific structural stressor types can be measured independently, and confer unique effects on neural function, which are valuable to understand and further explore.

1.2 Social drift

When discussing structural exposures, it is necessary to consider possible effects due to what is defined as "social drift." That is, vulnerable individuals may tend toward moving into more adverse neighborhood environments as emerging symptoms interfere with access to resources and day to day functioning. However, there is evidence that the effects of *stimulation* (March et al., 2008; Tandon, Nasrallah, & Keshavan, 2009), *discrepancy* (Termorshuizen, Smeets, Braam, & Veling, 2014), and *deprivation* exposures (Kirkbride, Jones, Ullrich, & Coid, 2012) on increased incidence of psychotic disorders remain even after accounting for social drift. Thus, while social drift may certainly be an influence in some cases, it is notable that many of the studies discussed in this review have accounted for social drift in their methods and study conceptualizations.

2. Converging mechanisms impacted by the three structural exposures

Chronic exposure to structural adverse environmental conditions, (whether these occur pre-natally, during early-postnatal periods, childhood, or adolescence through young adulthood) is likely to confer chronic stress, resulting in hypothalamic-pituitary-adrenal (HPA) dysregulation, a key factor in diathesis-stress conceptualizations of psychosis (Pruessner, Cullen, Aas, & Walker, 2017). Chronic stress is also causally linked to aberrant dopamine and glutamate transmission; this association is etiologically compelling, as dopamine dysregulation is closely linked with positive symptoms, and glutamate transmission has been associated with negative symptoms and cognitive dysfunction (Grace, 2016; Howes, McCutcheon, & Stone, 2015; Pruessner et al., 2017). Chronic inflammation resulting from chronic stress exposure is another putative mechanism conferring psychosis risk (Bergink, Gibney, & Drexhage, 2014). Dysregulation in these systems is liable to impact the brain in complex and multifaceted manners, which will be briefly delineated below.

2.1 HPA dysregulation

Chronic stress-induced prolonged HPA axis activation, leading to hyper secretion of glucocorticoids, can have detrimental effects on neural structure and function. Importantly, normative increases in cortisol secretion in adolescence and early adulthood coincide with the typical period of onset for psychotic disorders. With regards to the HPA axis, there is strong evidence that dysregulation due to chronic stress exposure could impact the prefrontal cortex (PFC) and hippocampus most prominently (Grace, 2016; Howes et al., 2015; Mittal & Walker, 2019; Pruessner et al., 2017; Walker, Mittal, & Tessner, 2008).

2.2 Dopamine dysregulation

Dopamine sensitization is both sensitive to chronic stress exposure broadly defined, and intricately related to psychotic disorder etiology. Chronic stress exposure leads to increased activity of dopamine neurons across the ventral tegmental area (VTA), PFC and nucleus accumbens (Grace, 2016). Grace and colleagues pose a key mechanism for downstream dopamine dysregulation involves interneurons, which enable brain information transfer and system coherence. Given that interneurons are among the last components to be incorporated during neurodevelopment, they are particularly susceptible to harm due to oxidative stress, most markedly early in the post-natal period. Thus, multiple hits beginning early in post-natal development and occurring additionally through adolescence and young adulthood (coinciding with normative increases in cortisol secretion) may lead to compounded interneuron loss during multiple critical periods, which would then confer increased risk of chronic dopamine dysregulation, and of development of psychotic disorders as a result. For instance, it seems that chronic stressors activate the dopamine system by acting on the hippocampus, which is also observed to be widely dysregulated in psychotic disorders; importantly, anterior hippocampus hyperactivity has been found to correlate both with presence of psychosis and with substantial decrease in interneurons in the hippocampus.

2.3 Glutamate dysregulation

Though the literature has been less prolific on the subject, studies have shown associations between chronic stress exposure and glutamate transmission, including glutamate release, receptors, clearance and metabolism (Popoli, Yan, McEwen, & Sanacora, 2012). In addition, abnormalities in hippocampal glutamatergic neurotransmission may be associated with elevated striatal dopamine, which could increase risk for developing a psychotic disorder (Howes et al., 2015; Stone et al., 2010).

2.4 Inflammation

Lastly, chronic inflammation resulting from chronic stress exposure has also been recently hypothesized as a mechanism through which chronic stress could confer increased risk for psychotic disorders (Bergink et al., 2014). Peripheral inflammation resulting from chronic early life and current structural stress exposure could lead to neuroinflammatory responses impacting cortico-amygdala function, as well as PFC structure, function and development (Nusslock & Miller, 2016).

2.5 Stimulation discrepancy deprivation model of psychosis

In the next sessions, the different types of structural exposures comprising the *Stimulation Discrepancy Deprivation* (SDD) model of psychosis will be described in detail (Fig. 1). The sections will include discussion of influences on different neural systems, as well as intermediary mechanisms through which these exposures could impact brain function and development.

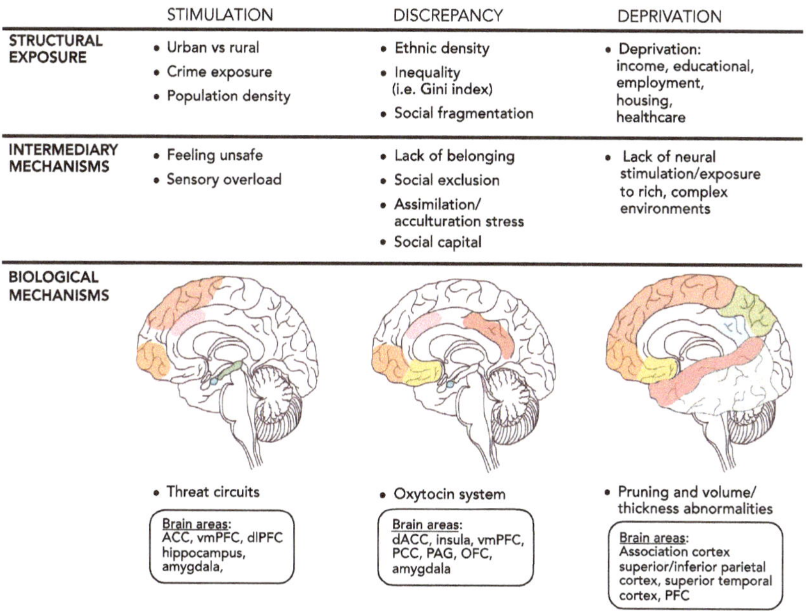

Fig. 1 The stimulation, discrepancy, and deprivation model: structural exposures, intermediary mechanisms, and biological mechanisms.

3. Stimulation structural exposure

3.1 Stimulation definition and relations to psychotic disorder risk

The first dimension is defined as "*stimulation*" neighborhood environments. Neighborhood environments in this dimension include urban/rural classifications, population density, and exposure to crime. Living in an urban environment has long been established as a risk factor for developing psychotic disorders (Kelly et al., 2010; Krabbendam & Van Os, 2005; March et al., 2008; Takei, Sham, O'Callaghan, Glover, & Murray, 1995; Vassos, Pedersen, Murray, Collier, & Lewis, 2012). There is evidence that the robustly observed effect applies across the continuum of rural to urban. Increasing levels of rurality have been associated with reduced risk for affective psychosis (Omer et al., 2016). Further, meta-analytic evidence suggests a linear relationship between increasing urbanicity and odds of risk for schizophrenia (Vassos et al., 2012), consistent with previous findings of a dose-response relationship (Pedersen & Mortensen, 2001). The association between urbanicity and psychotic experiences holds across the psychosis spectrum, and remains after accounting for confounds such as family socio-economic status, family psychiatric history, and substance use problems (Newbury et al., 2017). Experimental paradigms of busy urban environments have found urban "street exposure" results in significant increases in psychotic symptoms (Freeman et al., 2014). Relatedly, high population density, which characterizes urban areas, has been similarly associated with increased psychosis risk (Kirkbride et al., 2012). Emerging literature has also considered crime exposure and victimization as an additional exacerbating factor for increasing psychosis risk. For example, it has been found that adverse neighborhood social conditions and personal crime victimization combined exhibit a stronger association with psychotic experiences than either exposure alone (Newbury et al., 2017). Within urban areas specifically, associations have been found between high crime levels and incidence of schizophrenia (Bhavsar, Boydell, Murray, & Power, 2014).

3.2 Stimulation intermediary mechanisms

Structural exposures of urban environment, high population density, and high crime exposure share features through which they could negatively impact the individual and increase vulnerability for mental illness. The urban environment for one is often characterized by features such as significantly heightened levels of population density, which could act as visual stressors, offering chronic high levels of stimuli/attentional demands in an individual's environment (Gong, Palmer, Gallacher, Marsden, & Fone, 2016). In addition to population density, urban environments are also characterized by physical features such as limited "green" space, high noise levels, graffiti, increased rubbish, traffic, as well as hazardous waste sites (Weich et al., 2002). Taken together, these features could enact adaptation processes resulting in chronic heightened vigilance and physical arousal resulting from efforts to ensure safety and identify possible threats in the environment. Urban "street exposure" has indeed been linked in experimental paradigms to increased psychotic symptoms, as well as accompanying increases in paranoia and anxiety (Freeman et al., 2014). Epidemiologic and meta-analytic studies have also repeatedly observed associations between living in urban environments and increased psychological distress, along with risk for mental illness (including anxiety, depression, and psychosis) (Breslau, Marshall, Pincus, & Brown, 2014; Gong et al., 2016; Lederbogen et al., 2011; McKenzie, Murray, & Booth, 2013; Wilson-Genderson & Pruchno, 2013). Another facet that often characterizes urban environments is greater average crime rates (Jones & Fanek, 1997). Greater exposure to crime in an urban environment could exacerbate heightened chronic physical arousal and vigilance, leading to feelings of threat and lack of safety. This notion that high-crime urban conditions can lend themselves to chronic heightened physical arousal and vigilance is supported by investigations finding that perceptions of neighborhood safety along with exposure to neighborhood violence are important predictors of symptoms in otherwise healthy individuals (Wilson-Genderson & Pruchno, 2013).

3.3 Stimulation exposure neural mechanisms

The discussed components of neighborhood environments that are urban, have high population density, or have high exposure to crime could make them liable to alteration of neural threat systems. Chronic exposure to environments that could evoke feelings of threat or lack of safety in turn could affect the structure, function, and connectivity of threat circuit regions including the amygdala, hippocampus, and ventromedial prefrontal cortex (vmPFC) (McLaughlin, Sheridan, & Lambert, 2014). The literature offers compelling evidence that chronic exposure to threatening stimuli, especially during key developmental periods, can adversely impact these neural circuitry underlying fear conditioning and extinction (McLaughlin, Sheridan, & Lambert, 2014). Functions of the amygdala include detection of salient environmental stimuli, along with processing of emotion related cues (LeDoux, 2003; Tottenham & Sheridan, 2010; van Marle, Hermans, Qin, & Fernández, 2009). There is evidence that the amygdala plays an essential role in threat detection and identification (Adolphs, 2008; Isenberg et al., 1999; Öhman, 2005), and it is also key for acquiring and expressing learned fear (Harnett et al., 2019; Johansen, Cain, Ostroff, & LeDoux, 2011; LeDoux, 2003). Chronic exposure to highly stimulating and populated urban environments thus may over time impact novel fear learning. Over time, this could result in hypervigilance, along with a lowered threshold for experiencing fear as an incremental number of neutral cues are flagged as threatening; eventually leading to abnormal amygdala function (Cohen et al., 2013; van Marle et al., 2009).

The vmPFC is also critically involved in both fear inhibition and retention of extinction learning (Phelps, Delgado, Nearing, & LeDoux, 2004; Quirk, Russo, Barron, & Lebron, 2000). Importantly, the vmPFC also has direct projections to and down-regulates the amygdala (Milad & Quirk, 2012; Milad et al., 2007; Quirk et al., 2000; Quirk, Likhtik, Pelletier, & Paré, 2003).

For example, vmPFC activation during recall of extinction memory serves the role of inhibiting amygdala activity, thus dampening fear expression and facilitating extinction of learned feared stimuli (Knapska & Maren, 2009; Milad & Quirk, 2002). Another region that is implicated is the hippocampus. Early chronic threat exposure could impact hippocampal function and morphology through triggering shortening of dendritic spines and arborization (Brunson, Eghbal-Ahmadi, Bender, Chen, & Baram, 2001; Ivy et al., 2010). Early threat exposure has been shown to impact hippocampus-dependent facets of fear conditioning in animal models (Ishikawa et al., 2012; Matsumoto et al., 2008). Impairments in synaptic transmission within the vmPFC-hippocampus circuit have also been observed to result from impairments in fear extinction following prolonged early threat exposure (Toledo-Rodriguez & Sandi, 2007). Studies directly tying the structural exposures of urbanicity, population density and exposure to crime to threat neural circuitry have been scarce. Nonetheless, there is some evidence supporting these theories with regards to exposure to community violence. For example, a recent investigation of adolescents found an association between community violence exposure and lower hippocampal and amygdala volumes, as well as altered frontolimbic connectivity (Saxbe et al., 2018).

With regards to neural correlates of structural exposures in this domain, there is also robust evidence for the dorsolateral prefrontal cortex (dlPFC) and ACC being impacted by exposure to threat. For example, the lateral prefrontal cortex and ACC have both been shown to be implicated in processing of threat related stimuli (Bishop, Duncan, Brett, & Lawrence, 2004). The dlPFC specifically has also been shown to play a role in processing novel vs older stimuli while undergoing threat experiences (Balderston, Hsiung, Ernst, & Grillon, 2017). There is compelling evidence that living in urban regions may impact these threat circuitries. For instance, increased amygdala activity has been found in those living in urban regions, and activity in the perigenual anterior cingulate cortex (ACC), which regulates the amygdala, has been linked to urban upbringing (Lederbogen et al., 2011). Consistent with this finding, another investigation found urban upbringing related to lowered volumes in the right dlPFC and perigenual ACC (Haddad et al., 2014). Taken together, the existing literature allows for educated theorizing with regards to the specific effects of the urban/population density/crime exposure dimension. Future studies will benefit from enriching our understanding of this subject through linking specific structural exposures to neural mechanisms more directly.

4. Discrepancy structural exposure

4.1 Discrepancy definition and relations to psychotic disorder risk

The "*discrepancy*" dimension comprises neighborhood conditions including inequality, ethnic density, and high social fragmentation. Income inequality constitutes variation in income distributions within neighborhoods. High income inequality comprises great *discrepancy*, or heterogeneity, in incomes within a neighborhood, with low income inequality conversely meaning the neighborhood income distribution is relatively homogeneous. Studies have found an association between psychosis incidence and higher levels of income inequality. One investigation of 56 neighborhoods in London found that even after controlling for individual level factors and standardizing for age, sex ethnicity and socioeconomic status, greater income inequality explained substantial variation in psychosis risk (Kirkbride et al., 2012). Another investigation spanning 26 countries found a relationship between income inequality (assessed by the Gini index) and incidence of schizophrenia. Here also, the association between income inequality and schizophrenia incidence was not accounted for by urbanization, migration status, unemployment rate, population density, or gross domestic product per capita (Burns, Tomita, & Kapadia, 2014).

Also comprising part of the "*discrepancy*" dimension, ethnic density is defined as proportion of individuals in one's neighborhood that have a similar ethnic background. Higher ethnic density would entail more individuals in one's neighborhood that share a similar ethnic background. On the other hand, low ethnic density denotes a neighborhood in which there is a low proportion of individuals that share a similar ethnic background. Research has found living in lower ethnic density neighborhoods relates to increased incidence of psychosis in immigrant groups (Schofield et al., 2017; Veling et al., 2008). Compared to the general population, elevated rates of psychotic disorders have been consistently observed among ethnic minorities more generally (Lasalvia et al., 2014; Termorshuizen et al., 2014). This effect may be lessened, however, in areas with high ethnic density (Schofield et al., 2017; Termorshuizen et al., 2014; Veling et al., 2008).

Lastly, social fragmentation factors into the "*discrepancy*" dimension. Social fragmentation and its counter, social cohesion, are characterized by the stability and structure of a given social environmental context.

Social fragmentation can be operationalized through proportion of rental housing, frequency of changes in living arrangements among residents of a certain neighborhood (i.e., population turnover), proportion of single and/or divorced individuals in a neighborhood, as well as neighborhood disorder. High social cohesion, on the other hand, can be measured by aspects such as whether neighbors have shared values, trust each other, or are able to develop stable relationships within the neighborhood context (Crush, Arseneault, Jaffee, Danese, & Fisher, 2017). Research suggests that indexing social fragmentation distinctly from urban status and *deprivation* could be valuable. For example, studies found areas characterized by high social fragmentation had higher admission rates for psychosis, even after accounting for both *deprivation* and urban/rural living status (Allardyce et al., 2005). Furthermore, this investigation found evidence of a dose-response relationship between social fragmentation and first-ever admission rates for psychosis (Allardyce et al., 2005). Another investigation found that independently of individual characteristics, higher levels of neighborhood social fragmentation (indexed by proportion of single persons and divorced persons) predicted increased schizophrenia incidence, which was attributed to increased experiences of social isolation (Van Os, Driessen, Gunther, & Delespaul, 2000). Other studies operationalized social fragmentation in terms of population turnover, finding increased incidence of schizophrenia among areas with low social cohesion (Silver, Mulvey, & Swanson, 2002). In contrast, high social cohesion is a candidate protective factor against psychosis risk. For example, higher levels of neighborhood social cohesion were found to be protective against childhood psychotic symptoms in victimized children (Crush et al., 2017).

4.2 Discrepancy intermediary mechanisms

Income inequality, ethnic density, and social fragmentation share facets through which prolonged exposure could have a damaging impact. A sizeable portion of these shared facets can be conceptualized in terms of degrees of social capital. Social capital reflects the value of social networks to individuals and society (Bourdieu, 1986). Social capital can be defined as the extent and quality of an individual's social networks, mutual trust, and enacted social norms of reciprocity and support (Coleman, 1988). A key facet of social capital constitutes a sense of belonging to a certain group, or network of individuals (Butler & Muir, 2017; Nawyn, Gjokaj, Agbényiga, & Grace, 2012; Oliver & Cheff, 2014; Schellenberg, Lu, Schimmele, & Hou, 2018). Social capital has been intricately linked to health and life satisfaction (Elgar et al., 2011).

A wide body of epidemiological literature has shown that income inequality, which affects social capital, is similarly associated with poor mental and physical health and well-being (Elgar, Gariépy, Torsheim, & Currie, 2017; Karlsdotter, Martín, & del Amo González, 2012; Pickett & Wilkinson, 2015; Vilhjalmsdottir, Gardarsdottir, Bernburg, & Sigfusdottir, 2016). Income inequality at the neighborhood, state and country level, for one, has also been shown to elicit chronic feelings of distrust and lack of social cohesion, which are then associated with negative health outcomes (Bjornstrom, 2011; Fairbrother & Martin, 2013; Rözer & Volker, 2016). Further, emerging investigations have observed that early chronic exposure may be particularly impactful with regards to health and life satisfaction (Elgar et al., 2017). Thus, income inequality may impact feelings of trust, belonging, and social cohesion on a pervasive level.

There is evidence that ethnic density could impact mental health outcomes through a similar mechanism. Much like income inequality, low ethnic density creates a feeling of *discrepancy*, imposing roadblocks toward building social capital. Indeed, low ethnic density has been long proposed as a harmful influence for building social capital, which has in turn been linked to adverse physical and mental health outcomes (Eliacin, 2013; Yang, Lei, & Kurtulus, 2018). Low ethnic density has also been shown to deeply impact sense of belonging and community, which was related to psychiatric disorder incidence (Emerson, Minh, & Guhn, 2018; Finney & Jivraj, 2013; Pan & Carpiano, 2013).

Social fragmentation, like the previously discussed factors, is an additional key factor for impeding building of social capital. Social fragmentation could lead to experiences of social isolation, resulting in feelings of lack of trust and belonging, along with reduced social support (Eliacin, 2013; Van Os et al., 2000). Conversely, increases in the counter to social fragmentation, social cohesion, are associated with decreased psychological distress even after controlling for factors such as neighborhood socioeconomic status (Erdem, Prins, Voorham, Van Lenthe, & Burdorf, 2015). Thus, structural exposures in the *discrepancy* domain, putatively impeding the building of social capital could particularly impact the marked social deficits facet of psychotic disorders (Addington, McCleary, & Munroe-Blum, 1998).

4.3 Discrepancy neural mechanisms

Structural exposures in the "*discrepancy*" domain thus share overlapping components of impacted social capital, including lack of social cohesion, an impacted sense of trust and belonging, as well as less robust social networks and sense of support. There are several solid candidates as to biological systems that could be impacted by the above-mentioned exposures. One of these is the oxytocinergic system. The oxytocinergic system relates to social processes such as social attachment, empathy, and social threat (Heinrichs, Baumgartner, Kirschbaum, & Ehlert, 2003; Norman, Hawkley, Cole, Berntson, & Cacioppo, 2012). There is evidence of intranasal oxytocin attenuating amygdala-brainstem coupling in the face of threatening social stimuli (Kirsch et al., 2005). Studies have also found intranasal delivery of oxytocin to modulate evaluation of socially relevant faces through the amygdala (Gamer, Zurowski, & Büchel, 2010) and fusiform gyrus (Domes, Heinrichs, Michel, Berger, & Herpertz, 2007; Petrovic, Kalisch, Singer, & Dolan, 2008). Oxytocin has been linked to increased in-group trust (De Dreu et al., 2010; Kosfeld, Heinrichs, Zak, Fischbacher, & Fehr, 2005) as well as to elevated processing of positive social information (Di Simplicio, Massey-Chase, Cowen, & Harmer, 2009; Guastella, Mitchell, & Mathews, 2008). There is further evidence in the animal literature of increased oxytocin blunting social vigilance behavior, suggesting increased manifestations of trusting behavior (Ebitz, Watson, & Platt, 2013). Notably, there is evidence for a degree of specificity with regards to oxytocin modulating affective social processes. For example, while oxytocin has been found to decrease emotional arousal to threatening human stimuli, this was not the case for threatening animal stimuli, or for general positive/negative valence stimuli (Norman et al., 2011).

Speaking beyond oxytocin, neuroendocrine and autonomic nervous system alterations are other mechanisms through which lack of trust, belonging, and social capital could modulate adverse mental health outcomes (Norman et al., 2012). Neural regions involved in processing of social threat, rejection, and lack of belonging include the amygdala, orbitofrontal cortex, and ACC (Adolphs, 2009; Eisenberger, Lieberman, & Williams, 2003). These regions project to brainstem nuclei which regulate parasympathetic and sympathetic outflow (Critchley, 2005). Autonomic nervous system engagement can change based on high-level processing from these regions that also regulate processing of social threat, rejection, and lack of belonging. Given the responsivity of the autonomic nervous system to these socio-psychological contexts, chronic signaling from implicated regions could confer alterations in these circuits over time (Berntson, Cacioppo, Quigley, & Fabro, 1994). This notion is supported by investigations on individuals experiencing chronic loneliness, which exhibit alterations in autonomic reactivity across various contexts, leading to poorer physical and mental health outcomes over time (Cacioppo et al., 2002; Hawkley, Masi, Berry, & Cacioppo, 2006).

Studies have theorized social disconnection (both the perception/possibility of social disconnection and the experience of social disconnection/exclusion) to be processed by neural systems responsible for responding to harm and impending danger. This hypothesized "neural alarm system" includes the amygdala, as well as the dorsal ACC, anterior insula and periaqueductal gray (PAG), which are responsible for pain-related processing (Eisenberger & Cole, 2012). Chronic activation in these regions can trigger endocrine and autonomic responses that can negatively impact health over time. Conversely, vmPFC and posterior cingulate cortex (PCC) are involved in processing safety detection and fear, and increased activity in these regions may inhibit sympathetic and promote parasympathetic responses, which is protective for physical and mental health (Eisenberger & Cole, 2012). Dysregulation in these regions due to chronic stress exposure thus could possibly compromise these protective processes. Taken together, social neuroscience literature allows us to gain a fuller understanding of mechanisms through which the structural exposures in the "*discrepancy*" domain (income inequality, ethnic density, social fragmentation) could impact neurological systems over time. Nonetheless, future studies will be crucial in marrying the literature exploring psychosis incidence rates in light of these exposures, with the non-clinical literature exploring neural correlates of possible intermediary causal mechanisms (such as reduced social cohesion, sense of belonging, trust, and reduced social capital).

5. Deprivation structural exposure

5.1 Deprivation definition and relations to psychotic disorder risk

The "*deprivation*" dimension is characterized by environmental and neighborhood conditions with a lack of appropriate resources with regards to factors such as employment, income, education, health, living conditions, and barriers to housing and services. While exposure to these environments has been associated with a host of negative health outcomes, there is evidence that living in a high *deprivation* environment can confer increased risk for developing a psychotic disorder specifically (Bhavsar et al., 2014; Bhavsar, Fusar-Poli, & McGuire, 2018; Kirkbride et al., 2012; Lasalvia et al., 2014;

O'Donoghue et al., 2015; Omer et al., 2014). Two studies used an overall *deprivation* index including employment, income, education, health, *deprivation* in living conditions, as well as barriers to housing and services (Bhavsar et al., 2018; Kirkbride et al., 2012). These studies found consistent results, with the most deprived neighborhoods having a significant increase in incidence rate of psychosis risk status (Bhavsar et al., 2018; Kirkbride et al., 2012). Another investigation incorporated area income and employment *deprivation*, health *deprivation* and disability, education, skills and training *deprivation*, barriers to housing and services, and living environment *deprivation* (Bhavsar et al., 2014), finding a similar pattern of increased schizophrenia incidence among those living in high *deprivation* areas. Yet another study focused on material *deprivation*, finding greater rates of psychotic disorders in those exposed to material *deprivation* (Omer et al., 2014). It is key to note that though factors such as population density could relate to overall *deprivation* in certain regions, there is research suggesting that the association between high *deprivation* and schizophrenia incidence holds even after controlling for these factors (Bhavsar et al., 2014).

5.2 Deprivation intermediary mechanisms

While there are different types of *deprivation*, all share the aspect of a need or resource that is not being provided, possibly affecting adaptive neurodevelopment and ongoing neural function in an individual. That is, there is a lack of species- and age-typical complexity in environmental stimuli (McLaughlin, Sheridan, & Lambert, 2014). Thus, *deprivation* can be conceptualized as chronic exposure to environments that lack adequate stimulation for an individual. Chronic lack of needed educational, cognitive, economic, or health resources can lead to neural under-stimulation in certain key functions, which in turn could negatively impact neurodevelopment. This is especially the case during childhood, adolescence and early adulthood, which are marked by critical periods of neurodevelopment (Pechtel & Pizzagalli, 2011; Shaw et al., 2008). The literature has long observed an association between exposure to *deprivation* and health outcomes (Bak, Andersen, & Dokkedal, 2015). Prolonged exposure to *deprivation* has been associated with adverse mental health outcomes, including increased psychotic symptoms such as paranoia, as well as depressed symptoms (Wickham, Taylor, Shevlin, & Bentall, 2014). *Deprivation* exposure has also been associated with increased risk of hypertension, among other physical health outcomes (Matheson, Moineddin, & Glazier, 2008).

On the other hand, animal studies have shown that continued exposure to enriched, complex environments (the counter to environments high in *deprivation*) yields greater neurobiological resilience (Bardi et al., 2016; Lambert, Nelson, Jovanovic, & Cerdá, 2015). Neurobiological resilience was evidenced, for instance, through less observed anxiety in response to exposure to novel objects, less anxiety-typical behavior in response to a predator odor, as well as decreased amygdala activation following a water escape task designed to induce high stress levels. Lastly and importantly, exposure to *deprivation* (in the form of low socioeconomic status) has marked effects on cognitive performance for complex tasks, including language (Fernald, Marchman, & Weisleder, 2013; Weisleder & Fernald, 2013) executive function and memory (Evans & Schamberg, 2009; Farah et al., 2006; Hackman, Farah, & Meaney, 2010; Hackman, Gallop, Evans, & Farah, 2015; Kishiyama, Boyce, Jimenez, Perry, & Knight, 2009; Noble, McCandliss, & Farah, 2007). Markedly, these differences have been found to be mediated by a lack of exposure to enriching and complex activities in childhood (Bradley, Corwyn, McAdoo, & García Coll, 2001; Linver, Brooks-Gunn, & Kohen, 2002; Yeung, Linver, & Brooks-Gunn, 2002).

5.3 Deprivation neural mechanisms

Exposure to *deprivation* is likely to impact neural development through several key mechanisms. There is strong evidence of an impact of low environmental complexity on cortical development; an effect which has been observed in both the animal and human literature (Diamond, Rosenzweig, Bennett, Lindner, & Lyon, 1972; Leporé et al., 2010). Early and chronic exposure to low complexity environments may yield neural systems selectively designed to best process low complexity environments (McLaughlin, Sheridan, & Lambert, 2014). This, in turn, could yield long-term adverse mental and physical health outcomes when the individual inevitably has to adapt to a high-complexity environment and demands. The process through which environmental stimuli impacts neurodevelopment likely stems from central nervous system pruning of synaptic connections, deemed the "selective elimination" hypothesis (Changeux & Danchin, 1976; Huttenlocher, de Courten, Garey, & Van der Loos, 1982; Petanjek et al., 2011). There is an overabundance of synaptic connections that are initially created (i.e., synaptic proliferation), and these are pruned across development to accommodate and adapt to the being's needs. Pruning following synaptic proliferation ensues throughout childhood and adolescence (Huttenlocher, 1979; Rakic, Bourgeois, Eckenhoff, Zecevic, & Goldman-Rakic, 1986). Theoretically, this leads to an adaptive system whereby only the most environmentally helpful and efficient synaptic pathways remain. However, if the environment lacks adequate enrichment, chronic and early exposure to low complexity environments could foster excessive pruning, resulting in a lower number of synaptic connections, along with lower dendritic branching. Brain regions with protracted developmental trajectories, making them more sensitive to environmental influence, would be particularly likely to be impacted by exposure to *deprivation*. This includes the association cortex (Gogtay et al., 2004) and PFC (Pechtel & Pizzagalli, 2011). The association cortex refers to lateral and medial prefrontal, parietal, and temporal areas of the cortex involved in cognitive processing (Dumontheil, Thompson, & Duncan, 2011; Mountcastle, Lynch, Georgopoulos, Sakata, & Acuna, 1975; Pulvermüller & Garagnani, 2014).

Though the association cortex engages in stimuli processing across multiple sensory modalities, it is also necessary for higher-level cognitive functions including language, executive function, and social cognition (Badre & D'esposito, 2009; Culham & Kanwisher, 2001). Likewise, the PFC is chiefly involved in executive and affective regulatory functions (Pechtel & Pizzagalli, 2011). As alluded to earlier, the animal literature yields strong support for this interpretation. For example, global *deprivation* in rodents has been documented to result in decreases in dendritic arborization and spines throughout the cortex, as well as decreases in total brain volume (Bennett, Diamond, Krech, & Rosenzweig, 1964; Diamond et al., 1966; Globus, Rosenzweig, Bennett, & Diamond, 1973). These changes become more widespread and persistent the longer the exposure to the *deprivation* environment.

The human literature, though needing further studies in order to increase specificity of affected regions and degree of stability of the effect, paints a similar picture. Much of this literature has emphasized socioeconomic status. Though investigations of neighborhood socioeconomic status have been more scarce, low parental socioeconomic status has been associated to lower exposure to enriching cognitive experiences in the school and home environment (Bradley et al., 2001; Sirin, 2005). Conversely, cortical gray matter volume as well as cortical thickness has been found to be greater in students from higher-income backgrounds, with certain regions of the association cortex relating to greater academic performance (Mackey et al., 2015). Low socioeconomic status has been associated with decreased volume in the association cortex, particularly in PFC (Noble, Houston, Kan, & Sowell, 2012; Noble et al., 2007). Low socioeconomic status has also been associated with more diffuse patterns of activation in the association cortex while supporting performance on executive function and language tasks in children (Raizada & Kishiyama, 2010) and adults (Gianaros et al., 2008; Gianaros & Manuck, 2010). A perhaps more extreme *deprivation* exposure that has been covered in the literature is that of institutionalization, which often constitutes a clear scarcity of social and cognitive inputs, ranging from maternal care *deprivation*, lack of variation in daily routines and experiences, enriching cognitive stimuli (such as books or toys) to peer interaction opportunities (Gee et al., 2013; Nelson, Furtado, Fox, & Zeanah, 2009; Smyke et al., 2007; Tottenham, 2012; Tottenham et al., 2010). Here, too, findings have been consistent, whereby institutionalization has been associated with decreases in gray matter thickness and volume, along with cognitive dysfunction (McLaughlin et al., 2014; Mehta et al., 2009; Sheridan, Fox, Zeanah, McLaughlin, & Nelson, 2012). While further synthesis of the research is needed in order to solidify the theorized association between increased psychosis incidence and structural exposures related to *deprivation*, the literature in animal and non-clinical human populations offers us a theoretical framework of possible affected neural mechanisms.

6. Stimulation discrepancy deprivation model of psychosis summary

Partially inspired by diathesis-stress conceptualizations of psychosis, the SDD model distinguishes itself in highlighting the critical importance of systems-level considerations for assessing both pre-existing biological underlying vulnerabilities, and exposure to chronic stressors going into adolescence and young adulthood (Fig. 2). The SDD model is also novel in proposing three distinct types of structural exposures, along with neural underpinnings and intermediary mechanisms. We pose that exposure to structural stressors can bring about pre-existing neurobiological vulnerability through prolonged exposure during the pre-natal, early post-natal, and childhood periods. Structural stressors can then additionally contribute to increasing vulnerability throughout adolescence and young adulthood. Chronic exposure to structural stressors throughout adolescence and young adulthood can drive a cascade of HPA dysregulation, aberrant dopamine and glutamate neural transmission, as well as altered brain function and development, ultimately driving increased inflammation, cognitive and social

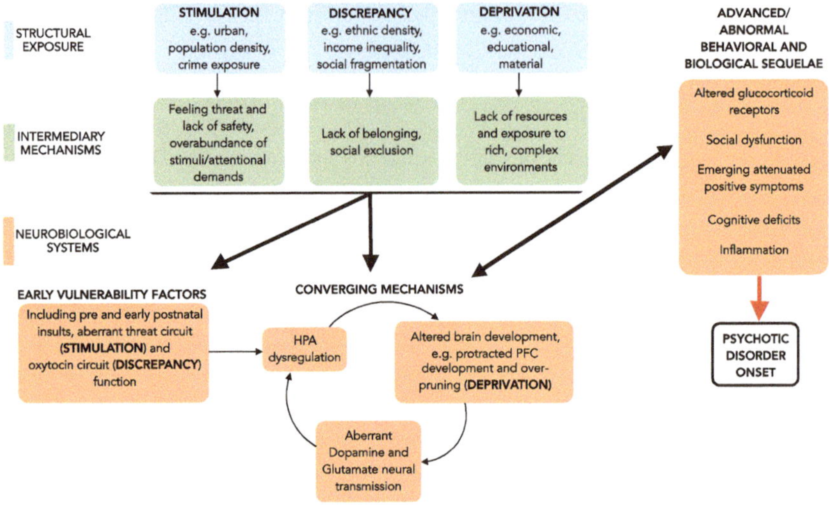

Fig. 2 The stimulation, discrepancy, and deprivation model as it relates to psychotic disorder onset.

dysfunction, altered glucocorticoid receptor function, among other factors specific to the three structural exposure types, eventually leading to onset of psychotic symptoms (Fig. 2). Structural stressors thus are key to consider in chronic stress models of psychosis, as they are liable to be critically relevant throughout the lifetime with regards to providing detrimental or protective influences on the individual.

7. Future directions

The present review drew on empirical evidence from the neuroscience literature, as well as from epidemiological evidence of structural exposures in order to propose intermediary mechanisms as well as neural underpinnings through which structural stressors could impact risk for developing a psychotic disorder. It will be necessary to directly test these theories in populations across the psychosis spectrum. Further, specifying sensitive and critical developmental periods during which these exposures could be particularly impactful is a critical future direction. Models testing both cumulative and distinct effects of these exposures would also be beneficial. Examining the interplay between individual and structural level stressors is also important in order to compare degrees of impact and decrease noise due to individual differences. Finally, further identifying and increasing understanding of buffering or protective environmental factors will aid in conceptualizing psychosis risk profiles. In conclusion, in the present review, we pose that system, or structural level factors deserve more attention within psychological research. It is crucial to enhance our understanding of manners in which structural factors may exert influence on their own, and through facilitating the expression and exacerbation of individual vulnerability (biological factors, genetic predispositions, individual behaviors, cognition and personality) (Glass & McAtee, 2006; Rose, 2001). Given their standalone influence and key role in facilitating expression of individual pre-existing vulnerability, targeting these system level conditions through epidemiological, public health approaches may ultimately exert a greater impact on an aggregate, societal level.

References

Addington, J., McCleary, L., & Munroe-Blum, H. (1998). Relationship between cognitive and social dysfunction in schizophrenia. *Schizophrenia Research*, *34*(1–2), 59–66.

Adolphs, R. (2008). Fear, faces, and the human amygdala. *Current Opinion in Neurobiology*, *18*(2), 166–172.

Adolphs, R. (2009). The social brain: Neural basis of social knowledge. *Annual Review of Psychology*, *60*, 693–716.

Allardyce, J., Gilmour, H., Atkinson, J., Rapson, T., Bishop, J., & McCreadie, R. (2005). Social fragmentation, deprivation and urbanicity: Relation to first-admission rates for psychoses. *The British Journal of Psychiatry*, *187*(5), 401–406.

Badre, D., & D'esposito, M. (2009). Is the rostro-caudal axis of the frontal lobe hierarchical? *Nature Reviews Neuroscience*, *10*(9), 659.

Bak, C. K., Andersen, P. T., & Dokkedal, U. (2015). The association between social position and self-rated health in 10 deprived neighbourhoods. *BMC Public Health*, *15*(1), 14.

Balderston, N. L., Hsiung, A., Ernst, M., & Grillon, C. (2017). Effect of threat on right dlPFC activity during behavioral pattern separation. *Journal of Neuroscience*, *37*(38), 9160–9171.

Bardi, M., Kaufman, C., Franssen, C., Hyer, M., Rzucidlo, A., Brown, M., et al. (2016). Paper or plastic? Exploring the effects of natural enrichment on behavioural and neuroendocrine responses in Long-Evans rats. *Journal of Neuroendocrinology*, *28*(5).

Bennett, E. L., Diamond, M. C., Krech, D., & Rosenzweig, M. R. (1964). Chemical and anatomical plasticity of brain. *Science*, *146*(3644), 610–619.

Bergink, V., Gibney, S. M., & Drexhage, H. A. (2014). Autoimmunity, inflammation, and psychosis: A search for peripheral markers. *Biological Psychiatry*, *75*(4), 324–331.

Berntson, G. G., Cacioppo, J. T., Quigley, K. S., & Fabro, V. T. (1994). Autonomic space and psychophysiological response. *Psychophysiology*, *31*(1), 44–61.

Bhavsar, V., Boydell, J., Murray, R., & Power, P. (2014). Identifying aspects of neighbourhood deprivation associated with increased incidence of schizophrenia. *Schizophrenia Research*, *156*(1), 115–121.

Bhavsar, V., Fusar-Poli, P., & McGuire, P. (2018). Neighbourhood deprivation is positively associated with detection of the ultra-high risk (UHR) state for psychosis in South East London. *Schizophrenia Research*, *192*, 371–376.

Bishop, S., Duncan, J., Brett, M., & Lawrence, A. D. (2004). Prefrontal cortical function and anxiety: Controlling attention to threat-related stimuli. *Nature Neuroscience*, 7(2), 184.

Bjornstrom, E. E. (2011). The neighborhood context of relative position, trust, and self-rated health. *Social Science & Medicine, 73*(1), 42–49.
Bourdieu, P. (1986). *The forms of capital*. New York, NY: Westview Press.
Bradley, R. H., Corwyn, R. F., McAdoo, H. P., & García Coll, C. (2001). The home environments of children in the United States part I: Variations by age, ethnicity, and poverty status. *Child Development, 72*(6), 1844–1867.
Breslau, J., Marshall, G. N., Pincus, H. A., & Brown, R. A. (2014). Are mental disorders more common in urban than rural areas of the United States? *Journal of Psychiatric Research, 56*, 50–55.
Bronfenbrenner, U. (1994). Ecological models of human development. *Readings on the Development of Children, 2*(1), 37–43.
Brunson, K. L., Eghbal-Ahmadi, M., Bender, R., Chen, Y., & Baram, T. Z. (2001). Long-term, progressive hippocampal cell loss and dysfunction induced by early-life administration of corticotropin-releasing hormone reproduce the effects of early-life stress. *Proceedings of the National Academy of Sciences of the United States of America, 98*(15), 8856–8861.
Burns, J. K., Tomita, A., & Kapadia, A. S. (2014). Income inequality and schizophrenia: Increased schizophrenia incidence in countries with high levels of income inequality. *International Journal of Social Psychiatry, 60*(2), 185–196.
Butler, R., & Muir, K. (2017). Young people's education biographies: Family relationships, social capital and belonging. *Journal of Youth Studies, 20*(3), 316–331.
Cacioppo, J. T., Hawkley, L. C., Crawford, L. E., Ernst, J. M., Burleson, M. H., Kowalewski, R. B., et al. (2002). Loneliness and health: Potential mechanisms. *Psychosomatic Medicine, 64*(3), 407–417.
Changeux, J.-P., & Danchin, A. (1976). Selective stabilisation of developing synapses as a mechanism for the specification of neuronal networks. *Nature, 264*(5588), 705.
Cohen, M. M., Jing, D., Yang, R. R., Tottenham, N., Lee, F. S., & Casey, B. (2013). Early-life stress has persistent effects on amygdala function and development in mice and humans. *Proceedings of the National Academy of Sciences of the United States of America, 110*(45), 18274–18278.
Coleman, J. S. (1988). Social capital in the creation of human capital. *American Journal of Sociology, 94*, S95–S120.
Critchley, H. D. (2005). Neural mechanisms of autonomic, affective, and cognitive integration. *Journal of Comparative Neurology, 493*(1), 154–166.
Crush, E., Arseneault, L., Jaffee, S. R., Danese, A., & Fisher, H. L. (2017). Protective factors for psychotic symptoms among poly-victimized children. *Schizophrenia Bulletin, 44*(3), 691–700.
Culham, J. C., & Kanwisher, N. G. (2001). Neuroimaging of cognitive functions in human parietal cortex. *Current Opinion in Neurobiology, 11*(2), 157–163.
De Dreu, C. K., Greer, L. L., Handgraaf, M. J., Shalvi, S., Van Kleef, G. A., Baas, M., et al. (2010). The neuropeptide oxytocin regulates parochial altruism in intergroup conflict among humans. *Science, 328*(5984), 1408–1411.
Di Simplicio, M., Massey-Chase, R., Cowen, P., & Harmer, C. (2009). Oxytocin enhances processing of positive versus negative emotional information in healthy male volunteers. *Journal of Psychopharmacology, 23*(3), 241–248.
Diamond, M. C., Law, F., Rhodes, H., Lindner, B., Rosenzweig, M. R., Krech, D., et al. (1966). Increases in cortical depth and glia numbers in rats subjected to enriched environment. *Journal of Comparative Neurology, 128*(1), 117–125.
Diamond, M. C., Rosenzweig, M. R., Bennett, E. L., Lindner, B., & Lyon, L. (1972). Effects of environmental enrichment and impoverishment on rat cerebral cortex. *Journal of Neurobiology, 3*(1), 47–64.

Domes, G., Heinrichs, M., Michel, A., Berger, C., & Herpertz, S. C. (2007). Oxytocin improves "mind-reading" in humans. *Biological Psychiatry*, *61*(6), 731–733.

Dumontheil, I., Thompson, R., & Duncan, J. (2011). Assembly and use of new task rules in fronto-parietal cortex. *Journal of Cognitive Neuroscience*, *23*(1), 168–182.

Ebitz, R. B., Watson, K. K., & Platt, M. L. (2013). Oxytocin blunts social vigilance in the rhesus macaque. *Proceedings of the National Academy of Sciences of the United States of America*, *110*(28), 11630–11635.

Eisenberger, N. I., & Cole, S. W. (2012). Social neuroscience and health: Neurophysiological mechanisms linking social ties with physical health. *Nature Neuroscience*, *15*(5), 669.

Eisenberger, N. I., Lieberman, M. D., & Williams, K. D. (2003). Does rejection hurt? An fMRI study of social exclusion. *Science*, *302*(5643), 290–292.

Elgar, F. J., Davis, C. G., Wohl, M. J., Trites, S. J., Zelenski, J. M., & Martin, M. S. (2011). Social capital, health and life satisfaction in 50 countries. *Health & Place*, *17*(5), 1044–1053.

Elgar, F. J., Gariépy, G., Torsheim, T., & Currie, C. (2017). Early-life income inequality and adolescent health and well-being. *Social Science & Medicine*, *174*, 197–208.

Eliacin, J. (2013). Social capital, narratives of fragmentation, and schizophrenia: An ethnographic exploration of factors shaping African-Caribbeans' social capital and mental health in a North London community. *Culture, Medicine and Psychiatry*, *37*(3), 465–487.

Emerson, S. D., Minh, A., & Guhn, M. (2018). Ethnic density of regions and psychiatric disorders among ethnic minority individuals. *International Journal of Social Psychiatry*, *64*(2), 130–144.

Erdem, Ö., Prins, R. G., Voorham, T. A., Van Lenthe, F. J., & Burdorf, A. (2015). Structural neighbourhood conditions, social cohesion and psychological distress in the Netherlands. *European Journal of Public Health*, *25*(6), 995–1001.

Evans, G. W., & Schamberg, M. A. (2009). Childhood poverty, chronic stress, and adult working memory. *Proceedings of the National Academy of Sciences of the United States of America*, *106*(16), 6545–6549.

Fairbrother, M., & Martin, I. W. (2013). Does inequality erode social trust? Results from multilevel models of US states and counties. *Social Science Research*, *42*(2), 347–360.

Farah, M. J., Shera, D. M., Savage, J. H., Betancourt, L., Giannetta, J. M., Brodsky, N. L., et al. (2006). Childhood poverty: Specific associations with neurocognitive development. *Brain Research*, *1110*(1), 166–174.

Fernald, A., Marchman, V. A., & Weisleder, A. (2013). SES differences in language processing skill and vocabulary are evident at 18 months. *Developmental Science*, *16*(2), 234–248.

Finney, N., & Jivraj, S. (2013). Ethnic group population change and neighbourhood belonging. *Urban Studies*, *50*(16), 3323–3341.

Freeman, D., Emsley, R., Dunn, G., Fowler, D., Bebbington, P., Kuipers, E., et al. (2014). The stress of the street for patients with persecutory delusions: A test of the symptomatic and psychological effects of going outside into a busy urban area. *Schizophrenia Bulletin*, *41*(4), 971–979.

Gamer, M., Zurowski, B., & Büchel, C. (2010). Different amygdala subregions mediate valence-related and attentional effects of oxytocin in humans. *Proceedings of the National Academy of Sciences of the United States of America*, *107*(20), 9400–9405.

Gee, D. G., Gabard-Durnam, L. J., Flannery, J., Goff, B., Humphreys, K. L., Telzer, E. H., et al. (2013). Early developmental emergence of human amygdala–prefrontal connectivity after maternal deprivation. *Proceedings of the National Academy of Sciences of the United States of America*, *110*(39), 15638–15643.

Gianaros, P. J., Horenstein, J. A., Hariri, A. R., Sheu, L. K., Manuck, S. B., Matthews, K. A., et al. (2008). Potential neural embedding of parental social standing. *Social Cognitive and Affective Neuroscience, 3*(2), 91–96.

Gianaros, P. J., & Manuck, S. B. (2010). Neurobiological pathways linking socioeconomic position and health. *Psychosomatic Medicine, 72*(5), 450.

Glass, T. A., & McAtee, M. J. (2006). Behavioral science at the crossroads in public health: Extending horizons, envisioning the future. *Social Science & Medicine, 62*(7), 1650–1671.

Globus, A., Rosenzweig, M. R., Bennett, E. L., & Diamond, M. C. (1973). Effects of differential experience on dendritic spine counts in rat cerebral cortex. *Journal of Comparative and Physiological Psychology, 82*(2), 175.

Gogtay, N., Giedd, J. N., Lusk, L., Hayashi, K. M., Greenstein, D., Vaituzis, A. C., et al. (2004). Dynamic mapping of human cortical development during childhood through early adulthood. *Proceedings of the National Academy of Sciences of the United States of America, 101*(21), 8174–8179.

Gong, Y., Palmer, S., Gallacher, J., Marsden, T., & Fone, D. (2016). A systematic review of the relationship between objective measurements of the urban environment and psychological distress. *Environment International, 96*, 48–57.

Grace, A. A. (2016). Dysregulation of the dopamine system in the pathophysiology of schizophrenia and depression. *Nature Reviews Neuroscience, 17*(8), 524–532.

Guastella, A. J., Mitchell, P. B., & Mathews, F. (2008). Oxytocin enhances the encoding of positive social memories in humans. *Biological Psychiatry, 64*(3), 256–258.

Hackman, D. A., Farah, M. J., & Meaney, M. J. (2010). Socioeconomic status and the brain: Mechanistic insights from human and animal research. *Nature Reviews Neuroscience, 11*(9), 651.

Hackman, D. A., Gallop, R., Evans, G. W., & Farah, M. J. (2015). Socioeconomic status and executive function: Developmental trajectories and mediation. *Developmental Science, 18*(5), 686–702.

Haddad, L., Schäfer, A., Streit, F., Lederbogen, F., Grimm, O., Wüst, S., et al. (2014). Brain structure correlates of urban upbringing, an environmental risk factor for schizophrenia. *Schizophrenia Bulletin, 41*(1), 115–122.

Harnett, N. G., Wheelock, M. D., Wood, K. H., Goodman, A. M., Mrug, S., Elliott, M. N., et al. (2019). Negative life experiences contribute to racial differences in the neural response to threat. *NeuroImage, 202*, 116086.

Hawkley, L. C., Masi, C. M., Berry, J. D., & Cacioppo, J. T. (2006). Loneliness is a unique predictor of age-related differences in systolic blood pressure. *Psychology and Aging, 21*(1), 152.

Heinrichs, M., Baumgartner, T., Kirschbaum, C., & Ehlert, U. (2003). Social support and oxytocin interact to suppress cortisol and subjective responses to psychosocial stress. *Biological Psychiatry, 54*(12), 1389–1398.

Howes, O., McCutcheon, R., & Stone, J. (2015). Glutamate and dopamine in schizophrenia: An update for the 21st century. *Journal of Psychopharmacology, 29*(2), 97–115.

Huttenlocher, P. R. (1979). Synaptic density in human frontal cortex-developmental changes and effects of aging. *Brain Research, 163*(2), 195–205.

Huttenlocher, P. R., de Courten, C., Garey, L. J., & Van der Loos, H. (1982). Synaptogenesis in human visual cortex—Evidence for synapse elimination during normal development. *Neuroscience Letters, 33*(3), 247–252.

Isenberg, N., Silbersweig, D., Engelien, A., Emmerich, S., Malavade, K., Beattie, B. a., et al. (1999). Linguistic threat activates the human amygdala. *Proceedings of the National Academy of Sciences of the United States of America, 96*(18), 10456–10459.

Ishikawa, S., Saito, Y., Yanagawa, Y., Otani, S., Hiraide, S., Shimamura, K. I., et al. (2012). Early postnatal stress alters extracellular signal-regulated kinase signaling in the corticolimbic system modulating emotional circuitry in adult rats. *European Journal of Neuroscience, 35*(1), 135–145.

Ivy, A. S., Rex, C. S., Chen, Y., Dubé, C., Maras, P. M., Grigoriadis, D. E., et al. (2010). Hippocampal dysfunction and cognitive impairments provoked by chronic early-life stress involve excessive activation of CRH receptors. *Journal of Neuroscience*, *30*(39), 13005–13015.

Johansen, J. P., Cain, C. K., Ostroff, L. E., & LeDoux, J. E. (2011). Molecular mechanisms of fear learning and memory. *Cell*, *147*(3), 509–524.

Jones, M., & Fanek, M. (1997). Crime in the urban environment. In *Paper presented at the proceedings of first international symposium on space syntax*.

Karlsdotter, K., Martín, J. J. M., & del Amo González, M. P. L. (2012). Multilevel analysis of income, income inequalities and health in Spain. *Social Science & Medicine*, *74*(7), 1099–1106.

Kelly, B. D., O'Callaghan, E., Waddington, J. L., Feeney, L., Browne, S., Scully, P. J., et al. (2010). Schizophrenia and the city: A review of literature and prospective study of psychosis and urbanicity in Ireland. *Schizophrenia Research*, *116*(1), 75–89.

Kirkbride, J. B., Jones, P. B., Ullrich, S., & Coid, J. W. (2012). Social deprivation, inequality, and the neighborhood-level incidence of psychotic syndromes in East London. *Schizophrenia Bulletin*, *40*(1), 169–180.

Kirsch, P., Esslinger, C., Chen, Q., Mier, D., Lis, S., Siddhanti, S., et al. (2005). Oxytocin modulates neural circuitry for social cognition and fear in humans. *Journal of Neuroscience*, *25*(49), 11489–11493.

Kishiyama, M. M., Boyce, W. T., Jimenez, A. M., Perry, L. M., & Knight, R. T. (2009). Socioeconomic disparities affect prefrontal function in children. *Journal of Cognitive Neuroscience*, *21*(6), 1106–1115.

Knapska, E., & Maren, S. (2009). Reciprocal patterns of c-Fos expression in the medial prefrontal cortex and amygdala after extinction and renewal of conditioned fear. *Learning & Memory*, *16*(8), 486–493.

Kosfeld, M., Heinrichs, M., Zak, P. J., Fischbacher, U., & Fehr, E. (2005). Oxytocin increases trust in humans. *Nature*, *435*(7042), 673.

Krabbendam, L., & Van Os, J. (2005). Schizophrenia and urbanicity: A major environmental influence—Conditional on genetic risk. *Schizophrenia Bulletin*, *31*(4), 795–799.

Lambert, K. G., Nelson, R. J., Jovanovic, T., & Cerdá, M. (2015). Brains in the city: Neurobiological effects of urbanization. *Neuroscience & Biobehavioral Reviews*, *58*, 107–122.

Lasalvia, A., Bonetto, C., Tosato, S., Zanatta, G., Cristofalo, D., Salazzari, D., et al. (2014). First-contact incidence of psychosis in North-Eastern Italy: Influence of age, gender, immigration and socioeconomic deprivation. *The British Journal of Psychiatry*, *205*(2), 127–134.

Lederbogen, F., Kirsch, P., Haddad, L., Streit, F., Tost, H., Schuch, P., et al. (2011). City living and urban upbringing affect neural social stress processing in humans. *Nature*, *474*(7352), 498.

LeDoux, J. (2003). The emotional brain, fear, and the amygdala. *Cellular and Molecular Neurobiology*, *23*(4–5), 727–738.

Leporé, N., Voss, P., Lepore, F., Chou, Y.-Y., Fortin, M., Gougoux, F., et al. (2010). Brain structure changes visualized in early-and late-onset blind subjects. *NeuroImage*, *49*(1), 134–140.

Linver, M. R., Brooks-Gunn, J., & Kohen, D. E. (2002). Family processes as pathways from income to young children's development. *Developmental Psychology*, *38*(5), 719.

Mackey, A. P., Finn, A. S., Leonard, J. A., Jacoby-Senghor, D. S., West, M. R., Gabrieli, C. F., et al. (2015). Neuroanatomical correlates of the income-achievement gap. *Psychological Science*, *26*(6), 925–933.

March, D., Hatch, S. L., Morgan, C., Kirkbride, J. B., Bresnahan, M., Fearon, P., et al. (2008). Psychosis and place. *Epidemiologic Reviews*, *30*(1), 84–100.

Matheson, F. I., Moineddin, R., & Glazier, R. H. (2008). The weight of place: A multilevel analysis of gender, neighborhood material deprivation, and body mass index among Canadian adults. *Social Science & Medicine*, *66*(3), 675–690.

Matsumoto, M., Togashi, H., Konno, K., Koseki, H., Hirata, R., Izumi, T., et al. (2008). Early postnatal stress alters the extinction of context-dependent conditioned fear in adult rats. *Pharmacology Biochemistry and Behavior*, *89*(3), 247–252.

McKenzie, K., Murray, A., & Booth, T. (2013). Do urban environments increase the risk of anxiety, depression and psychosis? An epidemiological study. *Journal of Affective Disorders*, *150*(3), 1019–1024.

McLaughlin, K. A., Sheridan, M. A., & Lambert, H. K. (2014). Childhood adversity and neural development: Deprivation and threat as distinct dimensions of early experience. *Neuroscience & Biobehavioral Reviews*, *47*, 578–591.

McLaughlin, K. A., Sheridan, M. A., Winter, W., Fox, N. A., Zeanah, C. H., & Nelson, C. A. (2014). Widespread reductions in cortical thickness following severe early-life deprivation: A neurodevelopmental pathway to attention-deficit/hyperactivity disorder. *Biological Psychiatry*, *76*(8), 629–638.

Mehta, M. A., Golembo, N. I., Nosarti, C., Colvert, E., Mota, A., Williams, S. C., et al. (2009). Amygdala, hippocampal and corpus callosum size following severe early institutional deprivation: The English and Romanian Adoptees study pilot. *Journal of Child Psychology and Psychiatry*, *50*(8), 943–951.

Milad, M. R., & Quirk, G. J. (2002). Neurons in medial prefrontal cortex signal memory for fear extinction. *Nature*, *420*(6911), 70.

Milad, M. R., & Quirk, G. J. (2012). Fear extinction as a model for translational neuroscience: Ten years of progress. *Annual Review of Psychology*, *63*, 129–151.

Milad, M. R., Wright, C. I., Orr, S. P., Pitman, R. K., Quirk, G. J., & Rauch, S. L. (2007). Recall of fear extinction in humans activates the ventromedial prefrontal cortex and hippocampus in concert. *Biological Psychiatry*, *62*(5), 446–454.

Mittal, V. A., & Walker, E. F. (2019). Advances in the neurobiology of stress and psychosis. *Schizophrenia Research*, *213*, 1–5.

Mountcastle, V. B., Lynch, J., Georgopoulos, A., Sakata, H., & Acuna, C. (1975). Posterior parietal association cortex of the monkey: Command functions for operations within extrapersonal space. *Journal of Neurophysiology*, *38*(4), 871–908.

Nawyn, S. J., Gjokaj, L., Agbényiga, D. L., & Grace, B. (2012). Linguistic isolation, social capital, and immigrant belonging. *Journal of Contemporary Ethnography*, *41*(3), 255–282.

Nelson, C. A., Furtado, E. A., Fox, N. A., & Zeanah, C. H. (2009). The deprived human brain: Developmental deficits among institutionalized Romanian children—And later improvements—Strengthen the case for individualized care. *American Scientist*, *97*(3), 222–229.

Newbury, J., Arseneault, L., Caspi, A., Moffitt, T. E., Odgers, C. L., & Fisher, H. L. (2017). Cumulative effects of neighborhood social adversity and personal crime victimization on adolescent psychotic experiences. *Schizophrenia Bulletin*, *44*(2), 348–358.

Noble, K. G., Houston, S. M., Kan, E., & Sowell, E. R. (2012). Neural correlates of socioeconomic status in the developing human brain. *Developmental Science*, *15*(4), 516–527.

Noble, K. G., McCandliss, B. D., & Farah, M. J. (2007). Socioeconomic gradients predict individual differences in neurocognitive abilities. *Developmental Science*, *10*(4), 464–480.

Norman, G. J., Cacioppo, J. T., Morris, J. S., Karelina, K., Malarkey, W. B., DeVries, A. C., et al. (2011). Selective influences of oxytocin on the evaluative processing of social stimuli. *Journal of Psychopharmacology*, *25*(10), 1313–1319.

Norman, G. J., Hawkley, L. C., Cole, S. W., Berntson, G. G., & Cacioppo, J. T. (2012). Social neuroscience: The social brain, oxytocin, and health. *Social Neuroscience*, 7(1), 18–29.

Nusslock, R., & Miller, G. E. (2016). Early-life adversity and physical and emotional health across the lifespan: A neuroimmune network hypothesis. *Biological Psychiatry*, *80*(1), 23–32.

O'Donoghue, B., Yung, A. R., Wood, S., Thompson, A., Lin, A., McGorry, P., et al. (2015). Neighbourhood characteristics and the rate of identification of young people at ultra-high risk for psychosis. *Schizophrenia Research*, *169*(1–3), 214–216.

Öhman, A. (2005). The role of the amygdala in human fear: Automatic detection of threat. *Psychoneuroendocrinology*, *30*(10), 953–958.

Oliver, V., & Cheff, R. (2014). The social network: Homeless young women, social capital, and the health implications of belonging outside the nuclear family. *Youth & Society*, *46*(5), 642–662.

Omer, S., Finnegan, M., Pringle, D. G., Kinsella, A., Fearon, P., Russell, V., et al. (2016). Socioeconomic status at birth and risk for first episode psychosis in rural Ireland: Eliminating the features of urbanicity in the Cavan-Monaghan First Episode Psychosis Study (CAMFEPS). *Schizophrenia Research*, *173*(1–2), 84–89.

Omer, S., Kirkbride, J. B., Pringle, D. G., Russell, V., O'Callaghan, E., & Waddington, J. L. (2014). Neighbourhood-level socio-environmental factors and incidence of first episode psychosis by place at onset in rural Ireland: The Cavan–Monaghan First Episode Psychosis Study [CAMFEPS]. *Schizophrenia Research*, *152*(1), 152–157.

Pan, S. W., & Carpiano, R. M. (2013). Immigrant density, sense of community belonging, and suicidal ideation among racial minority and white immigrants in Canada. *Journal of Immigrant and Minority Health*, *15*(1), 34–42.

Pechtel, P., & Pizzagalli, D. A. (2011). Effects of early life stress on cognitive and affective function: An integrated review of human literature. *Psychopharmacology*, *214*(1), 55–70.

Pedersen, C. B., & Mortensen, P. B. (2001). Evidence of a dose-response relationship between urbanicity during upbringing and schizophrenia risk. *Archives of General Psychiatry*, *58*(11), 1039–1046.

Petanjek, Z., Judaš, M., Šimić, G., Rašin, M. R., Uylings, H. B., Rakic, P., et al. (2011). Extraordinary neoteny of synaptic spines in the human prefrontal cortex. *Proceedings of the National Academy of Sciences of the United States of America*, *108*(32), 13281–13286.

Petrovic, P., Kalisch, R., Singer, T., & Dolan, R. J. (2008). Oxytocin attenuates affective evaluations of conditioned faces and amygdala activity. *Journal of Neuroscience*, *28*(26), 6607–6615.

Phelps, E. A., Delgado, M. R., Nearing, K. I., & LeDoux, J. E. (2004). Extinction learning in humans: Role of the amygdala and vmPFC. *Neuron*, *43*(6), 897–905.

Pickett, K. E., & Wilkinson, R. G. (2015). Income inequality and health: A causal review. *Social Science & Medicine*, *128*, 316–326.

Popoli, M., Yan, Z., McEwen, B. S., & Sanacora, G. (2012). The stressed synapse: The impact of stress and glucocorticoids on glutamate transmission. *Nature Reviews Neuroscience*, *13*(1), 22.

Pruessner, M., Cullen, A. E., Aas, M., & Walker, E. F. (2017). The neural diathesis-stress model of schizophrenia revisited: An update on recent findings considering illness stage and neurobiological and methodological complexities. *Neuroscience & Biobehavioral Reviews*, *73*, 191–218.

Pulvermüller, F., & Garagnani, M. (2014). From sensorimotor learning to memory cells in prefrontal and temporal association cortex: A neurocomputational study of disembodiment. *Cortex*, *57*, 1–21.

Quirk, G. J., Likhtik, E., Pelletier, J. G., & Paré, D. (2003). Stimulation of medial prefrontal cortex decreases the responsiveness of central amygdala output neurons. *Journal of Neuroscience*, *23*(25), 8800–8807.

Quirk, G. J., Russo, G. K., Barron, J. L., & Lebron, K. (2000). The role of ventromedial prefrontal cortex in the recovery of extinguished fear. *Journal of Neuroscience*, *20*(16), 6225–6231.

Raizada, R. D., & Kishiyama, M. M. (2010). Effects of socioeconomic status on brain development, and how cognitive neuroscience may contribute to leveling the playing field. *Frontiers in Human Neuroscience*, *4*, 3.

Rakic, P., Bourgeois, J.-P., Eckenhoff, M. F., Zecevic, N., & Goldman-Rakic, P. S. (1986). Concurrent overproduction of synapses in diverse regions of the primate cerebral cortex. *Science*, *232*(4747), 232–235.

Rose, G. (2001). Sick individuals and sick populations. *International Journal of Epidemiology*, *30*(3), 427–432.

Rözer, J. J., & Volker, B. (2016). Does income inequality have lasting effects on health and trust? *Social Science & Medicine*, *149*, 37–45.

Saxbe, D., Khoddam, H., Piero, L. D., Stoycos, S. A., Gimbel, S. I., Margolin, G., et al. (2018). Community violence exposure in early adolescence: Longitudinal associations with hippocampal and amygdala volume and resting state connectivity. *Developmental Science*, *21*(6), e12686.

Schellenberg, G., Lu, C., Schimmele, C., & Hou, F. (2018). The correlates of self-assessed community belonging in Canada: Social capital, neighbourhood characteristics, and rootedness. *Social Indicators Research*, *140*(2), 597–618.

Schofield, P., Thygesen, M., Das-Munshi, J., Becares, L., Cantor-Graae, E., Pedersen, C., et al. (2017). Ethnic density, urbanicity and psychosis risk for migrant groups—A population cohort study. *Schizophrenia Research*, *190*, 82–87.

Shaw, P., Kabani, N. J., Lerch, J. P., Eckstrand, K., Lenroot, R., Gogtay, N., et al. (2008). Neurodevelopmental trajectories of the human cerebral cortex. *Journal of Neuroscience*, *28*(14), 3586–3594.

Sheridan, M. A., Fox, N. A., Zeanah, C. H., McLaughlin, K. A., & Nelson, C. A. (2012). Variation in neural development as a result of exposure to institutionalization early in childhood. *Proceedings of the National Academy of Sciences of the United States of America*, *109*(32), 12927–12932.

Silver, E., Mulvey, E. P., & Swanson, J. W. (2002). Neighborhood structural characteristics and mental disorder: Faris and Dunham revisited. *Social Science & Medicine*, *55*(8), 1457–1470.

Sirin, S. R. (2005). Socioeconomic status and academic achievement: A meta-analytic review of research. *Review of Educational Research*, *75*(3), 417–453.

Smyke, A. T., Koga, S. F., Johnson, D. E., Fox, N. A., Marshall, P. J., Nelson, C. A., et al. (2007). The caregiving context in institution-reared and family-reared infants and toddlers in Romania. *Journal of Child Psychology and Psychiatry*, *48*(2), 210–218.

Stone, J. M., Howes, O. D., Egerton, A., Kambeitz, J., Allen, P., Lythgoe, D. J., et al. (2010). Altered relationship between hippocampal glutamate levels and striatal dopamine function in subjects at ultra high risk of psychosis. *Biological Psychiatry*, *68*(7), 599–602.

Takei, N., Sham, P. C., O'Callaghan, E., Glover, G., & Murray, R. M. (1995). Schizophrenia: Increased risk associated with winter and city birth—A case-control study in 12 regions within England and Wales. *Journal of Epidemiology and Community Health*, *49*(1), 106.

Tandon, R., Nasrallah, H. A., & Keshavan, M. S. (2009). Schizophrenia, "just the facts" 4. Clinical features and conceptualization. *Schizophrenia Research*, *110*(1–3), 1–23.

Termorshuizen, F., Smeets, H. M., Braam, A. W., & Veling, W. (2014). Neighborhood ethnic density and psychotic disorders among ethnic minority groups in Utrecht City. *Social Psychiatry and Psychiatric Epidemiology*, *49*(7), 1093–1102.

Toledo-Rodriguez, M., & Sandi, C. (2007). Stress before puberty exerts a sex-and age-related impact on auditory and contextual fear conditioning in the rat. *Neural Plasticity, 2007*.

Tottenham, N. (2012). Human amygdala development in the absence of species-expected caregiving. *Developmental Psychobiology*, *54*(6), 598–611.

Tottenham, N., Hare, T. A., Quinn, B. T., McCarry, T. W., Nurse, M., Gilhooly, T., et al. (2010). Prolonged institutional rearing is associated with atypically large amygdala volume and difficulties in emotion regulation. *Developmental Science*, *13*(1), 46–61.

Tottenham, N., & Sheridan, M. A. (2010). A review of adversity, the amygdala and the hippocampus: A consideration of developmental timing. *Frontiers in Human Neuroscience*, *3*, 68.

van Marle, H. J., Hermans, E. J., Qin, S., & Fernández, G. (2009). From specificity to sensitivity: How acute stress affects amygdala processing of biologically salient stimuli. *Biological Psychiatry*, *66*(7), 649–655.

Van Os, J., Driessen, G., Gunther, N., & Delespaul, P. (2000). Neighbourhood variation in incidence of schizophrenia: Evidence for person-environment interaction. *The British Journal of Psychiatry*, *176*(3), 243–248.

Vassos, E., Pedersen, C. B., Murray, R. M., Collier, D. A., & Lewis, C. M. (2012). Meta-analysis of the association of urbanicity with schizophrenia. *Schizophrenia Bulletin*, *38*(6), 1118–1123.

Veling, W., Susser, E., Van Os, J., Mackenbach, J. P., Selten, J.-P., & Hoek, H. W. (2008). Ethnic density of neighborhoods and incidence of psychotic disorders among immigrants. *American Journal of Psychiatry*, *165*(1), 66–73.

Vilhjalmsdottir, A., Gardarsdottir, R. B., Bernburg, J. G., & Sigfusdottir, I. D. (2016). Neighborhood income inequality, social capital and emotional distress among adolescents: A population-based study. *Journal of Adolescence*, *51*, 92–102.

Vos, T., Abajobir, A. A., Abate, K. H., Abbafati, C., Abbas, K. M., Abd-Allah, F., et al. (2017). Global, regional, and national incidence, prevalence, and years lived with disability for 328 diseases and injuries for 195 countries, 1990–2016: A systematic analysis for the Global Burden of Disease Study 2016. *The Lancet*, *390*(10100), 1211–1259.

Walker, E., Mittal, V., & Tessner, K. (2008). Stress and the hypothalamic pituitary adrenal axis in the developmental course of schizophrenia. *Annual Review of Clinical Psychology*, *4*, 189–216.

Weich, S., Blanchard, M., Prince, M., Burton, E., Erens, B., & Sproston, K. (2002). Mental health and the built environment: Cross–sectional survey of individual and contextual risk factors for depression. *The British Journal of Psychiatry*, *180*(5), 428–433.

Weisleder, A., & Fernald, A. (2013). Talking to children matters: Early language experience strengthens processing and builds vocabulary. *Psychological Science*, *24*(11), 2143–2152.

Wickham, S., Taylor, P., Shevlin, M., & Bentall, R. P. (2014). The impact of social deprivation on paranoia, hallucinations, mania and depression: The role of discrimination social support, stress and trust. *PLoS One*, *9*(8), e105140.

Wilson-Genderson, M., & Pruchno, R. (2013). Effects of neighborhood violence and perceptions of neighborhood safety on depressive symptoms of older adults. *Social Science & Medicine*, *85*, 43–49.

Yang, T.-C., Lei, L., & Kurtulus, A. (2018). Neighborhood ethnic density and self-rated health: Investigating the mechanisms through social capital and health behaviors. *Health & Place*, *53*, 193–202.

Yeung, W. J., Linver, M. R., & Brooks-Gunn, J. (2002). How money matters for young children's development: Parental investment and family processes. *Child Development*, *73*(6), 1861–1879.

CHAPTER FOUR

The influence of stress and early life adversity on addiction: Psychobiological mechanisms of risk and resilience

Mustafa al'Absi*
Department of Family Medicine and Biobehavioral Health, University of Minnesota Medical School, Duluth, MN, United States
*Corresponding author: e-mail address: malabsi@umn.edu

Contents

Abstract

Preclinical, clinical, and population research demonstrates that stress and early life adversity (ELA) increase vulnerability to initiate, maintain, and relapse in addiction. Individuals with addiction problems have higher prevalence of trauma exposure than nondrug users; and this association extends to young populations (e.g., adolescents). Mechanisms for these associations likely involve multiple systems, including changes in the mesolimbic reward functions, HPA axis stress response, and other stress- and reward-related pathways. Other brain morphological and functional changes are also

International Review of Neurobiology, Volume 152
ISSN 0074-7742
https://doi.org/10.1016/bs.irn.2020.03.012

likely involved and directly contribute to the neurohormonal and behavioral alterations observed during adulthood in those exposed to ELA. Stress-related risk is influenced by sex, genetic factors, and resilience, among other factors. Our heuristic model proposes that long-term effects of stress and ELA on the brain contribute to dysregulation of the stress response, emotional reactivity, reward systems, cognitive dysregulation, and delay discounting that lead to impulsive and high-risk behaviors, such as drug use and relapse.

1. Introduction and definitions

Accumulated research demonstrates the impact of acute and chronic stress on risk for addiction onset, maintenance, and relapse. This research has come from multiple disciplines, including psychology, neurobiology, psychiatry, and related areas of investigation; and it includes support from different approaches of investigation, including preclinical, clinical, community, and population studies (al'Absi, 2007; Anda et al., 2006; Dong et al., 2004; Dube et al., 2006; Dube, Felitti, Dong, Giles, & Anda, 2003; Lovallo, 2007; Sinha, 2008; Taha, Galea, Hien, & Goodwin, 2014). Evidence also comes from research in different regions of the world and from different populations. In all, this research converges in support of the significant roles of stress and early life adversity in addiction and other mental and medical health problems.

In this paper, we will review recent literature on the impact of stress and early life adversity on multiple neurobehavioral and neurobiological systems. We will develop an integrative model to guide our understanding of how stressful events increase risk for addictive disorders and lead to maintenance of, or hinder efforts to recover from, addiction. Changes that are caused by stress will be reviewed as potential mediators of the impact of stress and early life adversity on addictive processes. We will also review factors that may amplify or buffer the impact of stress and ELA on addiction risk and maintenance. An integration and a discussion of the existing literature within a heuristic model will be presented. Implication of this model for future research directions will also be provided.

First, we define important constructs and processes related to this review. In this review, we adopt the definition of stress as a psychological or physical factor that causes mental or physiological tension (Chrousos, 2009; Lovallo, 2004). We distinguish psychological from physical stressors because of the inherent physical changes caused by physical stressors, such as being exposed to freezing weather or to a toxic substance. Psychological stress produces its physiological, mental, and behavioral responses because of how we perceive events or situations in our environments rather than because of its physical attributes. It is important to draw a distinction between a stressor and a stress response. The latter represents a constellation of neurophysiological, biological, and behavioral changes that are produced as a result of one's perception of a stressful situation, as a way to cope with or address the situation. Review of specific stress response systems will be presented later.

In this review, we will focus on early life stress and adversity (ELA) as a model of chronic stress. ELA is defined as exposure to a range of environmental changes and adverse experiences, including exposure to violence domestically or in the community, maltreatment, parental loss, neglect, separation, and extreme poverty during childhood or adolescence. These experiences represent challenges for the individual's capacity to cope and require multiple behavioral and biological adaptive processes (Doucette et al., 2016; Duffy, McLaughlin, & Green, 2018). We focus primarily on ELA as a model of stress with long-term adverse effects on addiction and related behavioral health problems.

In addition to ELA, we will also address adult and ongoing stressful life events (SLE). SLE is defined here as events that may occur on an ongoing or periodic basis that are likely to trigger adjustment processes or responses that include both internal (systemic) and external social and personal activities. There is a focus in the current context on the negative characteristics of such events and their adverse effects on psychological and behavioral problems, particularly addictive behaviors and the manner by which SLE influence these behaviors (Dohrenwend, 2006).

Addiction represents a constellation of behavioral, biological, and psychological processes that contribute to impulsive behaviors and loss of control over use, obsessive thoughts about obtaining drugs, and continuation of use despite negative health and social consequences. Addiction is positively reinforced by acute effects, which are characterized by a sense of pleasure, or a high, produced by addictive substances. Both positive (euphoric) and negative (anxiolytic) reinforcement are features of addiction. Negative reinforcement tends to become the driving force as the disorder progresses. Abstinence from addictive substances is usually associated with craving, withdrawal syndrome, and other affective symptoms. Addiction or substance use disorders are defined by the DSM-V as a continuum from mild to severe and include both abuse and dependence. This version of the DSM also emphasizes use of the phrase "substance use disorders" instead of dependence or addiction. Because of our focus on addictive substances and behaviors, and to reflect the extant research domains, we will continue to use the term "addiction" in this paper.

2. Stress response neurobiological pathways

Exposure to acute stress activates multiple physiological and behavioral systems, including the hypothalamic pituitary adrenocortical (HPA) axis, the sympathetic-adrenal medullary (SAM) axis, and the endogenous opioid system (al'Absi & Arnett, 2000; Chrousos & Gold, 1992; Kudielka, Schommer, Hellhammer, & Kirschbaum, 2004; McEwen, 1997; Stratakis & Chrousos, 1995). The HPA axis plays a significant role in coordinating the neuroendocrine response to stress and in mediating effects of stress on health (al'Absi, 2007; McEwen, 1998; Sapolsky, 2000; Sapolsky, Romero, & Munck, 2000). The HPA axis includes three primary structures. (1) The neuronal cell bodies of the paraventricular nucleus (PVN) of the hypothalamus release corticotropin-releasing factor (CRF). (2) CRF is transported through the portal systems to the pituitary gland where it stimulates the release of adrenocorticotropic hormone (ACTH) and beta-endorphin. (3) ACTH travels through the system to the cortical part of the adrenal gland where it triggers the synthesis and release of corticosteroids (Dallman, 1993; Munck, Guyre, & Holbrook, 1984). Corticosteroids, primarily cortisol in humans, circulate back to the brain where they provide a negative feedback loop regulating the production of ACTH and CRF (Dallman, 1993; Munck et al., 1984).

The SAM response is mediated by production of epinephrine from the adrenal medulla. The SAM axis also stimulates the production of norepinephrine. Central activation of the sympathetic nervous system (SNS) also includes the locus coeruleus, a brainstem nucleus that is dense with catecholamine neurons. Stimulation of these neurons activates the adrenal medulla, leading to the synthesis and release epinephrine and norepinephrine into circulation, which go on to activate receptors in various tissues and systems of the body, including cardiac and smooth muscles, adipose tissues, endocrine glands, and the immune system (al'Absi, 2018; Coventry et al., 2001; Kiecolt-Glaser, McGuire, Robles, & Glaser, 2002; Lemieux & al'Absi, 2016; Thayer & Brosschot, 2005).

Endogenous opioids, which are opiate-like substances released by the brain, are important in regulating pain, mood, and reward (Drolet et al., 2001; Martin del Campo, Dowson, Herbert, & Paykel, 1994). Endogenous opioids are present in the amygdala, cingulate cortex, periaqueductal gray, and other brain regions that are also involved in regulating various stress response processes (Ableitner & Schulz, 1992; Apkarian, Bushnell, Treede, & Zubieta, 2005; Dunn, Swiergiel, & Palamarchouk, 2004; Gianoulakis, 2004; Pasternak, 2005; Thayer & Lane, 2000). Exposure to stress increases the release of endogenous opiates, which are involved in regulating the HPA and other physiological responses to stress (Akil et al., 1984; Drolet et al., 2001; Morris et al., 1990; Rushen, Schwartze, Ladewig, & Foxcroft, 1993; Wand & Schumann, 1998). Beyond their role in the stress response, endogenous opioids in the brainstem, limbic system, and prefrontal cortex interact with dopaminergic, GABAergic, and adrenergic systems (Barrot et al., 2000; Cowen & Lawrence, 1999; Marinelli, 2007; Nevo & Hamon, 1995; Piazza et al., 1996), implicating these factors in multiple processes that directly affect biological stress response systems as well as mood and drug reward. For example, inhibitory input from beta-endorphin neurons in the arcuate nucleus inhibits CRF neurons in the hypothalamus, an effect that is mediated by μ-opioid receptors (Hellbach et al., 1998; Henderson, 1983; Tsagarakis, Rees, Besser, & Grossman, 1990). Similarly, opiate neuronal innervation of neurons within the locus coeruleus reduces noradrenergic release (Christie, 1991; Singewald & Philippu, 1998; Valentino & Van Bockstaele, 2001) and input from arcuate nucleus to the nucleus accumbens and prefrontal cortex stimulates the mesolimbic dopaminergic pathway (Copeland, Neff, & Hadjiconstantinou, 2005; Koob & Kreek, 2007; Mizrahi et al., 2013; Picetti et al., 2013; Uhart & Wand, 2009; Volkow, Wang, Fowler, Tomasi, & Baler, 2012; Wang et al., 2005).

3. Addiction and brain reward pathways

Effects of addictive substances on the brain are influenced by the type of substance, course of use, and chronicity. While initial use is characterized by euphorigenic effects, continuing use leads to a compulsive pattern (Edwards & Koob, 2010; Erb, 2007; Koob, 2013, 2017, 2020). Multiple factors contribute to continuing engagement in addictive behaviors, ranging from genetic predisposition to the reinforcing effects of the behaviors. The brain's changes in response to these substances over time may underlie a motivation state caused by a complex set of adaptive processes that subsequently contribute to the maintenance of substance use by causing cravings, withdrawal syndrome, and subsequent relapse (al'Absi, 2006; Bruijnzeel, 2012; Gulliver, Hughes, Solomon, & Dey, 1995; Harris,

Wimmer, & Aston-Jones, 2005; Koob, 2020; Sinha & Jastreboff, 2013). This cycle of events is mediated by multiple neurobiological pathways, including the mesolimbic dopamine (DA) system.

It is well-established that many substances and other types of rewarding experiences induce their reinforcing effects by increasing the DA transmission and DA release. The influence of reward pathways is particularly salient during initiation and transition to compulsive substance use (Koob, 2013, 2017, 2020; Koob & Kreek, 2007; Koob & Le, 1997; Zorrilla, Logrip, & Koob, 2014). The DA system includes the ventral tegmental area (VTA) of the midbrain, which projects to the nucleus accumbens (NAc) as well as the limbic system and the orbitofrontal cortex. Within this network, the amygdala, hippocampus, and medial prefrontal cortex play a role in modifying reward system activity by sending excitatory projections to the NAc (Gilman et al., 2014; Le Merrer, Becker, Befort, & Kieffer, 2009; Makris et al., 2008; Pellissier, Gandía, Laboute, Becker, & Le Merrer, 2018). Beyond its role in drug use reinforcement, this pathway influences other neuronal functions, such as movement, memory, and attention (Arnsten & Goldman-Rakic, 1998; Gorwood et al., 2012; Hadjiconstantinou & Neff, 2011; Mittleman, Jones, & Robbins, 1991).

In addition to the DA system, research has demonstrated a role for the endogenous opioid system in addiction. Endogenous opioids regulate the reinforcing properties of addictive substances (Drolet et al., 2001; Martin del Campo et al., 1994); and brain structures that are involved in drug reward and reinforcement contain endogenous opioids and receptors (Ableitner & Schulz, 1992; Apkarian et al., 2005; Dunn et al., 2004; Gianoulakis, 2004; Pasternak, 2005; Thayer & Lane, 2000). These regions include the limbic system and brainstem. Indeed, preclinical studies show a direct interaction between central opioid activity and dopaminergic, GABAergic, and adrenergic pathways (Barrot et al., 2000; Cowen & Lawrence, 1999; Marinelli, 2007; Nevo & Hamon, 1995; Piazza et al., 1996), implicating this system is regulating reward and affective functions associated with addictive behaviors. Further, research has demonstrated the role of μ-opioid receptors in mediating reward effects and reinforcement. For example, μ-receptor agonists injected into the VTA or nucleus accumbens (NAc) facilitate a conditioned place preference (REF). The link between the opioid system and drug reward has also been demonstrated by genetic studies involving μ-opioid receptor knockout mice. These studies have shown that the knockout mice are insensitive to opiates and do not exhibit the expected self-administration behavior (Ayranci, Befort,

Lalanne, Kieffer, & Lutz, 2015; Belzeaux, Lalanne, Kieffer, & Lutz, 2018; Lissek et al., 2017; Lutz & Kieffer, 2013; Lutz, Reiss, Ouagazzal, & Kieffer, 2013). The reinforcing effects of other nonopioid substances are also reduced (Le Merrer et al., 2009; Lutz & Kieffer, 2013). These observations suggest a role for this system, through μ-receptor function, in regulating reward and reinforcement and, therefore, contributing to the risk for initiation and maintenance of addiction.

4. Stress and ELA as a risk factor for addiction

Evidence from multiple domains of research supports an influence of stress on all stages of addiction. This evidence demonstrates the need to approach addiction from multiple levels of analysis, including behavioral, affective, and physiological, to better reflect the complex nature of the influences of stress and ELA on addiction risk, maintenance, and relapse. Both preclinical and clinical evidence demonstrates the role of stressful events in facilitating initiation, maintenance of addictive behaviors, and relapse (al'Absi, 2018; al'Absi, Lemieux, Westra, & Allen, 2017; al'Absi, Nakajima, & Lemieux, 2018; Burke, Finn, McGuire, & Roche, 2017; Colich et al., 2019; Duffy et al., 2018; Lambert et al., 2019; Luijcks et al., 2016; McLaughlin et al., 2017; McLaughlin, DeCross, Jovanovic, & Tottenham, 2019). Beyond psychological and environmental stressors, negative affect, craving, and other symptoms associated with withdrawal may also be experienced as stressful (al'Absi, 2018; al'Absi et al., 2018; al'Absi, Nakajima, Allen, Lemieux, & Hatsukami, 2015; Bongard, al'Absi, Khalil, & Al Habori, 2011; Bruijnzeel, 2012; Koob, 2017, 2020; Lemieux & al'Absi, 2016; Potretzke, Nakajima, Cragin, & al'Absi, 2014; Slopen et al., 2013; Zorrilla et al., 2014) and may trigger substance use and relapse. Use in response to these symptoms, therefore, may be reinforced negatively by reduction or relief of these symptoms.

5. Mechanisms mediating effects of stress on addiction

5.1 Preclinical evidence on mechanisms mediating ELA and addiction

One of the potential mechanisms of ELA's effect on addictive behaviors is HPA dysregulation. Research with rhesus macaque monkeys has shown long-term effects of maternal separation during early life on HPA dysregulation, characterized by a flattened diurnal cortisol production

(Sánchez et al., 2005). These effects vary by sex, with only females showing increased cortisol. Similar effects were found with restraint stress (McCormick, Kehoe, Mallinson, Cecchi, & Frye, 2002). In addition to the impact of both handling and maternal separation on corticosterone and ACTH levels, these stressors also influence CRF mRNA expression and content in the PVN and they cause changes to glucocorticoid receptor levels in the hippocampus and hypothalamus. Preclinical studies show that after 24-h of separation from the dam (Smith, Kim, van Oers, & Levine, 1997), HPA hormones are increased and CRH mRNA levels decrease, effects that were not reversed by suppressing corticosterone release pharmacologically (Smith et al., 1997; van Bodegom, Homberg, & Henckens, 2017). On the other hand, handling and tactile stimulation or feeding reduced these effects (McCormick et al., 2002; Meaney et al., 1991, 1996). This illustrates the impact of maternal separation in stimulating the stress response (Pryce, Bettschen, Bahr, & Feldon, 2001) and the role of maternal behavior, such as nursing and licking, in regulating the HPA stress response in offspring (Kosten, Lee, & Kim, 2007).

It has been noted that the direction of these changes depends on age at separation, frequency, and duration (Lehmann, Stöhr, & Feldon, 2000). Handled rats exhibit decreased HPA responses to stress (Arborelius, Owens, Plotsky, & Nemeroff, 1999; Ladd et al., 2000; Meaney et al., 1993, 1996) and increased negative feedback sensitivity to glucocorticoids (Meaney et al., 1993, 1996). They also show lower CRF mRNA expression and content in PVN (Plotsky & Meaney, 1993) compared to nonhandled rats. On the other hand, maternal separation is associated with lowers GR receptor binding in hippocampus and hypothalamus as well as reduced CRF levels in hypothalamus, hippocampus, and amygdala (Heim, Ehlert, Rexhausen, Hanker, & Hellhammer, 1997; Liu & Nusslock, 2018; Plotsky & Meaney, 1993).

5.2 ELA and HPA blunted response in humans

Findings from animal models have been extended to prospective studies in humans which have shown that women who were abused as children show reduced adrenocortical activity during adolescence and early adulthood (Trickett, Noll, Susman, Shenk, & Putnam, 2010), and this has also been found in adults with history of childhood adversity (Cicchetti, Rogosch, Gunnar, & Toth, 2010; Gebremariam, Bjune, & Frich, 2011; Hostinar & Gunnar, 2015; Perry, DePasquale, Fisher, & Gunnar, 2019; Tarullo &

Gunnar, 2006). These patterns of dysregulation have also been documented in adolescent girls who were free from psychopathology, but who had experienced childhood maltreatment (Brown et al., 2009; Carpenter, Shattuck, Tyrka, Geracioti, & Price, 2011; MacMillan et al., 2009; Tyrka et al., 2008; Tyrka, Walters, Price, Anderson, & Carpenter, 2012). This dysregulation may be due to changes in central circuitry of the stress response, as suggested by blunted cortisol response to a dexamethasone/CRH test (Carpenter et al., 2009; Tyrka et al., 2008; Tyrka, Price, Marsit, Walters, & Carpenter, 2012).

While most studies do confirm blunted cortisol release and response as a long-term consequence of exposure to ELA, it should be noted that some studies have observed the opposite, including enhanced ACTH and cortisol responses to acute stress (Pruessner, Champagne, Meaney, & Dagher, 2004). Some authors have noted multiple factors that may explain this discrepancy, including presence of comorbid psychopathology, presence of high levels of ongoing stressors, as well as individual differences related to genetic risk (Enoch, 2011). There are indications of change in glucocorticoid receptor (GR) mRNA expression in certain areas of the brain, such as the hippocampus, in animals exposed to maternal neglect (Weaver, 2009), which may influence sensitivity of the GR-mediated feedback, leading to increased HPA activity. Nevertheless, support that this blunted stress response may serve as an intermediate marker of vulnerability to addiction has also been found in research targeting groups at risk for substance use, including those with family history of alcoholism (Lovallo, Acheson, Vincent, Sorocco, & Cohoon, 2018; Lovallo, Cohoon, Acheson, Sorocco, & Vincent, 2019), those with antisocial behavior (Lovallo, 2007, 2013; Lovallo, Farag, Sorocco, Cohoon, & Vincent, 2012; Snoek, Van Goozen, Matthys, Buitelaar, & van Engeland, 2004; van Goozen, Fairchild, Snoek, & Harold, 2007), and those with other risk factors (Oswald et al., 2006; Somaini et al., 2011; van Leeuwen et al., 2011).

5.3 Stress and mesolimbic DA functions

Mechanisms that contribute to the influence of ELA on addiction involve dysregulation of the mesolimbic DA functions (Koob & Bloom, 1988; Lovallo, Cohoon, Sorocco, et al., 2019; Meaney, Brake, & Gratton, 2002; Moffett et al., 2007; Pruessner et al., 2004; Sinha, 2008; Wise & Rompre, 1989). This has been supported by preclinical evidence showing, for example, that neonatal isolation increases DA responses to psychostimulants (Kosten & Ambrosio, 2002; Kosten, Miserendino, & Kehoe, 2000; Kosten, Sanchez, Zhang, & Kehoe, 2004; Kosten, Zhang, & Kehoe, 2003; Walters & Kosten, 2019). There is also evidence that exposure to stress stimulates the release of DA (Kalivas, Pierce, Cornish, & Sorg, 1998; Kalivas & Volkow, 2005; Moffett et al., 2007; Prasad, Sorg, Ulibarri, & Kalivas, 1995; Roth, Tam, Ida, Yang, & Deutch, 1988).

Preclinical research has also demonstrated the impact of stress and ELA on the mesolimbic DA response to amphetamine and cocaine administration, with greater stressed pups and juveniles exhibiting enhanced responses compared to control groups (Kosten et al., 2000, 2003, 2004; Kosten & Ambrosio, 2002; Walters & Kosten, 2019). Also, amphetamine increased extracellular DA in the NAc in adult rats with maternal separation experience relative to controls (Hall, Wilkinson, Humby, & Robbins, 1999). However, the opposite was found with restraint stress, where DA levels in the NAc was high in the nonhandled animals only (McCormick et al., 2002). It is worth noting that differences tended to be present in response to drugs rather than in basal levels (Hall et al., 1999; Zimmerberg & Brown, 1998), suggesting specific priming effects of ELA on the effects of drugs on the brain and behavior.

6. Moderators of the effects of stress and ELA on addiction

In the next section, we review important and well-supported factors that have been identified in the literature as determinants of the effects of stress and early life adversity on risk for addiction as well as on maintenance of addiction and relapse.

6.1 Genetic and epigenetic factors

Multiple areas of investigation have pointed-out the possibility that genetic variation may contribute to the long-term effects of ELA and may modify individuals' vulnerability to developing substance use and other behavioral problems during adulthood. For example, there are preclinical data suggesting that reduced gene transcription in the GR promoter region may result from maternal deprivation during childhood (Szyf, Weaver, & Meaney, 2007; Weaver et al., 2004). Research has also shown that the GR gene, NR3C1, mediates stress and drug effects. In one study, suppression of NR3C1 in mice DA receptive neurons reduced DA cell transmission and reduced cocaine self-administration (Ambroggi et al., 2009). Another experiment showed that inactivation of NR3C1 led to decreased stress-related behavioral responses (Ambroggi et al., 2009; Deroche-Gamonet et al., 2003).

Early life maternal separation in rats has been shown to induce long-term changes in brain-derived neurotrophic factor (BDNF) expression, with decreased BDNF in the hippocampus and increased BDNF in the VTA (Lippmann, Bress, Nemeroff, Plotsky, & Monteggia, 2007). There is also evidence that the impact of ELA may be mediated by long-term changes in BDNF expression. One study in rats showed that maternal separation decreased BDNF expression in the hippocampus but increased expression in the VTA (Lippmann et al., 2007). Altered BDNF gene expression has also been seen in the adult prefrontal cortex, resulting from the long-term effect of early life stress on BDNF DNA methylation (Roth, Lubin, Funk, & Sweatt, 2009). These changes may play a role in increased risk for addiction. This is supported by observations indicating a role for BDNF in drug reward (Vargas-Perez et al., 2009).

Research focusing on genetic factors in addiction has also demonstrated the influence of genes in shaping the stress response and their potential impact in the effects of stress on subsequent risk for drug addiction. For example, research has shown that cortisol response to stress is influence by the presence of the G allele of the μ-opioid receptor gene (*OPRM1* A118G) relative to those with AA homozygotes. This is characterized by blunted cortisol response in the *G allele carriers* due to enhance receptor binding bond opiate (Bond et al., 1998; Lovallo et al., 2017; Womersley, Martin, van der Merwe, Seedat, & Hemmings, 2018), leading to enhanced endogenous opioid regulation of the HPA system (Lovallo et al., 2015).

These findings are further supported by an established body of work demonstrating the role of gene-by-environment interactions as a key to the influence of stress on addiction, especially through their effects on stress response and reward regulation. For example, one study with human participants observed decreased levels of GR (*NR3C1*) mRNA and increased methylation of an *NR3C1* promoter in postmortem hippocampus of suicide victims who were exposed to high levels of ELA (McGowan et al., 2009). Evidence of an interaction effect with childhood maltreatment was also observed in research focusing on monoamine oxidase A gene (*MAOA*), which was found to influence risk for adult antisocial behavior (Caspi et al., 2002; Gallardo-Pujol, Andrés-Pueyo, & Maydeu-Olivares, 2013). The interactive effects of stress during adulthood were also observed with the serotonin transporter gene (*SLC6A4*) and its impact on risk for depression (Caspi et al., 2003; Suppli et al., 2017). Other conditions also show interaction of genes related to the stress response with ELA to increase vulnerability for PTSD, depression, and other behavioral problems (Caspi, Hariri, Holmes, Uher, & Moffitt, 2010; Dunn et al., 2019; Heim et al., 2009; Klengel et al., 2013; Roy, Gorodetsky, Yuan, Goldman, & Enoch, 2010; Suppli et al., 2017; van Rooij et al., 2016; Zannas et al., 2015).

6.2 Sex differences

Effects of stress and ELA on risk for addiction are modified by sex. Preclinical studies have demonstrated such effects, indicating greater effects of early stress in female rats compared to male rats (Brydges, Best, & Thomas, 2019; Holschneider, Guo, Mayer, & Wang, 2016; Najavits et al., 1998; Page, Opp, & Kozachik, 2016; Thomas & Becker, 2019; Walters & Kosten, 2019). For example, studies examining the impact of early life maternal separation on the HPA axis in adult rats have demonstrated sex differences (Calatayud & Belzung, 2001; Desbonnet, Garrett, Daly, McDermott, & Dinan, 2008; Litvin et al., 2010). While stress decreases GR receptor levels in the hippocampus of male rats, it increases receptors in female rats and may be related to greater maternal handling of male than female pups (Champagne, Diorio, Sharma, & Meaney, 2001; Sutanto, Rosenfeld, de Kloet, & Levine, 1996). These sex differences in maternal care may explain, in part, differences in the stress response seen in males and females (Beatty & Beatty, 1970; Kosten, Sanchez, Jatlow, & Kehoe, 2005).

Research in humans also demonstrates sex difference in the stress response. Men show larger cortisol responses to stress (al'Absi et al., 2004; al'Absi, Wittmers, Hatsukami, & Westra, 2008; Kirschbaum, Kudielka, Gaab, Schommer, & Hellhammer, 1999; Kudielka et al., 2004; Kudielka & Kirschbaum, 2005; Steptoe, Fieldman, Evans, & Perry, 1996) and women have larger HPA responses to opioid blockade (al'Absi et al., 2004; al'Absi, France, Harju, France, & Wittmers, 2006; Lovallo, King, et al., 2012; Lovallo et al., 2015; Lovallo, Farag, et al., 2012), suggesting enhanced endogenous opioid regulation of the stress response in women. Interestingly, differences in how μ-opioid receptor genes may influence this regulation also occur between men and women, with the *OPRM1* GA/GG genotypes showing enhanced regulation in women. In one study (Lovallo et al., 2015), *OPRM1* A118G polymorphism predicted blunted cortisol responses to stress in women carrying the G allele compared to AA carriers.

This difference was not observed in men (Lovallo et al., 2015). Specific to effects in females, research has noted that early life stress diminishes estrous stage effects in females; and stress effects on drug self-administration also generalize to other appetitive behaviors (Kosten et al., 2007). It is possible that ELA produces more profound effects in females and contributes to differential consequences during adulthood, such as eating disorders and depression, which may, in turn, contribute to substance use problems.

6.3 Resilience and factors that may buffer effects of stress on addiction

Beyond the adverse effects of stress and ELA on risk for psychopathology and substance use problems, recent research suggests the possibility that exposure to ELA may enhance responses to positive environmental changes in ways that enhance resiliency (Albott, Forbes, & Anker, 2018; McLaughlin, Conron, Koenen, & Gilman, 2010; Myers, McLaughlin, Wang, Blanco, & Stein, 2014). One study found that high ELA was associated with enhanced benefits from positive environmental changes, as indicated by increased reductions of psychiatric risk and substance use disorders in those with high compared to low ELA (Albott et al., 2018; Myers et al., 2014). Additional factors that could enhance resilience and buffer the impact of ELA on substance use and other psychological problems include good parenting, good peer relationships, and neighborhood safety (DuMont, Widom, & Czaja, 2007; Fergusson, Boden, & Horwood, 2008; Fergusson, Woodward, & Horwood, 1999; Jaffee, Caspi, Moffitt, Polo-Tomás, & Taylor, 2007; Lansford et al., 2006; Miller et al., 2018; Woodward, Fergusson, & Horwood, 2002).

7. Discussion and heuristic model

Observations indicate that ELA and stress-related brain changes lead to dysregulation of the stress response, impulsive behaviors, enhanced sensation seeking behaviors (al'Absi, 2018; Carpenter et al., 2011; Lovallo et al., 2017, 2018; Lovallo, Cohoon, Acheson, et al., 2019; Lovallo, Cohoon, Sorocco, et al., 2019; MacMillan et al., 2009; Ouellet-Morin et al., 2011; Voellmin et al., 2015), and exaggerated emotional reactivity (Briggs-Gowan et al., 2015, 2019; Dannlowski et al., 2013). These processes, in turn, increase risky behaviors, such as drug and alcohol use (Aldao & Nolen-Hoeksema, 2012; Aldao, Nolen-Hoeksema, & Schweizer, 2010; Corral-Frías et al., 2015, 2016; Yang, Bogiatzi, & Spence, 2015), cigarette smoking (Baker,

Brandon, & Chassin, 2004; Kassel et al., 2007; Kober et al., 2010), and other appetitive behaviors (Danese & Tan, 2014; Hemmingsson, Johansson, & Reynisdottir, 2014; Michopoulos et al., 2015; Volkow, Fowler, Wang, Swanson, & Telang, 2007; Wang et al., 2006; Wang, Volkow, Thanos, & Fowler, 2004).

Based on existing literature and studies reviewed above, we propose that exposure to stress and early life adversity triggers cellular and neurochemical changes that influence brain development and lead to long-term effects in stress response pathways, emotional regulation, and the reward system (see Fig. 1). These changes contribute to increased vulnerability to engage in high-risk behaviors, including maintenance of addictive behaviors (al'Absi, 2007; al'Absi et al., 2018; Duffy et al., 2018). Multiple neurobiological systems are influenced by early exposure to adversity and stress. These include brain cortical, subcortical, and peripheral systems. In the brain, multiple structures are affected within the limbic system and the cortex, including prefrontal, amygdala, and hippocampus structures involved in evaluating stress and threat. This network, in turn, influences triggering and organization of the stress response. Such a response is likely to be influenced by exposure to stress and adversity during childhood, leading to enhanced vigilance and emotional reactivity to environmental cues (Hein & Monk, 2017; Kelly et al., 2013; McCrory et al., 2013; Peckins et al., 2020; Pollak, Messner, Kistler, & Cohn, 2009; Scult, Knodt, Swartz, Brigidi, & Hariri, 2017; Sebastian et al., 2012, 2014; Swartz, Knodt, Radtke, & Hariri, 2015).

Another system that is influenced by early exposure to stress is the reward system; stress may lead to blunted responsivity of dopaminergic and other systems (Dillon et al., 2009; Hanson et al., 2016; Hanson, Hariri, & Williamson, 2015; Pizzagalli et al., 2009; Swartz et al., 2015), possibly leading to compensatory reward-seeking behaviors to overcome the reduced reward responsivity (Blum et al., 1996, 2000; Dichter, Damiano, & Allen, 2012; Kelley, Finley, & Schmeichel, 2019). This may represent a risk for drug use and may enhance reinforcement drawn for engaging in such behaviors. The impact of these two pathways may be moderated by factors, such as sex, genetic factors, and resilience.

ELA effects on stress and reward pathways may also impact functioning in brain regions that serve cognition and decision-making processes, potentially contributing to decisions that favor enhanced reactivity to threat and enhanced motivation for reward-stimulating behaviors (McKee et al., 2015; Sheridan, Peverill, Finn, & McLaughlin, 2017) as well as impulsive behaviors and increased delayed discounting (Bobova, Finn, Rickert, &

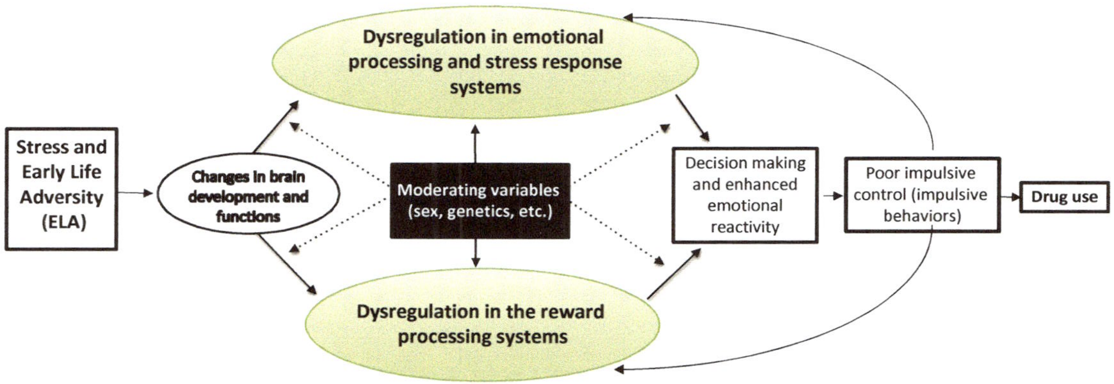

Fig. 1 Connection of stress and early life adversity with the risk for addiction.

Lucas, 2009; Müller et al., 2013; Ripke et al., 2012; Simmen-Janevska, Forstmeier, Krammer, & Maercker, 2015). This, in turn, increases chances for engaging in high-risk behaviors, including substance use and abuse.

8. Future directions

Notwithstanding extant literature on the impact of stress and ELA in addiction, several important gaps in our knowledge remain. One particular limitation in the existing literature is over reliance on cross-sectional studies to draw conclusions about the influence of stress and ELA on addiction. This demonstrates the need for prospective studies to address this link and avoid the pitfalls of correlational studies. Identifying ELA-related psychological and biological functions that can be targeted by future intervention research is also needed. It is also important to develop research that better distinguishes between ongoing ELA-related challenges that may not be seen as life threatening and those even at caused by exposure to specific high-intensity traumatic events. Methodological opportunities in assessment related to better capturing frequency, intensity, and contextual factors of stress and ELA also remain open for future research. Future research should also better assess and identify methods and factors that buffer the negative effects of ELA, including resilience-related psychosocial processes.

9. Conclusion

Stress and early life adversity contribute directly and indirectly to increased risk for drug addiction, maintenance, and relapse. Mechanisms mediating these effects include physiological, biological, behavioral, and psychosocial processes. Evidence reviewed here supports the roles of stress and ELA in addiction, with converging evidence obtained from research approaches that span both macro- (cross-sectional, population research) and micro-levels (molecular, genetic, epigenetic research). Biological systems involved in regulating reward and adapting to stress both centrally and peripherally provide substrates for the interaction between stress and addiction. We also explored multiple factors that may influence the impact of stress in addiction, including gender differences, genetic predisposition, and individual difference factors. Research is continuing to explore the link between many processes involved in these interactions, and new discoveries are likely to identify specific directions that can be followed to develop effective therapeutic and prevention strategies. Research on resilience, factors promoting resilience, and buffering the effects of stress and early life adversity provides tangible directions in this regard. There are several strategies that have been, or that have yet to be, explored as potential interventions to address the inherent challenges in treating addiction in light of the significant role that stress plays in maintaining addiction and in increasing risk of relapse. Research findings to date make a compelling case for multipronged approaches. Indeed, no single approach has been proven effective; and greater integrations of pharmacological, psychosocial, and behavioral intervention is an important and needed area of future research.

Acknowledgment

This paper was supported in in part by National Institutes of Health grants R01DA016351 and R01DA027232.

Declarations of interest

The author has no conflict of interest in this manuscript.

References

Ableitner, A., & Schulz, R. (1992). Neuroanatomical sites mediating the central actions of beta-endorphin as mapped by changes in glucose utilization: Involvement of mu opioid receptors. *The Journal of Pharmacology and Experimental Therapeutics*, *262*, 415–423.

Akil, H., Watson, S. J., Young, E., Lewis, M. E., Khachaturian, H., & Walker, J. M. (1984). Endogenous opioids: Biology and function. *Annual Review of Neuroscience*, 7, 223–255.

al'Absi, M. (2006). Hypothalamic-pituitary-adrenocortical responses to psychological stress and risk for smoking relapse. *International Journal of Psychophysiology*, *59*, 218–227. Netherlands.

al'Absi, M. (2007). *Stress and addiction: Biological and psychological mechanisms*. London: Academic Press/Elsevier.

al'Absi, M. (2018). Stress and addiction: When a robust stress response indicates resiliency. *Psychosomatic Medicine*, *80*, 2–16.

al'Absi, M., & Arnett, D. K. (2000). Adrenocortical responses to psychological stress and risk for hypertension. *Biomedicine & Pharmacotherapy*, *54*, 234–244.

al'Absi, M., France, C., Harju, A., France, J., & Wittmers, L. (2006). Adrenocortical and nociceptive responses to opioid blockade in hypertension-prone men and women. *Psychosomatic Medicine*, *68*, 292–298. United States.

al'Absi, M., Lemieux, A., Westra, R., & Allen, S. (2017). Early life adversity influences stress response association with smoking relapse. *Psychopharmacology*, *234*, 3375–3384.

al'Absi, M., Nakajima, M., Allen, S., Lemieux, A., & Hatsukami, D. (2015). Sex differences in hormonal responses to stress and smoking relapse: A prospective examination. *Nicotine & Tobacco Research*, *17*, 382–389.

al'Absi, M., Nakajima, M., & Lemieux, A. (2018). Impact of early life adversity on the stress biobehavioral response during nicotine withdrawal. *Psychoneuroendocrinology*, *98*, 108–118.

al'Absi, M., Wittmers, L. E., Ellestad, D., Nordehn, G., Kim, S. W., Kirschbaum, C., et al. (2004). Sex differences in pain and hypothalamic-pituitary-adrenocortical responses to opioid blockade. *Psychosomatic Medicine*, *66*, 198–206.

al'Absi, M., Wittmers, L. E., Hatsukami, D., & Westra, R. (2008). Blunted opiate modulation of hypothalamic-pituitary-adrenocortical activity in men and women who smoke. *Psychosomatic Medicine*, *70*, 928–935. United States.

Albott, C. S., Forbes, M. K., & Anker, J. J. (2018). Association of childhood adversity with differential susceptibility of transdiagnostic psychopathology to environmental stress in adulthood. *JAMA Network Open*, *1*, e185354.

Aldao, A., & Nolen-Hoeksema, S. (2012). When are adaptive strategies most predictive of psychopathology? *Journal of Abnormal Psychology*, *121*, 276–281.

Aldao, A., Nolen-Hoeksema, S., & Schweizer, S. (2010). Emotion-regulation strategies across psychopathology: A meta-analytic review. *Clinical Psychology Review*, *30*, 217–237.

Ambroggi, F., Turiault, M., Milet, A., Deroche-Gamonet, V., Parnaudeau, S., Balado, E., et al. (2009). Stress and addiction: Glucocorticoid receptor in dopaminoceptive neurons facilitates cocaine seeking. *Nature Neuroscience*, *12*, 247–249.

Anda, R. F., Felitti, V. J., Bremner, J. D., Walker, J. D., Whitfield, C., Perry, B. D., et al. (2006). The enduring effects of abuse and related adverse experiences in childhood. A convergence of evidence from neurobiology and epidemiology. *European Archives of Psychiatry and Clinical Neuroscience*, *256*, 174–186.

Apkarian, A. V., Bushnell, M. C., Treede, R. D., & Zubieta, J. K. (2005). Human brain mechanisms of pain perception and regulation in health and disease. *European Journal of Pain*, *9*, 463–484.

Arborelius, L., Owens, M. J., Plotsky, P. M., & Nemeroff, C. B. (1999). The role of corticotropin-releasing factor in depression and anxiety disorders. *The Journal of Endocrinology*, *160*, 1–12.

Arnsten, A. F., & Goldman-Rakic, P. S. (1998). Noise stress impairs prefrontal cortical cognitive function in monkeys: Evidence for a hyperdopaminergic mechanism. *Archives of General Psychiatry*, *55*, 362–368.

Ayranci, G., Befort, K., Lalanne, L., Kieffer, B. L., & Lutz, P. E. (2015). Dissociation of heroin-induced emotional dysfunction from psychomotor activation and physical dependence among inbred mouse strains. *Psychopharmacology*, *232*, 1957–1971.

Baker, T. B., Brandon, T. H., & Chassin, L. (2004). Motivational influences on cigarette smoking. *Annual Review of Psychology*, *55*, 463–491.

Barrot, M., Marinelli, M., Abrous, D. N., Rouge-Pont, F., Le Moal, M., & Piazza, P. V. (2000). The dopaminergic hyper-responsiveness of the shell of the nucleus accumbens is hormone-dependent. *The European Journal of Neuroscience*, *12*, 973–979.

Beatty, W. W., & Beatty, P. A. (1970). Hormonal determinants of sex differences in avoidance behavior and reactivity to electric shock in the rat. *Journal of Comparative and Physiological Psychology*, *73*, 446–455.

Belzeaux, R., Lalanne, L., Kieffer, B. L., & Lutz, P. E. (2018). Focusing on the opioid system for addiction biomarker discovery. *Trends in Molecular Medicine*, *24*, 206–220.

Blum, K., Braverman, E. R., Holder, J. M., Lubar, J. F., Monastra, V. J., Miller, D., et al. (2000). Reward deficiency syndrome: A biogenetic model for the diagnosis and treatment of impulsive, addictive, and compulsive behaviors. *Journal of Psychoactive Drugs*, *32*(Suppl.: i–iv), 1–112.

Blum, K., Sheridan, P. J., Wood, R. C., Braverman, E. R., Chen, T. J., Cull, J. G., et al. (1996). The D2 dopamine receptor gene as a determinant of reward deficiency syndrome. *Journal of the Royal Society of Medicine*, *89*, 396–400.

Bobova, L., Finn, P. R., Rickert, M. E., & Lucas, J. (2009). Disinhibitory psychopathology and delay discounting in alcohol dependence: Personality and cognitive correlates. *Experimental and Clinical Psychopharmacology*, *17*, 51–61.

Bond, C., LaForge, K. S., Tian, M., Melia, D., Zhang, S., Borg, L., et al. (1998). Single-nucleotide polymorphism in the human mu opioid receptor gene alters beta-endorphin binding and activity: Possible implications for opiate addiction. *Proceedings of the National Academy of Sciences of the United States of America, 95*, 9608–9613.

Bongard, S., al'Absi, M., Khalil, N. S., & Al Habori, M. (2011). Khat use and trait anger: Effects on affect regulation during an acute stressful challenge. *European Addiction Research, 17*, 285–291. Basel, Switzerland.

Briggs-Gowan, M. J., Estabrook, R., Henry, D., Grasso, D. G., Burns, J., McCarthy, K. J., et al. (2019). Parsing dimensions of family violence exposure in early childhood: Shared and specific contributions to emergent psychopathology and impairment. *Child Abuse & Neglect, 87*, 100–111.

Briggs-Gowan, M. J., Pollak, S. D., Grasso, D., Voss, J., Mian, N. D., Zobel, E., et al. (2015). Attention bias and anxiety in young children exposed to family violence. *Journal of Child Psychology and Psychiatry, 56*, 1194–1201.

Brown, R. A., Lejuez, C. W., Strong, D. R., Kahler, C. W., Zvolensky, M. J., Carpenter, L. L., et al. (2009). A prospective examination of distress tolerance and early smoking lapse in adult self-quitters. *Nicotine & Tobacco Research, 11*, 493–502.

Bruijnzeel, A. W. (2012). Tobacco addiction and the dysregulation of brain stress systems. *Neuroscience and Biobehavioral Reviews, 36*, 1418–1441.

Brydges, N. M., Best, C., & Thomas, K. L. (2019). Female HPA axis displays heightened sensitivity to pre-pubertal stress. *Stress, 23*, 1–11.

Burke, N. N., Finn, D. P., McGuire, B. E., & Roche, M. (2017). Psychological stress in early life as a predisposing factor for the development of chronic pain: Clinical and preclinical evidence and neurobiological mechanisms. *Journal of Neuroscience Research, 95*, 1257–1270.

Calatayud, F., & Belzung, C. (2001). Emotional reactivity in mice, a case of nongenetic heredity? *Physiology & Behavior, 74*, 355–362.

Carpenter, L. L., Shattuck, T. T., Tyrka, A. R., Geracioti, T. D., & Price, L. H. (2011). Effect of childhood physical abuse on cortisol stress response. *Psychopharmacology, 214*, 367–375.

Carpenter, L. L., Tyrka, A. R., Ross, N. S., Khoury, L., Anderson, G. M., & Price, L. H. (2009). Effect of childhood emotional abuse and age on cortisol responsivity in adulthood. *Biological Psychiatry, 66*, 69–75.

Caspi, A., Hariri, A. R., Holmes, A., Uher, R., & Moffitt, T. E. (2010). Genetic sensitivity to the environment: The case of the serotonin transporter gene and its implications for studying complex diseases and traits. *The American Journal of Psychiatry, 167*, 509–527.

Caspi, A., McClay, J., Moffitt, T. E., Mill, J., Martin, J., Craig, I. W., et al. (2002). Role of genotype in the cycle of violence in maltreated children. *Science, 297*, 851–854.

Caspi, A., Sugden, K., Moffitt, T. E., Taylor, A., Craig, I. W., Harrington, H., et al. (2003). Influence of life stress on depression: Moderation by a polymorphism in the 5-HTT gene. *Science, 301*, 386–389.

Champagne, F., Diorio, J., Sharma, S., & Meaney, M. J. (2001). Naturally occurring variations in maternal behavior in the rat are associated with differences in estrogen-inducible central oxytocin receptors. *Proceedings of the National Academy of Sciences of the United States of America, 98*, 12736–12741.

Christie, M. J. (1991). Mechanisms of opioid actions on neurons of the locus coeruleus. *Progress in Brain Research, 88*, 197–205.

Chrousos, G. P. (2009). Stress and disorders of the stress system. *Nature Reviews Endocrinology, 5*, 374–381.

Chrousos, G. P., & Gold, P. W. (1992). The concepts of stress and stress system disorders—Overview of physical and behavioral homeostasis. *JAMA-Journal of the American Medical Association, 267*, 1244–1252.

Cicchetti, D., Rogosch, F. A., Gunnar, M. R., & Toth, S. L. (2010). The differential impacts of early physical and sexual abuse and internalizing problems on daytime cortisol rhythm in school-aged children. *Child Development, 81*, 252–269.

Colich, N. L., Platt, J. M., Keyes, K. M., Sumner, J. A., Allen, N. B., & McLaughlin, K. A. (2019). Earlier age at menarche as a transdiagnostic mechanism linking childhood trauma with multiple forms of psychopathology in adolescent girls. *Psychological Medicine*, 1–9. https://doi.org/10.1017/S0033291719000953 (Epub ahead of print).

Copeland, B. J., Neff, N. H., & Hadjiconstantinou, M. (2005). Enhanced dopamine uptake in the striatum following repeated restraint stress. *Synapse, 57*, 167–174.

Corral-Frías, N. S., Nikolova, Y. S., Michalski, L. J., Baranger, D. A., Hariri, A. R., & Bogdan, R. (2015). Stress-related anhedonia is associated with ventral striatum reactivity to reward and transdiagnostic psychiatric symptomatology. *Psychological Medicine, 45*, 2605–2617.

Corral-Frías, N. S., Pizzagalli, D. A., Carré, J. M., Michalski, L. J., Nikolova, Y. S., Perlis, R. H., et al. (2016). COMT Val(158) Met genotype is associated with reward learning: A replication study and meta-analysis. *Genes, Brain, and Behavior, 15*, 503–513.

Coventry, T. L., Jessop, D. S., Finn, D. P., Crabb, M. D., Kinoshita, H., & Harbuz, M. S. (2001). Endomorphins and activation of the hypothalamo-pituitary-adrenal axis. *The Journal of Endocrinology, 169*, 185–193.

Cowen, M. S., & Lawrence, A. J. (1999). The role of opioid-dopamine interactions in the induction and maintenance of ethanol consumption. *Progress in Neuro-Psychopharmacology & Biological Psychiatry, 23*, 1171–1212.

Dallman, M. (1993). Stress update: Adaptation of the hypothalamic-pituitary-adrenal axis to chronic stress. *Trends in Endocrinology and Metabolism, 4*, 62–69.

Danese, A., & Tan, M. (2014). Childhood maltreatment and obesity: Systematic review and meta-analysis. *Molecular Psychiatry, 19*, 544–554.

Dannlowski, U., Kugel, H., Huber, F., Stuhrmann, A., Redlich, R., Grotegerd, D., et al. (2013). Childhood maltreatment is associated with an automatic negative emotion processing bias in the amygdala. *Human Brain Mapping, 34*, 2899–2909.

Deroche-Gamonet, V., Sillaber, I., Aouizerate, B., Izawa, R., Jaber, M., Ghozland, S., et al. (2003). The glucocorticoid receptor as a potential target to reduce cocaine abuse. *The Journal of Neuroscience, 23*, 4785–4790.

Desbonnet, L., Garrett, L., Daly, E., McDermott, K. W., & Dinan, T. G. (2008). Sexually dimorphic effects of maternal separation stress on corticotrophin-releasing factor and vasopressin systems in the adult rat brain. *International Journal of Developmental Neuroscience, 26*, 259–268.

Dichter, G. S., Damiano, C. A., & Allen, J. A. (2012). Reward circuitry dysfunction in psychiatric and neurodevelopmental disorders and genetic syndromes: Animal models and clinical findings. *Journal of Neurodevelopmental Disorders, 4*, 19.

Dillon, D. G., Holmes, A. J., Birk, J. L., Brooks, N., Lyons-Ruth, K., & Pizzagalli, D. A. (2009). Childhood adversity is associated with left basal ganglia dysfunction during reward anticipation in adulthood. *Biological Psychiatry, 66*, 206–213.

Dohrenwend, B. P. (2006). Inventorying stressful life events as risk factors for psychopathology: Toward resolution of the problem of intracategory variability. *Psychological Bulletin, 132*, 477–495.

Dong, M., Giles, W. H., Felitti, V. J., Dube, S. R., Williams, J. E., Chapman, D. P., et al. (2004). Insights into causal pathways for ischemic heart disease: Adverse childhood experiences study. *Circulation, 110*, 1761–1766.

Doucette, S., Levy, A., Flowerdew, G., Horrocks, J., Grof, P., Ellenbogen, M., et al. (2016). Early parent-child relationships and risk of mood disorder in a Canadian sample of offspring of a parent with bipolar disorder: Findings from a 16-year prospective cohort study. *Early Intervention in Psychiatry, 10*, 381–389.

Drolet, G., Dumont, E. C., Gosselin, I., Kinkead, R., Laforest, S., & Trottier, J. F. (2001). Role of endogenous opioid system in the regulation of the stress response. *Progress in Neuro-Psychopharmacology & Biological Psychiatry, 25*, 729–741.

Dube, S. R., Felitti, V. J., Dong, M., Giles, W. H., & Anda, R. F. (2003). The impact of adverse childhood experiences on health problems: Evidence from four birth cohorts dating back to 1900. *Preventive Medicine, 37*, 268–277.

Dube, S. R., Miller, J. W., Brown, D. W., Giles, W. H., Felitti, V. J., Dong, M., et al. (2006). Adverse childhood experiences and the association with ever using alcohol and initiating alcohol use during adolescence. *The Journal of Adolescent Health, 38* 444.e1-10.

Duffy, K. A., McLaughlin, K. A., & Green, P. A. (2018). Early life adversity and health-risk behaviors: Proposed psychological and neural mechanisms. *Annals of the New York Academy of Sciences, 1428*, 151–169.

DuMont, K. A., Widom, C. S., & Czaja, S. J. (2007). Predictors of resilience in abused and neglected children grown-up: The role of individual and neighborhood characteristics. *Child Abuse & Neglect, 31*, 255–274.

Dunn, E. C., Soare, T. W., Zhu, Y., Simpkin, A. J., Suderman, M. J., Klengel, T., et al. (2019). Sensitive periods for the effect of childhood adversity on DNA methylation: Results from a prospective, longitudinal study. *Biological Psychiatry, 85*, 838–849.

Dunn, A. J., Swiergiel, A. H., & Palamarchouk, V. (2004). Brain circuits involved in corticotropin-releasing factor-norepinephrine interactions during stress. *Annals of the New York Academy of Sciences, 1018*, 25–34.

Edwards, S., & Koob, G. F. (2010). Neurobiology of dysregulated motivational systems in drug addiction. *Future Neurology, 5*, 393–401.

Enoch, M. A. (2011). The role of early life stress as a predictor for alcohol and drug dependence. *Psychopharmacology, 214*, 17–31.

Erb, S. (2007). Neurobiology of stress and risk for relapse. In M. al'Absi (Ed.), *Stress and addiction: Biological and psychological mechanisms* (pp. 147–169). London: Academic Press/Elsevier.

Fergusson, D. M., Boden, J. M., & Horwood, L. J. (2008). The developmental antecedents of illicit drug use: Evidence from a 25-year longitudinal study. *Drug and Alcohol Dependence, 96*, 165–177.

Fergusson, D. M., Woodward, L. J., & Horwood, L. J. (1999). Childhood peer relationship problems and young people's involvement with deviant peers in adolescence. *Journal of Abnormal Child Psychology, 27*, 357–369.

Gallardo-Pujol, D., Andrés-Pueyo, A., & Maydeu-Olivares, A. (2013). MAOA genotype, social exclusion and aggression: An experimental test of a gene-environment interaction. *Genes, Brain, and Behavior, 12*, 140–145.

Gebremariam, M. K., Bjune, G. A., & Frich, J. C. (2011). Lay beliefs of TB and TB/HIV co-infection in Addis Ababa, Ethiopia: A qualitative study. *BMC Research Notes, 4*, 277.

Gianoulakis, C. (2004). Endogenous opioids and addiction to alcohol and other drugs of abuse. *Current Topics in Medicinal Chemistry, 4*, 39–50.

Gilman, J. M., Kuster, J K., Lee, S., Lee, M. J., Kim, B. W., Makris, N., et al. (2014). Cannabis use is quantitatively associated with nucleus accumbens and amygdala abnormalities in young adult recreational users. *The Journal of Neuroscience, 34*, 5529–5538.

Gorwood, P., Le Strat, Y., Ramoz, N., Dubertret, C., Moalic, J. M., & Simonneau, M. (2012). Genetics of dopamine receptors and drug addiction. *Human Genetics, 131*, 803–822.

Gulliver, S. B., Hughes, J. R., Solomon, L. J., & Dey, A. N. (1995). An investigation of self-efficacy, partner support and daily stresses as predictors of relapse to smoking in self-quitters. *Addiction, 90*, 767–772.

Hadjiconstantinou, M., & Neff, N. H. (2011). Nicotine and endogenous opioids: Neurochemical and pharmacological evidence. *Neuropharmacology, 60*, 1209–1220.

Hall, F. S., Wilkinson, L. S., Humby, T., & Robbins, T. W. (1999). Maternal deprivation of neonatal rats produces enduring changes in dopamine function. *Synapse*, *32*, 37–43.

Hanson, J. L., Albert, D., Iselin, A. M., Carré, J. M., Dodge, K. A., & Hariri, A. R. (2016). Cumulative stress in childhood is associated with blunted reward-related brain activity in adulthood. *Social Cognitive and Affective Neuroscience*, *11*, 405–412.

Hanson, J. L., Hariri, A. R., & Williamson, D. E. (2015). Blunted ventral striatum development in adolescence reflects emotional neglect and predicts depressive symptoms. *Biological Psychiatry*, *78*, 598–605.

Harris, G. C., Wimmer, M., & Aston-Jones, G. (2005). A role for lateral hypothalamic orexin neurons in reward seeking. *Nature*, *437*, 556–559. England.

Heim, C., Bradley, B., Mletzko, T. C., Deveau, T. C., Musselman, D. L., Nemeroff, C. B., et al. (2009). Effect of childhood trauma on adult depression and neuroendocrine function: Sex-specific moderation by CRH receptor 1 gene. *Frontiers in Behavioral Neuroscience*, *3*, 41.

Heim, C., Ehlert, U., Rexhausen, J., Hanker, J. P., & Hellhammer, D. H. (1997). Psychoendocrinological observations in women with chronic pelvic pain. *Annals of the New York Academy of Sciences*, *821*, 456–458.

Hein, T. C., & Monk, C. S. (2017). Research review: Neural response to threat in children, adolescents, and adults after child maltreatment—a quantitative meta-analysis. *Journal of Child Psychology and Psychiatry*, *58*, 222–230.

Hellbach, S., Gartner, P., Deicke, J., Fischer, D., Hassan, A. H., & Almeida OF (1998). Inherent glucocorticoid response potential of isolated hypothalamic neuroendocrine neurons. *The FASEB Journal*, *12*, 199–207.

Hemmingsson, E., Johansson, K., & Reynisdottir, S. (2014). Effects of childhood abuse on adult obesity: A systematic review and meta-analysis. *Obesity Reviews*, *15*, 882–893.

Henderson, G. (1983). Electrophysiological analysis of opioid action in the central nervous system. *British Medical Bulletin*, *39*, 59–64.

Holschneider, D. P., Guo, Y., Mayer, E. A., & Wang, Z. (2016). Early life stress elicits visceral hyperalgesia and functional reorganization of pain circuits in adult rats. *Neurobiology of Stress*, *3*, 8–22.

Hostinar, C. E., & Gunnar, M. R. (2015). Social support can buffer against stress and shape brain activity. *AJOB Neuroscience*, *6*, 34–42.

Jaffee, S. R., Caspi, A., Moffitt, T. E., Polo-Tomás, M., & Taylor, A. (2007). Individual, family, and neighborhood factors distinguish resilient from non-resilient maltreated children: A cumulative stressors model. *Child Abuse & Neglect*, *31*, 231–253.

Kalivas, P. W., Pierce, R. C., Cornish, J., & Sorg, B. A. (1998). A role for sensitization in craving and relapse in cocaine addiction. *Journal of Psychopharmacology*, *12*, 49–53.

Kalivas, P. W., & Volkow, N. D. (2005). The neural basis of addiction: A pathology of motivation and choice. *The American Journal of Psychiatry*, *162*, 1403–1413.

Kassel, J. D., Veilleux, J. C., Wardle, M. C., Yates, M. C., Greenstein, J. E., Evatt, D. P., et al. (2007). Negative affect and addiction. In M. al'Absi (Ed.), *Stress and addiction: Biological and psychological mechanisms* (pp. 171–190). London: Academic Press/Elsevier.

Kelley, N. J., Finley, A. J., & Schmeichel, B. J. (2019). After-effects of self-control: The reward responsivity hypothesis. *Cognitive, Affective, & Behavioral Neuroscience*, *19*, 600–618.

Kelly, P. A., Viding, E., Wallace, G. L., Schaer, M., De Brito, S. A., Robustelli, B., et al. (2013). Cortical thickness, surface area, and gyrification abnormalities in children exposed to maltreatment: Neural markers of vulnerability? *Biological Psychiatry*, *74*, 845–852.

Kiecolt-Glaser, J. K., McGuire, L., Robles, T. F., & Glaser, R. (2002). Psychoneuroimmunology: Psychological influences on immune function and health. *Journal of Consulting and Clinical Psychology*, *70*, 537–547.

Kirschbaum, C., Kudielka, B. M., Gaab, J., Schommer, N. C., & Hellhammer, D. H. (1999). Impact of gender, menstrual cycle phase, and oral contraceptives on the activity of the hypothalamus-pituitary-adrenal axis. *Psychosomatic Medicine*, *61*, 154–162.

Klengel, T., Mehta, D., Anacker, C., Rex-Haffner, M., Pruessner, J. C., Pariante, C. M., et al. (2013). Allele-specific FKBP5 DNA demethylation mediates gene-childhood trauma interactions. *Nature Neuroscience*, *16*, 33–41.

Kober, H., Mende-Siedlecki, P., Kross, E. F., Weber, J., Mischel, W., Hart, C. L., et al. (2010). Prefrontal-striatal pathway underlies cognitive regulation of craving. *Proceedings of the National Academy of Sciences of the United States of America*, *107*, 14811–14816.

Koob, G. F. (2013). Addiction is a reward deficit and stress surfeit disorder. *Frontiers in Psychiatry*, *4*, 72.

Koob, G. F. (2017). Antireward, compulsivity, and addiction: Seminal contributions of Dr. Athina Markou to motivational dysregulation in addiction. *Psychopharmacology*, *234*, 1315–1332.

Koob, G. F. (2020). Neurobiology of opioid addiction: Opponent process, hyperkatifeia, and negative reinforcement. *Biological Psychiatry*, *87*, 44–53.

Koob, G. F., & Bloom, F. E. (1988). Cellular and molecular mechanisms of drug dependence. *Science*, *242*, 715–723.

Koob, G., & Kreek, M. J. (2007). Stress, dysregulation of drug reward pathways, and the transition to drug dependence. *The American Journal of Psychiatry*, *164*, 1149–1159.

Koob, G. F., & Le, M. M. (1997). Drug abuse: Hedonic homeostatic dysregulation. *Science*, *278*, 52–58.

Kosten, T. A., & Ambrosio, E. (2002). HPA axis function and drug addictive behaviors: Insights from studies with Lewis and Fischer 344 inbred rats. *Psychoneuroendocrinology*, *27*, 35–69.

Kosten, T. A., Lee, H. J., & Kim, J. J. (2007). Neonatal handling alters learning in adult male and female rats in a task-specific manner. *Brain Research*, *1154*, 144–153.

Kosten, T. A., Miserendino, M. J., & Kehoe, P. (2000). Enhanced acquisition of cocaine self-administration in adult rats with neonatal isolation stress experience. *Brain Research*, *875*, 44–50.

Kosten, T. A., Sanchez, H., Jatlow, P. I., & Kehoe, P. (2005). Neonatal isolation alters the estrous cycle interactions on the acute behavioral effects of cocaine. *Psychoneuroendocrinology*, *30*, 753–761.

Kosten, T. A., Sanchez, H., Zhang, X. Y., & Kehoe, P. (2004). Neonatal isolation enhances acquisition of cocaine self-administration and food responding in female rats. *Behavioural Brain Research*, *151*, 137–149.

Kosten, T. A., Zhang, X. Y., & Kehoe, P. (2003). Chronic neonatal isolation stress enhances cocaine-induced increases in ventral striatal dopamine levels in rat pups. *Brain Research Developmental Brain Research*, *141*, 109–116.

Kudielka, B. M., & Kirschbaum, C. (2005). Sex differences in HPA axis responses to stress: A review. *Biological Psychology*, *69*, 113–132.

Kudielka, B. M., Schommer, N. C., Hellhammer, D. H., & Kirschbaum, C. (2004). Acute HPA axis responses, heart rate, and mood changes to psychosocial stress (TSST) in humans at different times of day. *Psychoneuroendocrinology*, *29*, 983–992.

Ladd, C. O., Huot, R. L., Thrivikraman, K. V., Nemeroff, C. B., Meancy, M. J., & Plotsky, P. M. (2000). Long-term behavioral and neuroendocrine adaptations to adverse early experience. *Progress in Brain Research*, *122*, 81–103.

Lambert, H. K., Peverill, M., Sambrook, K. A., Rosen, M. L., Sheridan, M. A., & McLaughlin, K. A. (2019). Altered development of hippocampus-dependent associative learning following early-life adversity. *Developmental Cognitive Neuroscience*, *38*, 100666.

Lansford, J. E., Malone, P. S., Stevens, K. I., Dodge, K. A., Bates, J. E., & Pettit, G. S. (2006). Developmental trajectories of externalizing and internalizing behaviors: Factors underlying resilience in physically abused children. *Development and Psychopathology*, *18*, 35–55.

Le Merrer, J., Becker, J. A., Befort, K., & Kieffer, B. L. (2009). Reward processing by the opioid system in the brain. *Physiological Reviews*, *89*, 1379–1412.

Lehmann, J., Stöhr, T., & Feldon, J. (2000). Long-term effects of prenatal stress experiences and postnatal maternal separation on emotionality and attentional processes. *Behavioural Brain Research*, *107*, 133–144.

Lemieux, A., & al'Absi, M. (2016). Stress psychobiology in the context of addiction medicine: From drugs of abuse to behavioral addictions. *Progress in Brain Research*, *223*, 43–62.

Lippmann, M., Bress, A., Nemeroff, C. B., Plotsky, P. M., & Monteggia, L. M. (2007). Long-term behavioral and molecular alterations associated with maternal separation in rats. *The European Journal of Neuroscience*, *25*, 3091–3098.

Lissek, T., Adams, M., Adelman, J., Ahissar, E., Akaaboune, M., Akil, H., et al. (2017). Building bridges through science. *Neuron*, *96*, 730–735.

Litvin, Y., Tovote, P., Pentkowski, N. S., Zeyda, T., King, L. B., Vasconcellos, A. J., et al. (2010). Maternal separation modulates short-term behavioral and physiological indices of the stress response. *Hormones and Behavior*, *58*, 241–249.

Liu, P. Z., & Nusslock, R. (2018). How stress gets under the skin: Early life adversity and glucocorticoid receptor epigenetic regulation. *Current Genomics*, *19*, 653–664.

Lovallo, W. R. (2004). *Stress & health: Biological and psychological interactions*. Thousand Oaks, CA: SAGE Knowledge.

Lovallo, W. (2007). Individual differences in response to stress and risk for addiction. In M. al'Absi (Ed.), *Stress and addiction: Biological and psychological mechanisms* (pp. 265–284). London: Academic Press/Elsevier.

Lovallo, W. R. (2013). Early life adversity reduces stress reactivity and enhances impulsive behavior: Implications for health behaviors. *International Journal of Psychophysiology*, *90*, 8–16.

Lovallo, W. R., Acheson, A., Vincent, A. S., Sorocco, K. H., & Cohoon, A. J. (2018). Early life adversity diminishes the cortisol response to opioid blockade in women: Studies from the Family Health Patterns project. *PLoS One*, *13*, e0205723.

Lovallo, W. R., Cohoon, A. J., Acheson, A., Sorocco, K. H., & Vincent, A. S. (2019). Blunted stress reactivity reveals vulnerability to early life adversity in young adults with a family history of alcoholism. *Addiction*, *114*, 798–806.

Lovallo, W. R., Cohoon, A. J., Sorocco, K. H., Vincent, A. S., Acheson, A., Hodgkinson, C. A., et al. (2019). Early-life adversity and blunted stress reactivity as predictors of alcohol and drug use in persons with COMT (rs4680) Val158Met genotypes. *Alcoholism, Clinical and Experimental Research*, *43*, 1519–1527.

Lovallo, W. R., Enoch, M. A., Acheson, A., Cohoon, A. J., Sorocco, K. H., Hodgkinson, C. A., et al. (2015). Cortisol stress response in men and women modulated differentially by the mu-opioid receptor gene polymorphism OPRM1 A118G. *Neuropsychopharmacology*, *40*, 2546–2554.

Lovallo, W. R., Enoch, M. A., Sorocco, K. H., Vincent, A. S., Acheson, A., Cohoon, A. J., et al. (2017). Joint impact of early life adversity and COMT Val158Met (rs4680) genotypes on the adult cortisol response to psychological stress. *Psychosomatic Medicine*, *79*, 631–637.

Lovallo, W. R., Farag, N. H., Sorocco, K. H., Cohoon, A. J., & Vincent, A. S. (2012). Lifetime adversity leads to blunted stress axis reactivity: Studies from the Oklahoma Family Health Patterns Project. *Biological Psychiatry*, *71*, 344–349.

Lovallo, W. R., King, A. C., Farag, N. H., Sorocco, K. H., Cohoon, A. J., & Vincent, A. S. (2012). Naltrexone effects on cortisol secretion in women and men in relation to a family history of alcoholism: Studies from the Oklahoma Family Health Patterns Project. *Psychoneuroendocrinology, 37*, 1922–1928.

Luijcks, R., Vossen, C. J., Roggeveen, S., van Os, J., Hermens, H. J., & Lousberg, R. (2016). Impact of early life adversity on EMG stress reactivity of the trapezius muscle. *Medicine (Baltimore), 95*, e4745.

Lutz, P. E., & Kieffer, B. L. (2013). The multiple facets of opioid receptor function: Implications for addiction. *Current Opinion in Neurobiology, 23*, 473–479.

Lutz, P. E., Reiss, D., Ouagazzal, A. M., & Kieffer, B. L. (2013). A history of chronic morphine exposure during adolescence increases despair-like behaviour and strain-dependently promotes sociability in abstinent adult mice. *Behavioural Brain Research, 243*, 44–52.

MacMillan, H. L., Georgiades, K., Duku, E. K., Shea, A., Steiner, M., Niec, A., et al. (2009). Cortisol response to stress in female youths exposed to childhood maltreatment: Results of the youth mood project. *Biological Psychiatry, 66*, 62–68.

Makris, N., Oscar-Berman, M., Jaffin, S. K., Hodge, S. M., Kennedy, D. N., Caviness, V. S., et al. (2008). Decreased volume of the brain reward system in alcoholism. *Biological Psychiatry, 64*, 192–202.

Marinelli, M. (2007). Dopaminergic reward pathways and effects of stress. In M. al'Absi (Ed.), *Stress and addiction: Biological and psychological mechanisms* (pp. 41–83). London: Academic Press/Elsevier.

Martin del Campo, A. F., Dowson, J. H., Herbert, J., & Paykel, E. S. (1994). Effects of naloxone on diurnal rhythms in mood and endocrine function: A dose-response study in man. *Psychopharmacology, 114*, 583–590.

McCormick, C. M., Kehoe, P., Mallinson, K., Cecchi, L., & Frye, C. A. (2002). Neonatal isolation alters stress hormone and mesolimbic dopamine release in juvenile rats. *Pharmacology, Biochemistry, and Behavior, 73*, 77–85.

McCrory, E. J., De Brito, S. A., Kelly, P. A., Bird, G., Sebastian, C. L., Mechelli, A., et al. (2013). Amygdala activation in maltreated children during pre-attentive emotional processing. *The British Journal of Psychiatry, 202*, 269–276.

McEwen, B. S. (1997). The brain is an important target of adrenal steroid actions. A comparison of synthetic and natural steroids. *Annals of the New York Academy of Sciences, 823*, 201–213.

McEwen, B. S. (1998). Protective and damaging effects of stress mediators. *New England Journal of Medicine, 338*, 171–179.

McGowan, P. O., Sasaki, A., D'Alessio, A. C., Dymov, S., Labonté, B., Szyf, M., et al. (2009). Epigenetic regulation of the glucocorticoid receptor in human brain associates with childhood abuse. *Nature Neuroscience, 12*, 342–348.

McKee, S. A., Potenza, M. N., Kober, H., Sofuoglu, M., Arnsten, A. F., Picciotto, M. R., et al. (2015). A translational investigation targeting stress-reactivity and prefrontal cognitive control with guanfacine for smoking cessation. *Journal of Psychopharmacology, 29*, 300–311.

McLaughlin, K. A., Conron, K. J., Koenen, K. C., & Gilman, S. E. (2010). Childhood adversity, adult stressful life events, and risk of past-year psychiatric disorder: A test of the stress sensitization hypothesis in a population-based sample of adults. *Psychological Medicine, 40*, 1647–1658.

McLaughlin, K. A., DeCross, S. N., Jovanovic, T., & Tottenham, N. (2019). Mechanisms linking childhood adversity with psychopathology: Learning as an intervention target. *Behaviour Research and Therapy, 118*, 101–109.

McLaughlin, K. A., Koenen, K. C., Bromet, E. J., Karam, E. G., Liu, H., Petukhova, M., et al. (2017). Childhood adversities and post-traumatic stress disorder: Evidence for stress sensitisation in the World Mental Health Surveys. *The British Journal of Psychiatry, 211*, 280–288.

Meaney, M. J., Bhatnagar, S., Larocque, S., McCormick, C., Shanks, N., Sharma, S., et al. (1993). Individual differences in the hypothalamic-pituitary-adrenal stress response and the hypothalamic CRF system. *Annals of the New York Academy of Sciences*, *697*, 70–85.

Meaney, M. J., Brake, W., & Gratton, A. (2002). Environmental regulation of the development of mesolimbic dopamine systems: A neurobiological mechanism for vulnerability to drug abuse? *Psychoneuroendocrinology*, *27*, 127–138.

Meaney, M. J., Diorio, J., Francis, D., Widdowson, J., LaPlante, P., Caldji, C., et al. (1996). Early environmental regulation of forebrain glucocorticoid receptor gene expression: Implications for adrenocortical responses to stress. *Developmental Neuroscience*, *18*, 49–72.

Meaney, M. J., Mitchell, J. B., Aitken, D. H., Bhatnagar, S., Bodnoff, S. R., Iny, L. J., et al. (1991). The effects of neonatal handling on the development of the adrenocortical response to stress: Implications for neuropathology and cognitive deficits in later life. *Psychoneuroendocrinology*, *16*, 85–103.

Michopoulos, V., Powers, A., Moore, C., Villarreal, S., Ressler, K. J., & Bradley, B. (2015). The mediating role of emotion dysregulation and depression on the relationship between childhood trauma exposure and emotional eating. *Appetite*, *91*, 129–136.

Miller, A. B., Sheridan, M. A., Hanson, J. L., McLaughlin, K. A., Bates, J. E., Lansford, J. E., et al. (2018). Dimensions of deprivation and threat, psychopathology, and potential mediators: A multi-year longitudinal analysis. *Journal of Abnormal Psychology*, *127*, 160–170.

Mittleman, G., Jones, G. H., & Robbins, T. W. (1991). Sensitization of amphetamine-stereotypy reduces plasma corticosterone: Implications for stereotypy as a coping response. *Behavioral and Neural Biology*, *56*, 170–182.

Mizrahi, R., Suridjan, I., Kenk, M., George, T. P., Wilson, A., Houle, S., et al. (2013). Dopamine response to psychosocial stress in chronic cannabis users: A PET study with [11C]-+-PHNO. *Neuropsychopharmacology*, *38*, 673–682.

Moffett, M. C., Vicentic, A., Kozel, M., Plotsky, P., Francis, D. D., & Kuhar, M. J. (2007). Maternal separation alters drug intake patterns in adulthood in rats. *Biochemical Pharmacology*, *73*, 321–330.

Morris, M., Salmon, P., Steinberg, H., Sykes, E. A., Bouloux, P., Newbould, E., et al. (1990). Endogenous opioids modulate the cardiovascular response to mental stress. *Psychoneuroendocrinology*, *15*, 185–192.

Müller, K. U., Mennigen, E., Ripke, S., Banaschewski, T., Barker, G. J., Büchel, C., et al. (2013). Altered reward processing in adolescents with prenatal exposure to maternal cigarette smoking. *JAMA Psychiatry*, *70*, 847–856.

Munck, A., Guyre, P. M., & Holbrook, N. J. (1984). Physiological functions of glucocorticoids in stress and their relation to pharmacological actions. *Endocrine Reviews*, *5*, 25–44.

Myers, B., McLaughlin, K. A., Wang, S., Blanco, C., & Stein, D. J. (2014). Associations between childhood adversity, adult stressful life events, and past-year drug use disorders in the National Epidemiological Study of Alcohol and Related Conditions (NESARC). *Psychology of Addictive Behaviors*, *28*, 1117–1126.

Najavits, L. M., Gastfriend, D. R., Barber, J. P., Reif, S., Muenz, L. R., Blaine, J., et al. (1998). Cocaine dependence with and without PTSD among subjects in the National Institute on Drug Abuse Collaborative Cocaine Treatment Study. *The American Journal of Psychiatry*, *155*, 214–219.

Nevo, I., & Hamon, M. (1995). Neurotransmitter and neuromodulatory mechanisms involved in alcohol abuse and alcoholism. *Neurochemistry International*, *26*, 305–336.

Oswald, L. M., Zandi, P., Nestadt, G., Potash, J. B., Kalaydjian, A. E., & Wand, G. S. (2006). Relationship between cortisol responses to stress and personality. *Neuropsychopharmacology*, *31*, 1583–1591.

Ouellet-Morin, I., Odgers, C. L., Danese, A., Bowes, L., Shakoor, S., Papadopoulos, A. S., et al. (2011). Blunted cortisol responses to stress signal social and behavioral problems among maltreated/bullied 12-year-old children. *Biological Psychiatry*, *70*, 1016–1023.

Page, G. G., Opp, M. R., & Kozachik, S. L. (2016). Sex differences in sleep, anhedonia, and HPA axis activity in a rat model of chronic social defeat. *Neurobiology of Stress*, *3*, 105–113.

Pasternak, G. W. (2005). Molecular biology of opioid analgesia. *Journal of Pain and Symptom Management*, *29*, S2–S9.

Peckins, M. K., Roberts, A. G., Hein, T. C., Hyde, L. W., Mitchell, C., Brooks-Gunn, J., et al. (2020). Violence exposure and social deprivation is associated with cortisol reactivity in urban adolescents. *Psychoneuroendocrinology*, *111*, 104426.

Pellissier, L. P., Gandía, J., Laboute, T., Becker, J. A. J., & Le Merrer, J. (2018). μ opioid receptor, social behaviour and autism spectrum disorder: Reward matters. *British Journal of Pharmacology*, *175*, 2750–2769.

Perry, N. B., DePasquale, C. E., Fisher, P. H., & Gunnar, M. R. (2019). Comparison of institutionally reared and maltreated children on socioemotional and biological functioning. *Child Maltreatment*, *24*, 235–243.

Piazza, P., Rouge-Pont, F., Deroche, V., Maccari, S., Simon, H., & Le Moal, M. (1996). Glucocorticoids have ste dependent stimulant effects on the mesencephalic dopaminergic transmission. *Neurobiology*, *93*, 8716–8720.

Picetti, R., Schlussman, S. D., Zhou, Y., Ray, B., Ducat, E., Yuferov, V., et al. (2013). Addictions and stress: Clues for cocaine pharmacotherapies. *Current Pharmaceutical Design*, *19*, 7065–7080. Netherlands.

Pizzagalli, D. A., Holmes, A. J., Dillon, D. G., Goetz, E. L., Birk, J. L., Bogdan, R., et al. (2009). Reduced caudate and nucleus accumbens response to rewards in unmedicated individuals with major depressive disorder. *The American Journal of Psychiatry*, *166*, 702–710.

Plotsky, P. M., & Meaney, M. J. (1993). Early, postnatal experience alters hypothalamic corticotropin- releasing factor (CRF) mRNA, median eminence CRF content and stress-induced release in adult rats. *Molecular Brain Research*, *18*, 195–200.

Pollak, S. D., Messner, M., Kistler, D. J., & Cohn, J. F. (2009). Development of perceptual expertise in emotion recognition. *Cognition*, *110*, 242–247.

Potretzke, S., Nakajima, M., Cragin, T., & al'Absi, M. (2014). Changes in circulating leptin levels during acute stress and associations with craving in abstinent smokers: A preliminary investigation. *Psychoneuroendocrinology*, *47*, 232–240.

Prasad, B. M., Sorg, B. A., Ulibarri, C., & Kalivas, P. W. (1995). Sensitization to stress and psychostimulants. Involvement of dopamine transmission versus the HPA axis. *Annals of the New York Academy of Sciences*, *771*, 617–625.

Pruessner, J. C., Champagne, F., Meaney, M. J., & Dagher, A. (2004). Dopamine release in response to a psychological stress in humans and its relationship to early life maternal care: A positron emission tomography study using [11C]raclopride. *The Journal of Neuroscience*, *24*, 2825–2831.

Pryce, C. R., Bettschen, D., Bahr, N. I., & Feldon, J. (2001). Comparison of the effects of infant handling, isolation, and nonhandling on acoustic startle, prepulse inhibition, locomotion, and HPA activity in the adult rat. *Behavioral Neuroscience*, *115*, 71–83.

Ripke, S., Hübner, T., Mennigen, E., Müller, K. U., Rodehacke, S., Schmidt, D., et al. (2012). Reward processing and intertemporal decision making in adults and adolescents: The role of impulsivity and decision consistency. *Brain Research*, *1478*, 36–47.

Roth, T. L., Lubin, F. D., Funk, A. J., & Sweatt, J. D. (2009). Lasting epigenetic influence of early-life adversity on the BDNF gene. *Biological Psychiatry*, *65*, 760–769.

Roth, R. H., Tam, S. Y., Ida, Y., Yang, J. X., & Deutch, A. Y. (1988). Stress and the mesocorticolimbic dopamine systems. *Annals of the New York Academy of Sciences*, *537*, 138–147.

Roy, A., Gorodetsky, E., Yuan, Q., Goldman, D., & Enoch, M. A. (2010). Interaction of FKBP5, a stress-related gene, with childhood trauma increases the risk for attempting suicide. *Neuropsychopharmacology*, *35*, 1674–1683.

Rushen, J., Schwartze, N., Ladewig, J., & Foxcroft, G. (1993). Opioid modulation of the effects of repeated stress on A.C.T.H., cortisol, prolactin, and growth hormone in pigs. *Physiology and Behavior, 53*, 923–928.

Sánchez, M. M., Noble, P. M., Lyon, C. K., Plotsky, P. M., Davis, M., Nemeroff, C. B., et al. (2005). Alterations in diurnal cortisol rhythm and acoustic startle response in non-human primates with adverse rearing. *Biological Psychiatry, 57*, 373–381.

Sapolsky, R. M. (2000). The possibility of neurotoxicity in the hippocampus in major depression: A primer on neuron death. *Biological Psychiatry, 48*, 755–765.

Sapolsky, R. M., Romero, L. M., & Munck, A. U. (2000). How do glucocorticoids influence stress responses? Integrating permissive, suppressive, stimulatory, and preparative actions. *Endocrine Reviews, 21*, 55–89.

Scult, M. A., Knodt, A. R., Swartz, J. R., Brigidi, B. D., & Hariri, A. R. (2017). Thinking and feeling: Individual differences in habitual emotion regulation and stress-related mood are associated with prefrontal executive control. *Clinical Psychological Science: A Journal of the Association for Psychological Science, 5*, 150–157.

Sebastian, C. L., McCrory, E. J., Cecil, C. A., Lockwood, P. L., De Brito, S. A., Fontaine, N. M., et al. (2012). Neural responses to affective and cognitive theory of mind in children with conduct problems and varying levels of callous-unemotional traits. *Archives of General Psychiatry, 69*, 814–822.

Sebastian, C. L., McCrory, E. J., Dadds, M. R., Cecil, C. A., Lockwood, P. L., Hyde, Z. H., et al. (2014). Neural responses to fearful eyes in children with conduct problems and varying levels of callous-unemotional traits. *Psychological Medicine, 44*, 99–109.

Sheridan, M. A., Peverill, M., Finn, A. S., & McLaughlin, K. A. (2017). Dimensions of childhood adversity have distinct associations with neural systems underlying executive functioning. *Development and Psychopathology, 29*, 1777–1794.

Simmen-Janevska, K., Forstmeier, S., Krammer, S., & Maercker, A. (2015). Does trauma impair self-control? Differences in delaying gratification between former indentured child laborers and nontraumatized controls. *Violence and Victims, 30*, 1068–1081.

Singewald, N., & Philippu, A. (1998). Release of neurotransmitters in the locus coeruleus. *Progress in Neurobiology, 56*, 237–267.

Sinha, R. (2008). Chronic stress, drug use, and vulnerability to addiction. *Annals of the New York Academy of Sciences, 1141*, 105–130.

Sinha, R., & Jastreboff, A. M. (2013). Stress as a common risk factor for obesity and addiction. *Biological Psychiatry, 73*, 827–835.

Slopen, N., Kontos, E. Z., Ryff, C. D., Ayanian, J. Z., Albert, M. A., & Williams, D. R. (2013). Psychosocial stress and cigarette smoking persistence, cessation, and relapse over 9–10 years: A prospective study of middle-aged adults in the United States. *Cancer Causes & Control, 24*, 1849–1863.

Smith, M. A., Kim, S. Y., van Oers, H. J., & Levine, S. (1997). Maternal deprivation and stress induce immediate early genes in the infant rat brain. *Endocrinology, 138*, 4622–4628.

Snoek, H., Van Goozen, S. H., Matthys, W., Buitelaar, J. K., & van Engeland, H. (2004). Stress responsivity in children with externalizing behavior disorders. *Development and Psychopathology, 16*, 389–406.

Somaini, L., Donnini, C., Manfredini, M., Raggi, M. A., Saracino, M. A., Gerra, M. L., et al. (2011). Adverse childhood experiences (ACEs), genetic polymorphisms and neurochemical correlates in experimentation with psychotropic drugs among adolescents. *Neuroscience and Biobehavioral Reviews, 35*, 1771–1778.

Steptoe, A., Fieldman, G., Evans, O., & Perry, L. (1996). Cardiovascular risk and responsivity to mental stress: The influence of age, gender and risk factors. *Journal of Cardiovascular Risk, 3*, 83–93.

Stratakis, C. A., & Chrousos, G. P. (1995). Neuroendocrinology and pathophysiology of the stress system. *Annals of the New York Academy of Sciences, 771*, 1–18.

Suppli, N. P., Bukh, J. D., Moffitt, T. E., Caspi, A., Johansen, C., Tjønneland, A., et al. (2017). Genetic variants in 5-HTTLPR, BDNF, HTR1A, COMT, and FKBP5 and risk for treated depression after cancer diagnosis. *Depression and Anxiety*, *34*, 845–855.

Sutanto, W., Rosenfeld, P., de Kloet, E. R., & Levine, S. (1996). Long-term effects of neonatal maternal deprivation and ACTH on hippocampal mineralocorticoid and glucocorticoid receptors. *Brain Research Developmental Brain Research*, *92*, 156–163.

Swartz, J. R., Knodt, A. R., Radtke, S. R., & Hariri, A. R. (2015). A neural biomarker of psychological vulnerability to future life stress. *Neuron*, *85*, 505–511.

Szyf, M., Weaver, I., & Meaney, M. (2007). Maternal care, the epigenome and phenotypic differences in behavior. *Reproductive Toxicology*, *24*, 9–19.

Taha, F., Galea, S., Hien, D., & Goodwin, R. D. (2014). Childhood maltreatment and the persistence of smoking: A longitudinal study among adults in the US. *Child Abuse & Neglect*, *38*, 1995–2006.

Tarullo, A. R., & Gunnar, M. R. (2006). Child maltreatment and the developing HPA axis. *Hormones and Behavior*, *50*, 632–639.

Thayer, J. F., & Brosschot, J. F. (2005). Psychosomatics and psychopathology: Looking up and down from the brain. *Psychoneuroendocrinology*, *30*, 1050–1058.

Thayer, J. F., & Lane, R. D. (2000). A model of neurovisceral integration in emotion regulation and dysregulation. *Journal of Affective Disorders*, *61*, 201–216.

Thomas, M. B., & Becker, J. B. (2019). Sex differences in prenatal stress effects on cocaine pursuit in rats. *Physiology & Behavior*, *203*, 3–9.

Trickett, P. K., Noll, J. G., Susman, E. J., Shenk, C. E., & Putnam, F. W. (2010). Attenuation of cortisol across development for victims of sexual abuse. *Development and Psychopathology*, *22*, 165–175.

Tsagarakis, S., Rees, L. H., Besser, M., & Grossman, A. (1990). Opiate receptor subtype regulation of CRF-41 release from rat hypothalamus in vitro. *Neuroendocrinology*, *51*, 599–605.

Tyrka, A. R., Price, L. H., Marsit, C., Walters, O. C., & Carpenter, L. L. (2012). Childhood adversity and epigenetic modulation of the leukocyte glucocorticoid receptor: Preliminary findings in healthy adults. *PLoS One*, *7*, e30148.

Tyrka, A. R., Walters, O. C., Price, L. H., Anderson, G. M., & Carpenter, L. L. (2012). Altered response to neuroendocrine challenge linked to indices of the metabolic syndrome in healthy adults. *Hormone and Metabolic Research*, *44*, 543–549.

Tyrka, A. R., Wier, L., Price, L. H., Ross, N., Anderson, G. M., Wilkinson, C. W., et al. (2008). Childhood parental loss and adult hypothalamic-pituitary-adrenal function. *Biological Psychiatry*, *63*, 1147–1154.

Uhart, M., & Wand, G. S. (2009). Stress, alcohol and drug interaction: An update of human research. *Addiction Biology*, *14*, 43–64.

Valentino, R. J., & Van Bockstaele, E. (2001). Opposing regulation of the locus coeruleus by corticotropin-releasing factor and opioids. Potential for reciprocal interactions between stress and opioid sensitivity. *Psychopharmacology*, *158*, 331–342.

van Bodegom, M., Homberg, J. R., & Henckens, M. J. A. G. (2017). Modulation of the hypothalamic-pituitary-adrenal axis by early life stress exposure. *Frontiers in Cellular Neuroscience*, *11*, 87.

van Goozen, S. H., Fairchild, G., Snoek, H., & Harold, G. T. (2007). The evidence for a neurobiological model of childhood antisocial behavior. *Psychological Bulletin*, *133*, 149–182.

van Leeuwen, A. P., Creemers, H. E., Greaves-Lord, K., Verhulst, F. C., Ormel, J., & Huizink, A. C. (2011). Hypothalamic-pituitary-adrenal axis reactivity to social stress and adolescent cannabis use: The TRAILS study. *Addiction*, *106*, 1484–1492.

van Rooij, S. J., Stevens, J. S., Ely, T. D., Fani, N., Smith, A. K., Kerley, K. A., et al. (2016). Childhood trauma and COMT genotype interact to increase hippocampal activation in resilient individuals. *Frontiers in Psychiatry*, *7*, 156.

Vargas-Perez, H., Ting-A Kee, R., Walton, C. H., Hansen, D. M., Razavi, R., Clarke, L., et al. (2009). Ventral tegmental area BDNF induces an opiate-dependent-like reward state in naive rats. *Science*, *324*, 1732–1734.

Voellmin, A., Winzeler, K., Hug, E., Wilhelm, F. H., Schaefer, V., Gaab, J., et al. (2015). Blunted endocrine and cardiovascular reactivity in young healthy women reporting a history of childhood adversity. *Psychoneuroendocrinology*, *51*, 58–67.

Volkow, N. D., Fowler, J. S., Wang, G. J., Swanson, J. M., & Telang, F. (2007). Dopamine in drug abuse and addiction: Results of imaging studies and treatment implications. *Archives of Neurology*, *64*, 1575–1579.

Volkow, N. D., Wang, G. J., Fowler, J. S., Tomasi, D., & Baler, R. (2012). Food and drug reward: Overlapping circuits in human obesity and addiction. *Current Topics in Behavioral Neurosciences*, *11*, 1–24.

Walters, H., & Kosten, T. A. (2019). Early life stress and the propensity to develop addictive behaviors. *International Journal of Developmental Neuroscience*, *78*, 156–169.

Wand, G. S., & Schumann, H. (1998). Relationship between plasma adrenocorticotropin, hypothalamic opioid tone, and plasma leptin. *The Journal of Clinical Endocrinology and Metabolism*, *83*, 2138–2142.

Wang, B., Shaham, Y., Zitzman, D., Azari, S., Wise, R. A., & You, Z. B. (2005). Cocaine experience establishes control of midbrain glutamate and dopamine by corticotropin-releasing factor: A role in stress-induced relapse to drug seeking. *The Journal of Neuroscience*, *25*, 5389–5396.

Wang, G. J., Volkow, N. D., Thanos, P. K., & Fowler, J. S. (2004). Similarity between obesity and drug addiction as assessed by neurofunctional imaging: A concept review. *Journal of Addictive Diseases*, *23*, 39–53.

Wang, G. J., Yang, J., Volkow, N. D., Telang, F., Ma, Y., Zhu, W., et al. (2006). Gastric stimulation in obese subjects activates the hippocampus and other regions involved in brain reward circuitry. *Proceedings of the National Academy of Sciences of the United States of America*, *103*, 15641–15645.

Weaver, I. C. (2009). Shaping adult phenotypes through early life environments. *Birth Defects Research Part C, Embryo Today*, *87*, 314–326.

Weaver, I. C., Cervoni, N., Champagne, F. A., D'Alessio, A. C., Sharma, S., Seckl, J. R., et al. (2004). Epigenetic programming by maternal behavior. *Nature Neuroscience*, *7*, 847–854.

Wise, R. A., & Rompre, P. P. (1989). Brain dopamine and reward. *Annual Review of Psychology*, *40*, 191–225.

Womersley, J. S., Martin, L. I., van der Merwe, L., Seedat, S., & Hemmings, S. M. J. (2018). Hypothalamic-pituitary-adrenal axis variants and childhood trauma influence anxiety sensitivity in South African adolescents. *Metabolic Brain Disease*, *33*, 601–613.

Woodward, L. J., Fergusson, D. M., & Horwood, L. J. (2002). Deviant partner involvement and offending risk in early adulthood. *Journal of Child Psychology and Psychiatry*, *43*, 177–190.

Yang, C., Bogiatzi, C., & Spence, J. D. (2015). Risk of stroke at the time of carotid occlusion. *JAMA Neurology*, *72*, 1261–1267.

Zannas, A. S., Arloth, J., Carrillo-Roa, T., Iurato, S., Röh, S., Ressler, K. J., et al. (2015). Lifetime stress accelerates epigenetic aging in an urban, African American cohort: Relevance of glucocorticoid signaling. *Genome Biology*, *16*, 266.

Zimmerberg, B., & Brown, R. C. (1998). Prenatal experience and postnatal stress modulate the adult neurosteroid and catecholaminergic stress responses. *International Journal of Developmental Neuroscience*, *16*, 217–228.

Zorrilla, E. P., Logrip, M. L., & Koob, G. F. (2014). Corticotropin releasing factor: A key role in the neurobiology of addiction. *Frontiers in Neuroendocrinology*, *35*, 234–244.

CHAPTER FIVE

Stress, cortisol and suicide risk

Daryl B. O'Connor[a,*], Nicola Gartland[a], Rory C. O'Connor[b]
[a]School of Psychology, University of Leeds, Leeds, United Kingdom
[b]Suicidal Behavior Research Laboratory, Institute of Health & Wellbeing, University of Glasgow, Glasgow, United Kingdom
*Corresponding author: e-mail address: d.b.oconnor@leeds.ac.uk

Contents

Abstract

Suicide is a global health issue accounting for at least 800,000 deaths per annum. Numerous models have been proposed that differ in their emphasis on the role of psychological, social, psychiatric and neurobiological factors in explaining suicide risk. Central to many models is a stress-diathesis component which states that suicidal behavior is the result of an interaction between acutely stressful events and a susceptibility to suicidal behavior (a diathesis). This article presents an overview of studies that demonstrate that stress and dysregulated hypothalamic-pituitary-adrenal (HPA) axis activity, as measured by cortisol levels, are important additional risk factors for suicide. Evidence for other putative stress-related suicide risk factors including childhood trauma, impaired executive function, impulsivity and disrupted sleep are considered together with the impact of family history of suicide, perinatal and epigenetic influences on suicide risk.

International Review of Neurobiology, Volume 152
ISSN 0074-7742
https://doi.org/10.1016/bs.irn.2019.11.006

1. Introduction

Every 40s a person dies by suicide somewhere in the world (WHO, 2014). Suicide is a leading cause of mortality and is a major global health issue. It is estimated that 800,000 people die by suicide each year and there are 25 million nonfatal suicide attempts annually. As a result, for many decades, researchers have been exploring the causes of suicidal behavior with an aim to identify targets for suicide prevention. Numerous models have been proposed that differ in their emphasis on the role of psychological, social, psychiatric and neurobiological factors in predicting risk of suicide (Mann, Waternaux, Haas, & Malone, 1999; O'Connor, 2011; O'Connor & Nock, 2014; O'Connor & Kirtley, 2018; van Heeringen & Mann, 2014; van Orden et al., 2010). However, central to many models is a stress-diathesis component which states that suicidal behavior is a result of an interaction between acutely stressful events and a susceptibility to suicidal behavior (a diathesis). Research findings are accruing from post-mortem, neuroimaging and in vivo studies that a trait diathesis is manifested in dysregulation of hypothalamic-pituitary-adrenal (HPA) axis stress response activity as well as in impairments of the serotonergic and noradrenergic neurotransmitter systems, in structural brain abnormalities and via epigenetic pathways (Mann, 2013; Turecki, Ernst, Jollant, Labonté, & Mechawa, 2012; van Heeringen, Bijttebier, & Godfrin, 2011; van Heeringen & Mann, 2014). Indeed, evidence is emerging to suggest that biomarkers of a trait diathesis following serious stressful and traumatic psychosocial events, independent of psychiatric co-morbidities, may be useful predictors of suicide risk (van Heeringen & Mann, 2014). However, surprisingly, a relatively small body of work has explored the role of stress and its concomitant biomarker, cortisol, in the context of suicide and suicide vulnerability. The aim of this chapter is to provide an overview of studies that have investigated the role of stress and cortisol in the context of suicide risk together with studies that have examined other putative stress-related risk factors including childhood trauma, impaired executive function, impulsivity, disrupted sleep and perinatal and epigenetic influences on suicide risk.

The study of stress has a long history. Scientific interest dates back to the First World War, when soldiers were found to exhibit "shellshock," an extreme reaction to the trauma of battle that was subsequently acknowledged to be a manifestation of post-traumatic stress disorder (Lazarus, 1999). Since this time, stress has become part of everyday vernacular, and there has been a marked increase in media coverage of stress, and as a result, this has led to increased research and public awareness. In terms of research, over many decades we have learned that when we experience stress, the HPA axis is activated and releases cortisol from the adrenal glands. Once released, cortisol has several important functions such as increasing access to energy stores, increasing protein and fat mobilization, as well as regulating the magnitude and duration of inflammatory responses (Sapolsky, Romero, & Munck, 2000). As such, cortisol is the primary effector hormone of the HPA axis stress response system. The HPA axis is regulated by a negative feedback system, whereby the hypothalamus and the pituitary gland have receptors that detect changes in cortisol levels. For example, cortisol secretion will be inhibited when circulating levels rise or it will be stimulated when levels fall. However, if the HPA axis is repeatedly activated, this will trigger increased cortisol output, thereby exposing bodily tissues to excessive concentrations of the hormone (McEwen, 1998, 2000; Miller, Chen, & Zhou, 2007). Over time, such repetitive activation may contribute to tissue damage and future ill health by placing excessive pressure on various bodily systems including the HPA axis (known as allostatic load; McEwen, 1998). In addition, in the longer-term repeated activation may lead to dysregulation of the HPA axis as evidenced by flattened patterns of cortisol secretion across the day (including in the morning as well as in response to stressors). Indeed, in the context of suicide, evidence is accumulating to suggest a link between dysregulation of the HPA axis following chronic exposure to stress and vulnerability to suicide. An important issue this chapter will return to soon, but first a brief overview of the role of stress in the leading models of suicide.

1.1 Stress-diathesis models of suicidal behavior

Stress-diathesis models have a long history in the field of suicide research (see O'Connor, Ferguson, Green, O'Carroll, & O'Connor, 2016; O'Connor, Cleare, Eschle, Wetherall, & Kirtley, 2016). More than 30 years ago, Schotte and Clum (1987) put forward their distress-stress-hopelessness model of suicidal behavior. Therein, they posited and found evidence that impaired social problem-solving, a specific cognitive vulnerability factor acted as a diathesis; it was associated with suicide risk in the presence of stress. Since then, there has been an exponential growth in studies which have investigated how a range of different diatheses are associated with suicide risk under particular circumstances. Some of these diatheses are biological, others are cognitive in nature, and others still are personality factors.

For example, there is a considerable body of research illustrating how distinct components of perfectionism increase one's risk of suicidal thinking and behavior in the presence of stress (O'Connor, 2007).

Another diathesis-stress model, developed by Mann and colleagues, was the clinical model of suicidal behavior (Mann et al., 1999) where risk is posited to vary as a function of the interaction between psychiatric disorder (stressor) and a trait-like diathesis (e.g., impulsivity). This clinical model has been especially influential within psychiatry and clinical medicine. Whereas psychiatric disorder was a key element within the Mann et al. model, in 2008. Wenzel and Beck (2008) put forward a cognitive model of suicidal behavior which focuses on psychological treatment for suicidal behavior. Similar to the clinical model, it has adopted a distress-stress framework; however, on this occasion, the model is psychological in orientation and is characterized by three main constructs: (i) dispositional vulnerability factors, (ii) cognitive processes associated with psychiatric problems and (iii) cognitive processes associated with suicidal behavior. When it was published, this latter model was noteworthy as it systematically identified theoretical components which could be targeted in the delivery of cognitive therapy for suicidal patients. More recently still there has been greater recognition of the heterogeneity of suicide risk and the identification of suicidal sub-types (Bernanke, Stanley, & Oquendo, 2017) in the context of the relationship between stress and suicide risk. To this end, Bernanke et al. (2017) have proposed two distinct phenotypes of suicidal behavior, with one being stress-responsive (governed by the cortisol system) and the other being non-stress responsive (associated with the serotonin system). As these two phenotypes have been suggested as only two of potentially numerous suicidal subtypes, more research is required to better describe the complexity of suicide risk in terms of diathesis-stress responses.

1.2 The integrated motivational–volitional model of suicidal behavior

Building upon the work of Mann et al. (1999) and Williams (1997), O'Connor (2011) published the integrated motivational–volitional (IMV) model of suicidal behavior in 2011 and refined in 2018 (O'Connor & Kirtley, 2018). The aim of this model was to bring together the disparate constructs from existing models of suicide and integrate them into a single overarching theoretical framework. At its core, the IMV model is a diathesis-stress model which tracks the development of suicide risk across three phases (see Fig. 1). The first phase, the pre-motivational phase, outlines

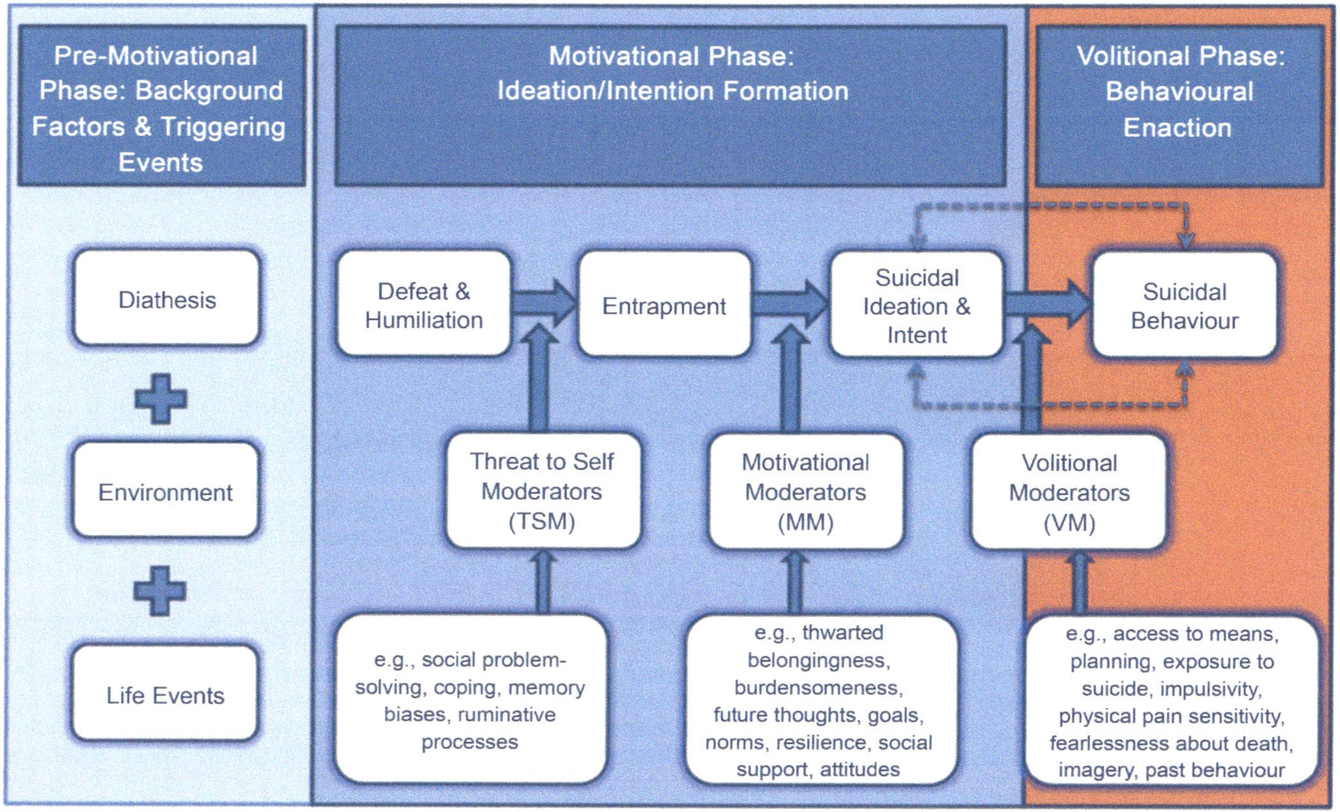

Fig. 1 Integrated motivational–volitional model of suicidal behavior (O'Connor, 2011; O'Connor & Kirtley, 2018).

the context in which suicidal thinking and suicidal behavior emerge. In this phase, it is posited that vulnerabilities interact with life stress and environmental influences to increase the likelihood that suicidal thinking may occur. However, the presence of vulnerabilities and stress are not sufficient to explain the increase in suicide risk on their own. According to the model, in phase 2 (the motivational phase) suicidal thinking is more likely to emerge if an individual is trapped by feelings of defeat, humiliation and loss. Needless to say, a stressful life event is often a key driver to feelings of defeat or humiliation from which the individual is endeavoring to escape. Defeat and entrapment are part of the final common pathway to the emergence of suicidal thinking. The final phase of the IMV model, the volitional phase, is concerned with the transition from thinking about suicide to acting upon one's thoughts of suicide, i.e., attempting suicide/dying by suicide. In this behavioral enaction phase, an individual is more likely to attempt suicide if volitional phase factors are also present. These volitional phase factors include having access to the means of suicide, being impulsive, being exposed to the suicidal behavior of others and having higher levels of fearlessness about death (O'Connor & Kirtley, 2018). In addition, although stress is not a key driver to the emergence of suicidal thoughts (beyond defeat and entrapment), it may be important in behavioral enaction (O'Connor, Rasmussen, & Hawton, 2012). Across a series of studies, as predicted by the IMV model, we have shown that the presence of such factors differentiate individuals who think about suicide or self-harm from those who engage in suicidal behavior or self-harm (Branley-Bell et al., 2019; Mars et al., 2019; Wetherall et al., 2018). In short, the IMV model is a useful model to consider the role of stress in the context of suicide risk.

2. Cortisol and suicide risk

Broadly speaking previous research on HPA axis, cortisol and suicidal behavior has focused in three main areas: (1) assessing HPA axis functioning through pharmacological manipulation of the stress system (Mann & Currier, 2007; Pompili et al., 2010) using the Dexamethasone Suppression Test (DST; Carroll, Martin, & Davies, 1968), (2) exploring naturally fluctuating cortisol levels and suicidal behavior, and (3) investigating HPA axis functioning following acute laboratory stressors in vulnerable and non-vulnerable groups.

2.1 The dexamethasone suppression test (DST) and suicide risk

For many decades researchers have been concentrating scientific effort in identifying clinical and biological predictors of suicide. In particular, during the early 1980s, studies were emerging to suggest that death by suicide may be associated with HPA axis hyperactivity and that a useful clinical tool to detect HPA axis hyperactivity was the dexamethasone suppression test (DST). The DST usually involves participants receiving oral administration of the synthetic glucocorticoid dexamethasone (e.g., 1 mg) on one morning (say at 11 am) and then plasma cortisol levels being assessed the following day in the morning (8 am) and afternoon (4 pm). Failure to suppress cortisol is evidence for HPA axis hyperactivity (due to glucocorticoid receptor insensivity) and has been found, in a number of studies, to predict completed suicide in different groups vulnerable to suicide (Coryell & Schlesser, 1981; Coryell, Young, & Carroll, 2006; Jokinen & Nordstrom, 2008, 2009; Norman, Brown, Miller, Keitner, & Overholser, 1990). An early example comes from a study by Coryell and Schlesser (1981) in patients with major depressive disorder. These authors showed that the risk estimate of suicide was around 27% in a group of patients who failed to suppress cortisol levels following DST compared to only 3% in patients who exhibited cortisol suppression. Similarly, Jokinen and Nordstrom (2008), in a 17-year follow-up study of elderly hospitalized mood disorder patients, found that DST non-suppression doubled the suicide risk and that patients who had completed suicide had higher post-dexamethasone serum cortisol levels compared to survivors.

Other existing evidence from prospective DST studies suggests that HPA hyperactivity is more consistently associated with completed suicide compared to suicide attempt (Mann & Currier, 2007). For example, in a sample of patients who met the criteria for major depressive disorder, mania, or schizoaffective disorder, Coryell and Schlesser (2001) reported that there was a 14-fold higher risk of suicide in individuals who failed to exhibit suppression of their cortisol levels in the DST compared to individuals who did exhibit suppression. However, the evidence for a clear relationship between HPA hyperactivity, as assessed using the DST, and suicide attempt is mixed. Some studies have shown that DST suppression status is unable to distinguish between individuals who will attempt suicide and those who will not. Yet, as outlined above, other research findings have demonstrated DST non-suppression is associated with a higher rate of suicide attempts (see Mann & Currier, 2007, for a review). In their review of biological predictors of suicidal behavior in individuals with mood disorders,

Mann and Currier (2007) suggest that an important reason why non-suppression on the DST is predictive of completed suicide may be because it is also associated with "a failure to respond to antidepressant treatment or a tendency for early relapse such as shortly after discharge" (p. 10).

Nevertheless, a recent meta-analysis of biological risk factors for suicidal behaviors was inconclusive with regards to the prediction of future suicide behaviors (Chang et al., 2016). Of the small number of tests included (4 for suicide attempt, 8 for completed suicide), the results showed that DST suppression significantly predicted completed suicide (Odds Ratio = 1.75 [1.05–2.90]), but did not significantly predict suicide attempt (Odds Ratio = 1.49 [0.58–3.82]). Moreover, there was also some evidence of publication bias suggesting that if three missing cases (below the mean) were included, the weighted mean odds ratio would have been non-significant.

Therefore, taken together, while DST research has contributed enormously to knowledge regarding HPA axis dysregulation and suicide vulnerability, findings remain inconsistent and contradictory (Chang et al., 2016; McGirr, Diaconu, Berlim, & Turecki, 2011). Pharmacological manipulation has also been criticized as it may not adequately mimic the size of the endogenous HPA response to naturally occurring stressors (Burke, Davis, Otte, & Mohr, 2005). In addition, more recent studies have begun to explore other aspects of the cortisol response, such as the diurnal cortisol rhythm (including morning and afternoon/evening cortisol levels; e.g., O'Connor, Green, Ferguson, O'Carroll, & O'Connor, 2018) and cortisol reactivity to stressors (e.g., McGirr et al., 2010) in order to improve understanding of the role of the stress response system and the HPA axis in the context of suicide behaviors.

2.2 Naturally fluctuating cortisol and suicidal behavior

The second broad area of research investigating the HPA axis and suicidal behavior has focused on exploring the relationship between naturally fluctuating (or baseline) cortisol levels and suicide behaviors. However, before outlining this research, it is important to note that cortisol has a distinct pattern over any 24 h period. The diurnal pattern of cortisol production is characterized by two distinct components: the peak levels after awakening (i.e., the cortisol awakening response, CAR) and the diminishing levels throughout the rest of the day (i.e., the diurnal slope). As will be shown later, evidence is beginning to converge to suggest that lower (or blunted) CAR and a flatter cortisol slope across the day are associated with more negative health outcomes (e.g., Adam et al., 2017; O'Connor et al., 2020, 2009).

Similar to the findings from the DST studies, research that has explored the associations between naturally fluctuating cortisol and different aspects of suicide behaviors have yielded inconsistent findings. For example, Westrin, Ekman, and Träskman-Bendz (1999) found elevated cortisol levels in patients who had recently attempted suicide compared to healthy controls, and Chatzittofis et al. (2013) found higher cortisol levels in (medication free) individuals who had attempted suicide compared to healthy volunteers. In contrast, Lindqvist, Isaksson, Träskman-Bendz, and Brundin (2008) and Lindqvist, Traskman-Bendz, and Vang (2008) found that cortisol levels were significantly lower in individuals who had attempted suicide compared to controls and more recently McGirr et al. (2011) also showed patients with depressive disorders exhibited lower levels of cortisol. A number of methodological factors may account for these mixed findings including the timing of the cortisol sampling (morning vs afternoon/evening), study quality, absence of a control comparison group and age of the sample. Given these disparate findings, O'Connor, Ferguson, et al., 2016; O'Connor, Cleare, et al., 2016 conducted a meta-analysis of all existing studies that has compared participants with at least one prior suicide *attempt* with a comparison group with no suicide attempt history in order: (i) to estimate the strength and variability of the association between naturally fluctuating cortisol levels and suicidal behavior and (ii) to identify moderators of this relationship. The systematic literature identified 27 studies ($N=2226$; 779 suicide attempters and 1447 non-attempters) that met the inclusion criteria. Overall, there was no significant effect of suicide group on cortisol. However, significant associations between cortisol and suicide attempts were observed as a function of age (see Fig. 2). In studies where the mean age of the sample was below 40 years the association was positive (i.e., higher cortisol was associated with suicide attempts; $r=0.234$, $P<0.001$), and where the mean age was 40 or above the association was negative (i.e., lower cortisol was associated with suicide attempts; $r=-0.129$, $P<0.001$).

The authors concluded that these meta-analytic findings confirm that HPA axis activity, as indicated by age-dependent variations in naturally occurring cortisol levels, are associated with suicide attempt. Moreover, these findings suggest that a reversal in the association between cortisol and suicide attempt occurs when the average age of the sample is around 40 years (or older). This is not to imply that for any individual the shift would happen at 40 years, this is on average (and was the mean age for the sample of studies that were included in this meta-analysis). Nonetheless, what these analyses do show is that for older people the association is negative and

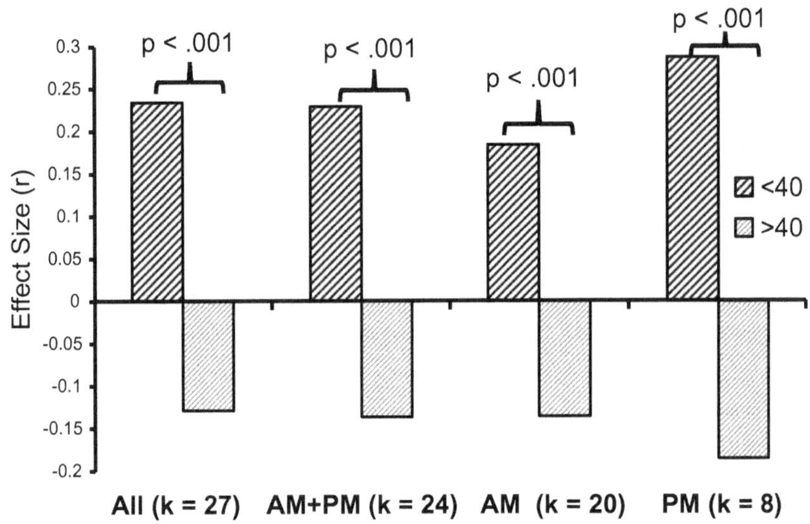

Fig. 2 Summary of moderating effects of age showing that where the average age of the sample was below 40 years the association was positive, whereas it was negative for samples where the average age was equal to or above 40 years. Note: Effect size = association between naturally fluctuating cortisol levels and suicide attempt history; AM = samples collected before midday, PM = samples collected in the afternoon or evening.

for younger people it is positive and that the relationship between cortisol and suicide attempts is more nuanced and complicated than past research has recognized. Furthermore, these results may have implications for research studies (reviewed earlier) that have assessed HPA axis functioning using pharmacological manipulation of the stress system such as the DST and raises the possibility that age may also moderate cortisol suppression following DST manipulation.

An important question remains unanswered by the findings of this meta-analysis. How might the reversal in the association be explained? It is likely that some of the variability will be accounted for by differences in study design, participants, suicide attempt measures and cortisol measurement. However, the findings are broadly consistent with McEwen's notion of allostatic load, whereby if the HPA axis is repeatedly activated (by stress) the immune, cardiovascular and the endocrine systems are potentially exposed to excessive demands that over time can lead to dysregulation of these systems (McEwen, 1998, 2000). Moreover, in the context of suicide vulnerability, naturally fluctuating cortisol levels may provide an index (or proxy) for the amount of stress exposure that vulnerable individuals have encountered (O'Connor et al., 2009). This view is also consistent with Fries, Hesse, Hellhammer, and Hellhammer (2005) account of the development of hypocortisolism, which suggests that the latter phenomenon occurs after a prolonged period of hyperactivity of the HPA axis due to chronic stress. Therefore, it would follow that in individuals who are older (40 years or older) and who have likely been exposed to stressful and traumatic events over a more sustained period, their HPA axis is more likely to have become dysregulated leading to lower secretion of cortisol levels. Such patterns have been observed in older Holocaust survivors with PTSD compared to those without PTSD (Yehuda et al., 1995). In contrast, younger individuals (less than 40 years), who have been exposed to serious stressful and psychosocial events, are likely to continue to exhibit an adaptive HPA axis stress response in the short to medium term (by releasing high levels of naturally fluctuating cortisol in response to their adverse and stressful environment).

The findings from this meta-analysis are also important because they demonstrate that both types of observations (hypereactivity and hyporeactivity) may be valid and true in terms of the relationship between cortisol levels and suicide attempt, but may be accounted for by age-dependent exposure to stress over time. However, much more work is required to understand how naturally fluctuating cortisol-suicide vulnerability relations change prospectively. Future research ought to improve the quality of their studies in this area by utilizing longitudinal designs (over many years) that incorporate assessments of suicidal behavior using clinical interviews or validated scales and ensure cortisol is measured at numerous time points across the day (morning, afternoon, evening) over multiple days (cf., Gartland, O'Connor, Lawton, & Bristow, 2014) to capture the full profile of cortisol, in doing so, use appropriate measurement (e.g., accuracy of sampling, accounting for variables known to influence cortisol, etc).

2.3 Cortisol reactivity to laboratory stress and suicide behavior

The third broad area investigating the HPA axis and suicidal behavior involves studies that have examined whether cortisol reactivity to a laboratory stress task can differentiate individuals who have a history of suicide attempt or ideation compared to individuals who have no such history (e.g., Giletta et al., 2015; McGirr et al., 2010; O'Connor, Green, Ferguson, O'Carroll, & O'Connor, 2017). A leading study in this area was conducted by McGirr et al. (2010). These authors investigated whether dysregulation of the HPA axis to a laboratory stressor was a heritable risk factor for suicidal behavior. A sample of first-degree relatives of individuals who had died by suicide and matched controls were compared on their cortisol reactivity to a well-established psychosocial stressor known as the Trier Social Stress Test, a public speaking task and mental arithmetic task in front of a judgmental/negative audience (TSST; Kirschbaum, Pirke, & Hellhammer, 1993). The results showed that the first-degree relatives exhibited a blunted (i.e., lower) cortisol response to stress. The authors have argued that their findings indicate that blunted cortisol reactivity to stress may represent a trait marker (or phenotype) of suicide risk.

More recently, two studies have used acute laboratory stressors to examine HPA axis responses to stress in vulnerable, at risk groups (Melhem et al., 2016; O'Connor et al., 2017). Melhem et al. (2016) examined cortisol responses to stress (i.e., the TSST) in a large sample of adult offspring of parents with mood disorder. This study found that an offspring suicide attempter group exhibited the lowest levels of total cortisol output during the stressor compared to an offspring with suicide-related behavior but never attempted suicide group, a non-suicidal offspring group and a healthy control group. Moreover, the suicide attempter group also showed the lowest baseline cortisol levels pre-TSST, but, contrary to expectations, there were no significant differences between groups on their measure of cortisol reactivity to stress.

A second study, conducted by O'Connor et al. (2017), aimed to investigate whether cortisol reactivity to the Maastricht Acute Stress Test (MAST, Smeets et al., 2012) differentiated individuals who had previously made a suicide attempt from those who had thought about suicide (a suicide ideation group) and control participants. The MAST stress protocol was designed to be both physiologically and psychologically challenging by combining an uncontrollable physical stressor (i.e., a cold pressor challenge) with a social-evaluative (i.e., mental arithmetic) component (Smeets et al., 2012). The results showed that participants who had made a previous suicide attempt exhibited significantly lower cortisol response to the MAST compared to participants in the ideator and control groups (see Fig. 3). Furthermore, participants who made an attempt within the past year exhibited a blunted cortisol response compared to participants with a more distant history of attempt. In addition, lower levels of cortisol in response to the MAST were associated with higher levels of suicidal ideation at 1-month follow-up in the suicide attempters group.

In the O'Connor et al. (2017) study, the finding that participants who attempted suicide within the last 12 months appear to exhibit a blunted

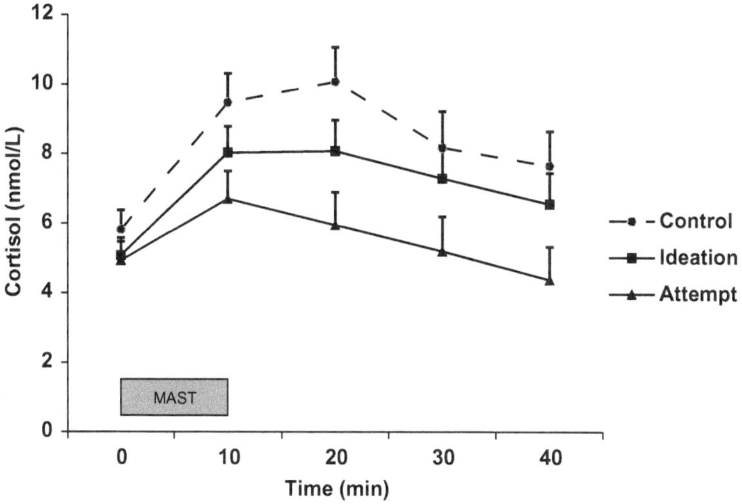

Fig. 3 Effects of suicide vulnerability group on cortisol reactivity to stress ($n = 145$). Note: Error bars represent the standard error of the mean. MAST was delivered between 0 and 10 min (see above) and cortisol measures taken at the beginning of the stress task (0 min), at +10, +20, +30 and +40 min post-task.

cortisol response to the laboratory stressor, compared to those with a lifetime history of suicide attempt, is a noteworthy observation. It is important because it suggests, in this study at least, that the cortisol response to stress may have returned to close to normal in the lifetime history group, although, these levels remain lower than in the control and ideator groups. This latter finding is promising as it is consistent with the notion that psychological and pharmacological intervention may yield benefits over time and help facilitate (partial) recovery of the HPA axis stress response system reflecting the higher cortisol levels in the lifetime history group. Therefore, an obvious next step would be for researchers to utilize longitudinal designs to explore whether dysregulation of cortisol reactivity to stress is restored over time and to investigate if the HPA axis has the potential to return to normal following psychological (e.g., stress management interventions) and/or pharmacological intervention.

Taken together, the results from recent laboratory based cortisol reactivity studies suggest that blunted or lower HPA axis activity may increase risk for suicide attempt among vulnerable individuals. The findings also indicate that the HPA axis stress response system may have become dysregulated in individuals who have tried to take their own lives and as such may increase future suicide risk by impairing their ability to cope and adapt to acute and non-acute stressors.

3. Childhood trauma—Cortisol—Suicide risk

Childhood trauma has been identified as an important variable in the etiology of suicide risk. For example, Marshall, Galea, Wood, and Kerr (2013) found high levels of moderate and severe childhood trauma being associated with suicide attempt in a prospective cohort study of illicit drug users. In particular, they showed that severe sexual, physical and emotional childhood abuse conferred a substantial increased repeated suicide risk in adulthood. In another study, Sachiapone, Carli, Cuomo, and Roy (2007) found that high levels of childhood trauma were associated with suicide attempt in patients with unipolar depression. Similarly, a large longitudinal population-based study in the Netherlands (Enns et al., 2006) found that childhood neglect, psychological abuse and physical abuse were strongly associated with new onset suicide ideation and suicide attempt over a 3-year follow-up. More recently, O'Connor et al. (2018) found that 78.7% of participants with a history of suicide attempt reported exposure to at least one type of childhood trauma that was classified as moderate or severe compared to 37.7% and 17.8% in an ideation and control group, respectively.

3.1 Childhood trauma and the hypothalamic-pituitary-adrenal axis

Research has begun to focus on the links between childhood trauma and altered dynamics of the HPA axis. In the context of depression, Heim et al. (2000) and Heim, Newport, Mletzko, Miller, and Nemeroff (2008) have shown associations between childhood trauma and dysregulated HPA axis and to persistent sensitization of the stress response system. Childhood trauma effects on depression have also been explained by changes in glucocorticoid resistance, increased central corticotropin-releasing factor (CRF) activity, immune activation, and reduced hippocampal volume. In contrast, the results are less clear relating childhood trauma to cortisol activity (e.g., cortisol reactivity to stress). A study by Carpenter et al. (2007) showed decreased cortisol levels in response to a laboratory stressor in childhood maltreated men who were never depressed. In a later study, the same team also found that women who reported childhood physical abuse displayed a blunted cortisol response to the TSST compared to women without physical abuse (Carpenter, Shattuck, Tyrka, Geracioti, & Price, 2011). These findings contradict earlier work by Heim et al. (2000) who showed that women who had a history of childhood abuse, with and without major depression, exhibited increased cortisol to an acute laboratory stressor. However, more broadly, there is also converging evidence to suggest that early life adversity is associated with blunted cortisol reactivity to stress (e.g., Lovallo, Farag, Sorocco, Cohoon, & Vincent, 2012). For example, Lovallo et al. (2012) using data from the Oklahoma Family Health Patterns Project showed that experience of adversity predicted reduced cortisol response to an acute laboratory stress challenge.

Similarly, two recent studies by O'Connor et al. (2020, 2018) provide further support linking childhood trauma with blunted, or lower cortisol levels in response to stress and in naturalistic settings. In a laboratory-based study investigating the effects of childhood trauma on cortisol reactivity to an acute stressor and on resting cortisol levels, O'Connor et al. (2018) found that higher levels of trauma were associated with blunted cortisol reactivity to stress and lower resting cortisol levels. In particular, individuals who reported more than one moderate or severe type of childhood trauma exhibited the lowest cortisol levels in response to stress (see Fig. 4) and at rest. In a second study, O'Connor et al. (2020), investigated for the first time, whether childhood trauma and daily stressors and emotions were associated with diurnal cortisol levels (i.e., cortisol levels following waking and the decline in cortisol levels across the rest of the day) over a 7-day study in individuals vulnerable to suicide. The results showed that participants

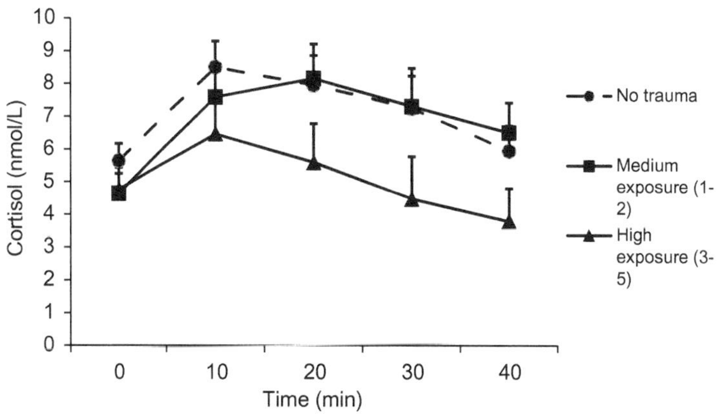

Fig. 4 Effects of childhood trauma levels on cortisol reactivity to stress in a combined attempt and ideation group ($n = 100$). The results show that individuals exposed to the highest number of childhood traumas release the lowest levels of cortisol in response to acute laboratory stressor.

with a history of suicide attempt (a suicide attempt group) or previously had thoughts of ending their life (an ideation group) released significantly lower cortisol upon awakening (CAR) and had a tendency toward flatter wake-peak to 12h (WP-12) cortisol slopes compared to individuals with no history of attempt or ideation. Moreover, childhood trauma was found to be associated with significantly lower CAR and a tendency toward flatter WP-12 cortisol slope and it had an indirect effect on suicide vulnerability group membership via lower daily CAR levels. The latter finding is particularly important as it shows, for the first time, that the effects of childhood trauma has indirect, as well as, direct effects on suicide vulnerability through lower levels of daily CAR.

Taken together, the studies reviewed are important as they suggest that the experience of childhood trauma may predispose individuals to vulnerability to suicide in adulthood by leading to diminished HPA axis activity during awakening (and possibly a tendency toward a flatter diurnal profile across the day) as well as during stress. These findings are in keeping with a recent large scale meta-analysis by Adam et al. (2017) that showed flatter cortisol cycles were common to a wide range of mental and physical health outcomes. Moreover, these results are also consistent with the development of hypocortisolism posited by Fries et al. (2005), as outlined earlier, which suggests that hypocortisolism occurs after a prolonged period of hyperactivity of the HPA axis due to chronic stress. Moreover, we have previously suggested that Lovallo's (2013) conceptual model of addiction linking adverse life experiences in childhood and adolescence to adverse health outcomes in adulthood should be extended to suicide risk (O'Connor et al., 2018). Lovallo (2013) has argued that adverse life experiences cause modifications in frontolimbic brain function which may then lead directly to: (1) reduced stress reactivity, (2) altered cognition (characterized by a shift in focus to more short-term goals and impulsive response selection) and (3) unstable affect regulation. Lovallo (2013) has also suggested that these three negative consequences influence the development of a more impulsive behavioral style that may increase risk of addiction and the engagement in poor health behaviors. We believe that exhibiting a low or blunted CAR may be another negative consequence of the modification of brain function (Boehringer et al., 2015).

4. Possible mechanisms linking stress and suicide risk

The association between stress and suicide has been described in the models of suicide risk, and the role of the HPA axis in this relationship has been outlined above. However, there are likely to be multiple interrelated mechanisms that link stress and suicide. We will look at a few more possible pathways here that may help to answer the question: How does the experience of stress influence subsequent suicidal behavior, sometimes decades later?

4.1 Executive function and impulsivity

One possible mechanism tying stress to suicide behavior is executive function. Executive function is a broad term for a range of cognitive processes which manage and control thoughts, emotions and actions. These functions are required whenever we must pay attention to a task or are effortfully pursuing a goal; they help us to concentrate, consider possible courses of action, and make informed decisions. There are three core executive functions (Diamond, 2013): inhibition (this includes both the self-control of behavior, as well was stopping interferences to thought necessary for selective attention); working memory (keeping information temporarily available for processing); and cognitive flexibility (the ability to switch from thinking about one concept to another, and also the ability to adapt thoughts or behaviors based on changes in the environment). The personality trait "impulsivity" is related to executive function as it is characterized by behaviors which reflect impaired self-regulation. At the behavioral level, this might include poor planning, premature responding without considering the consequences of one's actions, taking risks and an inability to delay gratification. These behaviors are suggested to originate from deficits in working memory, self-regulation of affect-motivation-arousal, internalization of speech and behavioral analysis that affords hindsight, forethought, and goal-directed action (Barkley, 1997; Gvion & Apter, 2011). Impulsivity tends to lead to the underestimation of potential consequences of actions, has been shown to be positively associated with suicide risk (Brezo, Paris, & Turecki, 2006; Gvion & Apter, 2011; McGirr et al., 2009). Dysfunctional executive decision-making, such as cognitive rigidity, has also been suggested to result in suicidal mental states (Marzuk, Hartwell, Leon, & Portera, 2005; for review, see Bredemeier & Miller, 2015).

There is evidence to support the suggestion that both distal and proximal stress can have an effect on executive function. Greater levels of adverse life experience (such as physical and sexual abuse, separation from parents, and a family history of substance abuse) has been shown to predict lower working memory function, greater impulsive decision-making, and lower mental age (Lovallo, 2013; Lovallo et al., 2013). As rates of early adversity are high in individuals who have attempted suicide (O'Connor et al., 2018), a mediated pathway is plausible where stressful experiences early in life alter cognitive function and that these altered thought processes can increase the risk of suicidal behaviors throughout the lifecourse.

However, a moderation pathway is also possible. Some aspects of executive control are considered heritable (Swan & Carmelli, 2002). McGirr et al. (2010) compared relatives of suicide completers with matched controls, and found no difference in baseline measures of executive function. However, performance on the Word-Color Inhibition Test and Trail Making Test were differentially affected by a controlled laboratory stressor (the Trier Social Stress Test). Relatives of suicide completers failed to improve on executive function tests after the TSST, specifically on switching and inhibition conditions. This indicates a level of cognitive inflexibility in these individuals, but only after stress induction. McGirr and colleagues argue that cognitive inflexibility and a decreased ability to inhibit inappropriate action in response to real-life stressors could be potential factors that increase the risk of suicidal behavior. These findings provide a possible moderation mechanism for the stress diathesis hypothesis, where "at risk" individuals respond to stress with cognitions which could increase their risk of suicidal behavior.

4.2 Family history

Suicidal behavior aggregates in families (Brent, Bridge, Johnson, & Connolly, 1996; Brent et al., 2002; Kim et al., 2005; McGirr et al., 2009, 2011). In a register-based case control study, Mittendorfer-Rutz, Rasmussen, and Wasserman (2008) demonstrated that individuals whose sibling had attempted suicide were nearly 3.5 times more likely to attempt suicide themselves; a maternal suicide attempt carried a 2.7 times greater risk and paternal suicide attempt carried a 1.9 times greater risk. Genetic transmission of personality traits such as impulsivity has been one factor suggested to account for familial aggregation of suicidal behavior in families (Mittendorfer-Rutz et al., 2008). However, it is difficult to tease out the relationships between stress, family history of suicide and suicidal behavior because a family history of suicide can be a substantial source of stress in itself, as well as providing a potentially direct genetic/hereditary pathway to suicidal behavior.

There is evidence that in a sample of depressed outpatients, a family history of suicide was associated with lower plasma cortisol levels (McGirr et al., 2011). This effect was independent of psychopathology and the individual's previous suicide attempts, and suggests an overall down-regulation of HPA-axis activity in this group. However, from this we cannot determine where in the diurnal rhythm of cortisol the levels are reduced. Different points of this rhythm are implicated in different aspects of HPA functionality and reactivity; research focusing on HPA stress reactivity has used cortisol saliva samples alongside acute laboratory stressors to determine whether differences in cortisol stress reactivity exist between these two groups.

Evidence from first-degree family members of suicide completers shows low (or blunted) salivary cortisol responses to an acute laboratory stressor, compared to controls (McGirr et al., 2010). O'Connor et al. (2017) also found that having a family history of suicide was associated with the lowest cortisol response to an acute laboratory stressor. Melhem et al. (2016) found that offspring of parents who had attempted suicide exhibited lower levels of total cortisol output during an acute laboratory stressor compared to offspring of parents with mood disorder but had not attempted suicide, but they did not find significant differences in cortisol reactivity to the task between these groups. These studies tentatively suggest that having a family history of suicide is associated with blunted HPA axis stress reactivity, and are consistent with the idea that dysregulation of the stress response system may be a heritable risk factor for suicidal behavior. Again, this dysregulation of the stress system would act as a moderation pathway, where the relationship between stress and suicidal behavior is strengthened due to the dysregulated HPA axis response to stress in these individuals. Indeed, in a sample of suicide attempters, lower levels of cortisol in response to acute stress have been shown to predict higher levels of suicidal ideation 1 month later (O'Connor et al., 2017). Therefore, HPA axis dysregulation may exacerbate the stress-suicide relationship such that the link between stress and suicide is stronger for those with a family history of suicide.

However, it cannot be determined whether this dysregulation occurs through a genetic commonality between family members or whether it is an effect of the stress of losing a close family member to suicide. Do members of families with a history of suicide have a shared genetic vulnerability or diathesis, which increases the risk of suicide in the face of stress? Or has the stressful experience of losing a close relative to suicide caused the dysregulation of the HPA axis in these individuals? Further research is warranted, but will require carefully controlled groups to compare those highly affected by family history of suicide and those less affected, for example, where the parental suicide attempt was before the child was born. Adoption studies could also provide valuable insights.

4.3 Perinatal influences and epigenetics

Familial influences on suicide risk may also work through adverse in-utero and perinatal conditions. In a recent systematic review, Orri et al. (2019) assessed family and parental characteristics during pregnancy and around the time of birth in relation to suicide, suicide attempt and suicide ideation throughout the lifespan. Factors associated with higher suicide risk included high birth order, teenage mothers, single mothers, low maternal and paternal education level, fetal growth and small for gestational age. Only one study in this review directly measured maternal stress, in the form of bereavement (Class et al., 2014). This Swedish population-based study of over 2,000,000 offspring found that the death of a first degree relative of the mother during the first postnatal year increased the risk of suicide attempt and completed suicide in offspring. Orri and colleagues also argue that factors such as teenage mothers, single mothers, and low socioeconomic position at birth may reflect a wider adverse psychosocial environment which would be associated with greater levels of maternal stress both during pregnancy and the perinatal period. While other psychosocial mechanisms are likely to be at work, there is some evidence that maternal stress may influence fetal brain development through epigenetic (non-genetic influences on gene expression) or gene-by-environment interaction mechanisms.

There is evidence that prenatal anxiety is associated with higher levels of waking cortisol in children at age 10, suggesting that prenatal experiences can influence HPA activity in offspring (O'Connor et al., 2005). Turecki et al. (2012) outline a model to explain increased risk of suicide in individuals exposed to early-life adversity through HPA axis dysregulation. Early life stress is proposed to increase methylation (addition of a methyl group to a DNA nucleotide) of hippocampal glucocorticoid receptor (GR) genes, which disrupts the GR gene expression. One of the studies that support this suggestion demonstrated increased methylation of the GR promoter gene in adolescent children in cases where their mothers were exposed to intimate partner violence during pregnancy (Radtke et al., 2011). The methylation status of the GR gene in the mothers was not affected by intimate partner violence, but the prenatal stress experienced appears to have had a long-lasting impact on the gene expression of their children. This may provide a mechanism to explain findings that prenatal stress alters HPA axis activity later in life (O'Connor et al., 2005). These hippocampal receptors play a crucial role in the negative feedback loop controlling cortisol levels, and thus alterations in the number and sensitivity of these receptors influences the body's ability to regulate the amount of circulating cortisol. This, in turn, is proposed to lead to the development of emotional, behavioral and cognitive phenotypes (e.g., chronic anxiety, impulsivity) and cognitive alterations (e.g., executive function, as discussed above) which are associated with increased suicide risk. For comprehensive reviews of the animal and human research into the epigenetics of stress in the early years of life, see Roy and Dwivedi (2017), Turecki et al. (2012), and Turecki and Meaney (2016). For a review of the evidence linking epigenetic changes with suicidal behavior, see Labonté and Turecki (2010).

Epigenetic research has also suggested that the neuropeptide oxytocin, which is sensitive to environmental stress, may be implicated in the transmission of maternal stress during pre- and postnatal periods (Toepfer et al., 2017). Further investigation for the role of epigenetic mechanisms in suicidal behavior comes from research assessing GR gene expression in post-mortem hippocampi obtained from suicide completers with a history of childhood abuse, suicide completers with no history of childhood abuse, and controls (Labonté et al., 2012; McGowan et al., 2009). These studies demonstrate reduced hippocampal GR gene expression in suicide victims with a history of childhood abuse in comparison to suicide victims with no history of abuse and controls, while there was no difference between the non-abused and control groups. This suggests that while childhood abuse is associated with epigenetic changes in gene expression which influences HPA function, but does not provide evidence for a link between these changes and risk of suicidal behavior.

Using a whole genome-wide approach to investigate DNA methylation in the hippocampi of suicide completers, Labonté et al. (2013) confirmed their previous findings that promoter DNA methylation levels are greater in suicide completers compared to controls. Interestingly, they report increased levels of methylation in promoters of four specific genes known to be involved in cognitive processes related to executive function (e.g., learning, working memory, behavior). Therefore, taken with previous findings that childhood abuse is related to DNA methylation, this provides initial evidence for an epigenetic mechanism where stress in childhood leads to changes in hippocampal gene expression which could cause impairments in executive function that increase the risk of suicide.

4.4 Sleep

Another mechanism by which stress could potentially affect suicide behaviors is through disruption to sleep. There is strong evidence that insomnia and nightmares are associated with increased suicide risk

(Bernert, Kim, Iwata, & Perlis, 2015; Nadorff, Nazem, & Fiske, 2011, 2013; Pigeon, Pinquart, & Conner, 2012). However, causality and third variable effects are hard to establish. Research to date has mainly focused on psychological mediators of this effect, identifying defeat and entrapment, as well as emotional regulation, social isolation and negative appraisals as mediators of this effect (Russell, Rasmussen, & Hunter, 2018; for review, see Littlewood, Kyle, Pratt, Peters, & Gooding, 2017). However, it has also been suggested that disturbances in sleep may lead to cognitive impairments and impulsive decision-making (Porras-Segovia et al., 2019). Interestingly, evidence suggests that the duration of sleep disturbance (e.g., for how long an individual has been experiencing nightmares) is a significant factor in risk of suicide, where longer durations are associated with increased risk (Golding, Nadorff, Winer, & Ward, 2015; Nadorff et al., 2013).

The main issue with this body of literature is that it consists predominantly of cross-sectional studies which cannot determine the direction of the relationships between sleep and suicidal behaviors. A recently ecological momentary assessment study by Littlewood et al. (2019) addressed this issue and demonstrated a unidirectional relationship between sleep disturbance and suicidal thoughts. Objectively and subjectively determined short sleep duration, and poor sleep quality predicted more severe next-day suicidal thoughts; there was no relationship between suicidal ideation and sleep duration or quality the following night. This study establishes the causal direction of this day-to-day relationship, which is a valuable step in our understanding. However, reciprocal and bidirectional relationships are still possible along longer time scales and merit investigation.

High levels of stress are associated with both chronic insomnia symptoms and recurrent short sleep duration (Abell, Shipley, Ferrie, Kivimäki, & Kumari, 2016). Stress effects sleep not only in terms of sleep quality, but it also disrupts the EEG spectral profile of sleep in both healthy participants and patients with chronic insomnia (Ackermann, Cordi, La Marca, Seifritz, & Rasch, 2019; Hall et al., 2000). In insomnia patients, the tendency to experience stress-related intrusive thoughts is associated with poorer subjective sleep quality, and higher levels of subjective stress burden are associated with decreases in delta activity which is an indication of hyperarousal during sleep (Hall et al., 2000). Therefore, different aspects of stress may influence distinct characteristics of sleep. In another study with healthy participants, an acute laboratory-based stressor was used to investigate the immediate effects of psychosocial stress on napping; this form of acute stress increased sleep latency, but also reduced slow wave activity and enhanced alpha activity (Ackemann et al., 2019).

Sleep disruption also influences HPA axis activity. While some older research did not find associations between measures of sleep quality or insomnia with salivary cortisol, more recent research with improved methodologies have confirmed an effect of shorter sleep and poor sleep quality on diurnal cortisol (Castro-Diehl et al., 2015). Cross-sectional research has also found that those who report frequent nightmares show a blunted CAR on a working day, but not on a leisure day (Nagy et al., 2015). In an impressive 10-year follow-up in the Whitehall II study, Abell et al. (2016) provide evidence that recurrent short sleep (measured at three timepoints during the 10-year period) was associated with a flatter diurnal cortisol pattern, characterized by higher levels of cortisol later in the day. A steeper CAR was also observed in those who reported insomnia symptoms at all three timepoints and those reporting short sleep twice, compared to those who did not report sleep problems at any time point. These findings have yet to be related to suicide risk, but given the accumulating evidence for different aspects of HPA axis dysregulation in suicide risk and the potential pathways between stress, sleep, and cortisol, this seems a promising avenue for future research.

However, the inter-relations of these variables are complex and potentially reciprocal and therefore are not easy to disentangle. While it is possible that there is a direct pathway from stress to suicidal behavior via sleep disturbance, it is likely that any such mechanism will also interact with the other variables we have mentioned here. For example, disruption to sleep patterns could influence executive function, and there is evidence that familial risk for insomnia can be measured through HPA axis dysregulation in response to stress (Drake, Cheng, Almeida, & Roth, 2017). Disruption to sleep also has been hypothesized as a stressor in itself, contributing to allostatic load (McEwen, 2006). This idea is consistent with the findings suggesting that duration of insomnia or nightmares is predictive of suicidal risk, as the cumulative effects of ongoing sleep disturbance leads to wear and tear on bodily systems. Research into the interconnections of these mechanisms could provide vital and effective insights for the development of interventions, through the identification of vulnerable populations and provision of targeted tools to reduce the risk of suicide in vulnerable populations.

5. General conclusion

This article has presented an overview of studies that demonstrate that stress and dysregulated hypothalamic-pituitary-adrenal (HPA) axis activity, as measured by cortisol levels, are important additional risk factors for suicide. It has also highlighted the IMV model of suicide as a useful framework to understand suicide risk. Evidence for other stress-related putative suicide risk factors including childhood trauma, impaired executive function, impulsivity and disrupted sleep have been shown to play an important role together with family history of suicide, perinatal and epigenetic influences on suicide risk. In order to further improve our understanding of the precise pathways through which stress and HPA axis dysregulation contribute to suicide, there is a need for future research to investigate simultaneously the impact of distal and proximal determinants of suicidal behavior.

References

Abell, J. G., Shipley, M. J., Ferrie, J. E., Kivimäki, M., & Kumari, M. (2016). Recurrent short sleep, chronic insomnia symptoms and salivary cortisol: A 10-year follow-up in the Whitehall II study. *Psychoneuroendocrinology*, *68*, 91–99. https://doi.org/10.1016/j.psyneuen.2016.02.021.

Ackermann, S., Cordi, M., La Marca, R., Seifritz, E., & Rasch, B. (2019). Psychosocial stress before a nap increases sleep latency and decreases early slow-wave activity. *Frontiers in Psychology*, *10*, 20. https://doi.org/10.3389/fpsyg.2019.00020.

Adam, E. K., Quinn, M. E., Tavernier, R., McQuillan, M. T., Dahlke, K. A., & Gilbert, K. E. (2017). Diurnal cortisol slopes and mental and physical health outcomes: A systematic review and meta-analysis. *Psychoneuroendocrinology*, *83*, 25–41.

Barkley, R. A. (1997). Behavioral inhibition, sustained attention, and executive functions: Constructing a unifying theory of ADHD. *Psychological Bulletin*, *121*, 65–94.

Bernanke, J. A., Stanley, B. H., & Oquendo, M. A. (2017). Toward fine-grained phenotyping of suicidal behavior: The role of suicidal subtypes. *Molecular Psychiatry*, *22*, 1080–1081.

Bernert, R. A., Kim, J. S., Iwata, N. G., & Perlis, M. L. (2015). Sleep disturbances as an evidence-based suicide risk factor. *Current Psychiatry Reports*, *17*, 1–9.

Boehringer, A., Tost, H., Haddad, L., Lederbogen, F., Wust, S., Schwarz, E., et al. (2015). Neural correlates of the cortisol awakening responses in humans. *Neuropsychopharmacology*, *40*, 2278–2285.

Branley-Bell, D., O'Connor, D. B., Green, J. A., Ferguson, E., O'Carroll, R. E., & O'Connor, R. C. (2019). Distinguishing suicide ideation from suicide attempts: Further test of the integrated motivational-volitional model of suicidal behaviour. *Journal of Psychiatric Research*, *117*, 100–107.

Bredemeier, K., & Miller, I. W. (2015). Executive function and suicidality: A systematic qualitative review. *Clinical Psychology Review*, *40*, 170–183. https://doi.org/10.1016/j.cpr.2015.06.005.

Brent, D. A., Bridge, J., Johnson, B. A., & Connolly, J. (1996). Suicidal behavior runs in families. A controlled family study of adolescent suicide victims. *Archives of General Psychiatry*, *53*(12), 1145–1152. https://doi.org/10.1001/archpsyc.1996.01830120085015.

Brent, D. A., Oquendo, M., Birmaher, B., Greenhill, L., Kolko, D., Stanley, B., et al. (2002). Familial pathways to early-onset suicide attempt: Risk for suicidal behavior in offspring of mood-disordered suicide attempters. *Archives of General Psychiatry*, *59*(9), 801–807. https://doi.org/10.1001/archpsyc.59.9.801.

Brezo, J., Paris, J., & Turecki, G. (2006). Personality traits as correlates of suicidal ideation, suicide attempts, and suicide completions: A systematic review. *Acta Psychiatrica Scandinavica*, *113*, 180–206. https://doi.org/10.1111/j.1600-0447.2005.00702.x.

Burke, H. M., Davis, M. C., Otte, C., & Mohr, D. C. (2005). Depression and cortisol responses to psychological stress: A meta-analysis. *Psychoneuroendocrinology*, *30*, 846–856.

Carpenter, L. L., Carvalho, J. P., Tyrka, A. R., Wier, L. M., Mello, A. F., Mello, M. F., et al. (2007). Decreased adrenocorticotropic hormone and cortisol responses to stress in healthy adults reporting significant childhood maltreatment. *Biological Psychiatry*, *62*, 1080–1087.

Carpenter, L. L., Shattuck, T. T., Tyrka, A. R., Geracioti, T. D., & Price, L. H. (2011). Effect of childhood physical abuse on cortisol stress response. *Psychopharmacology*, *214*, 367–375.

Carroll, B., Martin, F., & Davies, B. (1968). Resistance to suppression by dexamethasone of plasma 11-OHCS levels in severe depressive illness. *British Medical Journal*, *3*, 285.

Castro-Diehl, C., Diez Roux, A. V., Redline, S., Seeman, T., Shrager, S. E., Shea, S., et al. (2015). Association of sleep duration and quality with alterations in the hypothalamic-pituitary-adrenocortical axis: The multi-ethnic study of atherosclerosis. *Journal of Clinical Endocrinology and Metabolism*, *100*, 3149–3158.

Chang, B. P., Franklin, J. C., Ribeiro, J. D., Fox, K. R., Bentley, K. H., Kleinman, E. M., et al. (2016). Biological risk factors for suicidal behaviors: A meta-analysis. *Translational Psychiatry*, *6*, e887.

Chatzittofis, A., Nordström, P., Hellström, C., Arver, S., Asberg, M., & Jokinen, J. (2013). CSF 5-HIAA, cortisol and DHEAS levels in suicide attempters. *European Neuropsychopharmacology*, *23*, 1280–1287.

Class, Q. A., Abel, K. M., Khashan, A. S., Ricket, M. E., Dalman, C., Larsson, H., et al. (2014). Offspring psychopathology following preconception, prenatal, and postnatal maternal bereavement stress. *Psychological Medicine*, *44*(1), 71–84. https://doi.org/10.1017/S0033291713000780.

Coryell, W., & Schlesser, M. A. (1981). Suicide and the dexamethasone suppression test in unipolar depression. *American Journal of Psychiatry*, *138*, 1120–1121.

Coryell, W., & Schlesser, M. A. (2001). The dexamethasone suppression test and suicide prediction. *American Journal of Psychiatry*, *158*, 748–753.

Coryell, W., Young, E., & Carroll, B. (2006). Hyperactivity of the hypothalamic-pituitary-adrenal axis and mortality in major depressive disorder. *Psychiatry Research*, *142*, 99–104.

Diamond, A. (2013). Executive functions. *Annual Review of Psychology*, *64*, 135–168. https://doi.org/10.1146/annurev-psych-113011-143750.

Drake, C. L., Cheng, P., Almeida, D. M., & Roth, T. (2017). Familial risk for insomnia is associated with abnormal cortisol response to stress. *Sleep*, *40*(10). zsx143. https://doi.org/10.1093/sleep/zsx143.

Enns, M. W., Cox, B. J., Afifi, T. O., de Graff, R., ten Have, M., & Sareen, J. (2006). Childhood adversities and risk for suicide ideation and attempts: A longitudinal population-based study. *Psychological Medicine*, *36*, 1769–1778.

Fries, E., Hesse, J., Hellhammer, J., & Hellhammer, D. H. (2005). A new view of hypocortisolism. *Psychoneuroendocrinology*, *30*, 1010–1016.

Gartland, N., O'Connor, D. B., Lawton, R., & Bristow, M. (2014). Exploring day-to-day dynamics of daily stressor appraisals, physical symptoms and the cortisol awakening response. *Psychoneuroendocrinology*, *50*, 130–138.

Giletta, M., Calhoun, C. D., Hastings, P. D., Rudolph, K. D., Nock, M. K., & Prinstein, M. J. (2015). Multi-level risk factors for suicide ideation among at-risk adolescent females: The role of hypothalamic-pituitary-adrenal axis responses to stress. *Journal of Abnormal Child Psychology*, *43*, 807–820.

Golding, S., Nadorff, M. R., Winer, E. S., & Ward, K. C. (2015). Unpacking sleep and suicide in older adults in a combined online sample. *Journal of Clinical Sleep Medicine*, *11*(12), 1385–1392.

Gvion, Y., & Apter, A. (2011). Aggression, impulsivity, and suicide behavior: A review of the literature. *Archives of Suicide Research*, *15*, 93–112. https://doi.org/10.1080/13811118.2011.565265.

Hall, M., Buysse, D. J., Nowell, P. D., Nofzinger, E. A., Houck, P., Reynolds, C. F., et al. (2000). Symptoms of stress and depression as correlates of sleep in primary insomnia. *Psychosomatic Medicine, 62*, 227–230. https://doi.org/10.1097/00006842-200003000-00014.
Heim, C., Newport, D. J., Heit, S., Graham, Y. P., Wilcox, M., Bonsall, R., et al. (2000). Pituitary–adrenal and autonomic responses to stress in women after sexual and physical abuse in childhood. *JAMA, 284*, 592–597.
Heim, C., Newport, D. J., Mletzko, T., Miller, A. H., & Nemeroff, C. B. (2008). The link between childhood trauma and depression: Insights from HPA axis studies in humans. *Psychoneuroendocrinology, 33*, 693–710.
Jokinen, J., & Nordstrom, P. (2008). HPA axis hyperactivity as suicide predictor in elderly mood disorder inpatients. *Psychoneuroendocrinology, 33*, 1387–1393.
Jokinen, J., & Nordstrom, P. (2009). HPA axis hyperactivity and attempted suicide in young adult mood disorder inpatients. *Journal of Affective Disorders, 116*, 117–120.
Kim, C. D., Seguin, M., Therrien, N., Riopel, G., Chawky, N., Lesage, A. D., et al. (2005). Familial aggregation of suicidal behavior: A family study of male suicide completers from the general population. *American Journal of Psychiatry, 162*(5), 1017–1019. https://doi.org/10.1176/appi.ajp.162.5.1017.
Kirschbaum, C., Pirke, K. M., & Hellhammer, D. H. (1993). The 'Trier Social Stress Test'—A tool for investigating psychobiological stress responses in a laboratory setting. *Neuropsychobiology, 28*, 76–81.
Labonté, B., Suderman, M., Maussion, G., Lopez, J. P., Navarro-Sánchez, L., Yerko, V., et al. (2013). Genome-wide methylation changes in the brains of suicide completers. *American Journal of Psychiatry, 170*, 511–520.
Labonté, B., & Turecki, G. (2010). The epigenetics of suicide: Explaining the biological effects of early life environmental adversity. *Archives of Suicide Research, 14*, 291–310. https://doi.org/10.1080/13811118.2010.524025.
Labonté, B., Yerko, V., Gross, J., Mechawar, N., Meaney, M. J., Szyf, M., et al. (2012). Differential glucocorticoid receptor exon 1_B, 1_C, and 1_H expression and methylation in suicide completers with a history of childhood abuse. *Biological Psychiatry, 72*, 41–48. https://doi.org/10.1016/j.biopsych.2012.01.034.
Lazarus, R. S. (1999). *Stress and emotion: A new synthesis*. London: Springer.
Lindqvist, D., Isaksson, A., Träskman-Bendz, L., & Brundin, L. (2008). Salivary cortisol and suicidal behavior—A follow-up study. *Psychoneuroendocrinology, 33*, 1061–1068.
Lindqvist, D., Traskman-Bendz, L., & Vang, F. (2008). Suicidal intent and the HPA-axis characteristics of suicide attempters with major depressive disorder and adjustment disorder. *Archives of Suicide Research, 12*, 197–207.
Littlewood, D. L., Kyle, S. D., Carter, L., Peters, S., Pratt, D., & Gooding, P. (2019). Short sleep duration and poor sleep quality predict next-day suicidal ideation: An ecological momentary assessment study. *Psychological Medicine, 49*, 403–411. https://doi.org/10.1017/S0033291718001009.
Littlewood, D., Kyle, S. D., Pratt, D., Peters, S., & Gooding, P. (2017). Examining the role of psychological factors in the relationship between sleep problems and suicide. *Clinical Psychology Review, 54*, 1–16. https://doi.org/10.1016/j.cpr.2017.03.009.
Lovallo, W. R. (2013). Early life adversity reduces stress reactivity and enhances impulsive behavior: Implications for health behaviors. *International Journal of Psychophysiology, 90*(1), 8–16. https://doi.org/10.1016/j.ijpsycho.2012.10.006.
Lovallo, W. R., Farag, N. H., Sorocco, K. H., Cohoon, A. J., & Vincent, A. S. (2012). Lifetime adversity leads to blunted stress axis reactivity: Studies from the Oklahoma Family Health Patterns Project. *Biological Psychiatry, 71*, 344–349.
Lovallo, W. R., Farag, N. H., Sorocco, K. H., Acheson, A., Cohoon, A. J., & Vincent, A. S. (2013). Early life adversity contributes to impaired cognition and impulsive behaviour:

Studies from the Oklahoma family health patterns project. *Alcoholism: Clinical and Experimental Research*, *37*(4), 616–623. https://doi.org/10.1111/acer.12016.
Mann, J. J. (2013). The serotonergic system in mood disorders and suicidal behaviour. *Philosophical Transactions of the Royal Society London B: Biological Sciences*, *368*, 20120537.
Mann, J. J., & Currier, D. (2007). A review of prospective studies of biologic predictors of suicidal behavior in mood disorders. *Archives of Suicide Research*, *11*, 3–16.
Mann, J. J., Waternaux, C., Haas, G. L., & Malone, K. M. (1999). Towards a clinical model of suicidal behaviour in psychiatric patients. *American Journal of Psychiatry*, *156*, 181–189.
Mars, B., Heron, L., Klonsky, D., Moran, P., O'Connor, R. C., Tilling, K., et al. (2019). What distinguishes adolescents with suicidal thoughts from those who have attempted suicide? A population-based birth cohort study. *Journal of Child Psychology and Psychiatry*, *60*, 91–99.
Marshall, B. D. L., Galea, S., Wood, E., & Kerr, T. (2013). Longitudinal associations between types of childhood trauma and suicidal behaviour among substance users: A cohort study. *American Journal of Public Health*, *103*, e69–e75.
Marzuk, P. M., Hartwell, N., Leon, A. C., & Portera, L. (2005). Executive functioning in depressed patients with suicidal ideation. *Acta Psychiatrica Scandinavica*, *112*, 294–301.
McEwen, B. S. (1998). Protective and damaging effects of stress mediators. *New England Journal of Medicine*, *338*, 171–179.
McEwen, B. S. (2000). Allostasis and allostatic load: Implications for neuropsychopharmacology. *Neuropsychopharmacology*, *22*, 108–124.
McEwen, B. S. (2006). Protective and damaging effects of stress mediators: Central role of the brain. *Dialogues in Clinical Neuroscience*, *8*(4), 367–381.
McGirr, A., Alda, M., Séguin, M., Cabot, S., Lesage, A., & Tureki, G. (2009). Familial aggregation of suicide explained by cluster B traits: A three-group family study of suicide controlling for major depressive disorder. *American Journal of Psychiatry*, *166*(10), 1124–1134. https://doi.org/10.1176/appi.ajp.2009.08111744.
McGirr, A., Diaconu, G., Berlim, M. T., Pruessner, J. C., Sablé, R., Cabot, S., et al. (2010). Dysregulation of the sympathetic nervous system, hypothalamic-pituitary-adrenal axis and executive function in individuals at risk for suicide. *Journal of Psychiatry & Neuroscience*, *35*(6), 399–408.
McGirr, A., Diaconu, G., Berlim, M. T., & Turecki, G. (2011). Personal and family history of suicidal behavior is associated with lower peripheral cortisol in depressed outpatients. *Journal of Affective Disorders*, *131*, 368–373.
McGowan, P. O., Sasaki, A., B'Alessio, A. C., Dymov, S., Labonte, B., Szyf, M., et al. (2009). Epigenetic regulation of the glucocorticoid receptor in human brain associates with childhood abuse. *Nature Neuroscience*, *12*, 342–348.
Melhem, N. M., Keilp, J. G., Porta, G., Oquendo, M. A., Burke, A., Stanley, B., et al. (2016). Blunted HPA axis activity in suicide attempters compared to those at high risk for suicidal behavior. *Neuropsychopharmacology*, *41*, 1447–1456. https://doi.org/10.1038/npp.2015.309.
Miller, G. E., Chen, E., & Zhou, E. S. (2007). If it goes up, must it come down? Chronic stress and the hypothalamic-pituitary-adrenocortical axis in humans. *Psychological Bulletin*, *133*, 25–45.
Mittendorfer-Rutz, E., Rasmussen, F., & Wasserman, D. (2008). Familial clustering of suicidal behaviour and psychopathology in young suicide attempters: A register-based nested case control study. *Social Psychiatry and Psychiatric Epidemiology*, *43*, 28–36. https://doi.org/10.1007/s00127-007-0266-0.
Nadorff, M. R., Nazem, S., & Fiske, A. (2011). Insomnia symptoms, nightmares, and suicidal ideation in a college student sample. *Sleep*, *34*(1), 93–98.

Nadorff, M. R., Nazem, S., & Fiske, A. (2013). Insomnia symptoms, nightmares, and suicide risk: Duration of sleep disturbance matters. *Suicide & Life-Threatening Behavior, 43*(2), 139–149.

Nagy, T., Salavecz, G., Simor, P., Purebl, G., Bódizs, R., Dockray, S., et al. (2015). Frequent nightmares are associated with blunted cortisol awakening response in women. *Physiology & Behavior, 147*, 233–237. https://doi.org/10.1016/j.physbeh.2015.05.001.

Norman, W. H., Brown, W., Miller, I. W., Keitner, G. I., & Overholser, J. (1990). The dexamethasone suppression test and completed suicide. *Acta Psychiatrica Scandinavica, 81*, 120–125.

O'Connor, R. C. (2007). The relations between perfectionism and suicide risk: A systematic review. *Suicide and Life-threatening Behavior, 37*, 698–714.

O'Connor, R. C. (2011). Towards an integrated motivational-volitional of suicidal behaviour. In R. O'Connor, S. Platt, & J. Gordon (Eds.), *International handbook of suicide prevention: Research, policy and practice* (pp. 181–198). Wiley Blackwell.

O'Connor, T. G., Ben-Shlomo, Y., Heron, J., Golding, J., Adams, D., & Glover, V. (2005). Prenatal anxiety predicts individual differences in cortisol in pre-adolescent children. *Biological Psychiatry, 58*, 211–217. https://doi.org/10.1016/j.biopsych.2005.03.032.

O'Connor, D. B., Branley-Bell, D., Green, J., Ferguson, E., O'Carroll, R., & O'Connor, R. C. (2020). Effects of childhood trauma, daily stress and emotions on daily cortisol levels in individuals vulnerable to suicide. *Journal of Abnormal Psychology,* in press. https://doi.org/10.1037/abn0000482.

O'Connor, D. B., Ferguson, E., Green, J., O'Carroll, R. E., & O'Connor, R. C. (2016). Cortisol and suicidal behavior: A meta-analysis. *Psychoneuroendocrinology, 63*, 370–379.

O'Connor, D. B., Green, J., Ferguson, E., O'Carroll, R. E., & O'Connor, R. C. (2017). Cortisol reactivity and suicidal behavior: Investigating the role of hypothalamic-pituitary-adrenal (HPA) axis responses to stress in suicide attempters and ideators. *Psychoneuroendocrinology, 75*, 183–191.

O'Connor, D. B., Green, J. A., Ferguson, E., O'Carroll, R. E., & O'Connor, R. C. (2018). Effects of childhood trauma on cortisol levels in suicide attempters and ideators. *Psychoneuroendocrinology, 88*, 9–16.

O'Connor, D. B., Hendrickx, H., Dadd, T., Talbot, D., Mayes, A., Elliman, T., et al. (2009). Cortisol awakening rise in middle-aged women in relation to chronic psychological stress. *Psychoneuroendocrinology, 34*, 1486–1494.

O'Connor, R. C., & Nock, M. (2014). The psychology of suicidal behaviour. *Lancet Psychiatry, 1*, 73–85.

O'Connor, R. C., Rasmussen, S., & Hawton, K. (2012). From thoughts to action: Distinguishing adolescents who think about self-harm from those who engage in self-harm. *British Journal of Psychiatry, 200*, 330–335.

O'Connor, R. C., Cleare, S., Eschle, S., Wetherall, K., & Kirtley, O. J. (2016). The integrated motivational-volitional model of suicidal behaviour: An update. In R. C. O'Connor & J. Pirkis (Eds.), *The international handbook of suicide prevention* (2nd ed.). Chichester: Wiley-Blackwell, https://doi.org/10.1098/rstb.2017.0268.

O'Connor, R. C., & Kirtley, O. J. (2018). The integrated motivational-volitional model of suicidal behaviour. *Philosophical Transactions of the Royal Society B, 373*. 20170268.

Orri, M., Gunnell, D., Richard-Devantoy, S., Bolanis, D., Boruff, J., Turecki, G., et al. (2019). In-utero and perinatal influences on suicide risk: A systematic review and meta-analysis. *Lancet Psychiatry, 6*(6), 477–492.

Pigeon, W. R., Pinquart, M., & Conner, K. (2012). Meta-analysis of sleep disturbance and suicidal thoughts and behaviors. *Journal of Clinical Psychiatry, 73*, e1160–e1167.

Pompili, M., Serafini, G., Innamorati, M., Moller-Leimkuhler, A. M., Giupponi, G., Girardi, P., et al. (2010). The hypothalamic-pituitary-adrenal axis and serotonin abnormalities: A selective overview for the implications of suicide prevention. *European Archives of Psychiatry and Clinical Neuroscience*, *260*, 583–600.

Porras-Segovia, A., Pérez-Rodríguez, M. M., López-Esteban, P., Courtet, P., Barrigón, M. L., López-Castromán, J., et al. (2019). Contribution of sleep deprivation to suicidal behaviour: A systematic review. *Sleep Medicine Reviews*, *44*, 37–47. https://doi.org/10.1016/j.smrv.2018.12.005.

Radtke, K. M., Ruf, M., Gunter, H. M., Dohrmann, K., Schauer, M., Meyer, A., et al. (2011). Transgenerational impact of intimate partner violence on methylation in the promoter of the glucocorticoid receptor. *Translational Psychiatry*, *1*, e21. https://doi.org/10.1038/tp.2011.21.

Roy, B., & Dwivedi, Y. (2017). Understanding epigenetic architecture of suicide neurobiology: A critical perspective. *Neuroscience and Biobehavioral Reviews*, *72*, 10–27. https://doi.org/10.1016/j.neubiorev.2016.10.031.

Russell, K., Rasmussen, S., & Hunter, S. C. (2018). Insomnia and nightmares as markers of risk for suicidal ideation in young people: Investigating the role of defeat and entrapment. *Journal of Clinical Sleep Medicine*, *14*(5), 775–784. https://doi.org/10.5664/jcsm.7104.

Sachiapone, M., Carli, V., Cuomo, C., & Roy, A. (2007). Childhood trauma and suicide attempts in patients with unipolar depression. *Depression and Anxiety*, *24*, 268–272.

Sapolsky, R. M., Romero, L. M., & Munck, A. U. (2000). How do glucocorticoids influence stress responses? Integrating permissive, suppressive, stimulatory, and preparative actions. *Endocrine Reviews*, *21*, 55–89.

Schotte, D. E., & Clum, C. A. (1987). Problem-solving skills in psychiatric patients. *Journal of Consulting and Clinical Psychology*, *55*, 49–54.

Smeets, T., Cornelisse, S., Quaedflieg, C., Meyer, T., Jelicic, M., & Merckelbach, H. (2012). Introducing the Maastricht Acute Stress Test (MAST): A quick and non-invasive approach to elicit robust autonomic and glucocorticoid stress responses. *Psychoneuroendocrinology*, *37*, 1998–2008.

Swan, G. E., & Carmelli, D. (2002). Evidence for genetic mediation of executive control: A study of aging male twins. *Journal of Gerontology: Psychological Sciences*, *57B*(2), 133–143.

Toepfer, P., Heim, C., Entringer, S., Binder, E., Wadhwa, P., & Buss, C. (2017). Oxytocin pathways in the intergenerational transmission of maternal early life stress. *Neuroscience & Biobehavioral Reviews*, *73*, 293–308. https://doi.org/10.1016/j.neubiorev.2016.12.026.

Turecki, G., Ernst, C., Jollant, F., Labonté, B., & Mechawa, N. (2012). The neurodevelopmental origins of suicidal behavior. *Trends in Neurosciences*, *35*(1), 14–23. https://doi.org/10.1016/j.tins.2011.11.008.

Turecki, G., & Meaney, M. J. (2016). Effects of the social environment and stress on glucocorticoid receptor gene methylation: A systematic review. *Biological Psychiatry*, *79*, 87–96. https://doi.org/10.1016/j.biopsych.2014.11.022.

van Heeringen, C., Bijttebier, S., & Godfrin, K. (2011). Suicidal brains: A review of functional and structural brain studies in association with suicidal behaviour. *Neuroscience and Biobehavioral Reviews*, *35*, 688–698.

van Heeringen, K., & Mann, J. J. (2014). The neurobiology of suicide. *Lancet Psychiatry*, *1*, 63–72.

van Orden, K. A., Witte, T. K., Cukrowicz, K. C., Braithwaite, S. R., Selby, E. A., & Joiner, T. E., Jr. (2010). The interpersonal theory of suicide. *Psychological Review*, *117*, 575–600.

Wenzel, A., & Beck, A. T. (2008). A cognitive model of suicidal behaviour: Theory and treatment. *Applied and Preventive Psychology*, *12*, 189–201.

Westrin, A., Ekman, R., & Träskman-Bendz, L. (1999). Alterations of corticotropin releasing hormone (CRH) and neuropeptide Y (NPY) plasma levels in mood disorder patients with a recent suicide attempt. *European Neuropsychopharmacology*, *9*, 205–211.

Wetherall, K., Cleare, S., Eschle, S., Ferguson, E., O'Connor, D. B., O'Carroll, R., et al. (2018). From ideation to action: Differentiating between those who think about suicide and those who attempt suicide in a national study of young adults. *Journal of Affective Disorders*, *241*, 475–483.

Williams, J. M. G. (1997). Cry of pain. In *World Health Organization (WHO). Preventing suicide: A global imperative, 2014, Geneva, Switzerland*, Penguin: Oxford.

Yehuda, R., Kahana, B., Binder-Brynes, K., Southwick, S. M., Mason, J. W., & Giller, E. L. (1995). Low urinary cortisol excretion in Holocaust survivors with posttraumatic stress disorder. *American Journal of Psychiatry*, *152*, 982–986.

CHAPTER SIX

Stress and cortisol in Parkinson's disease

Daniel J. van Wamelen[a,b,*], **Yi-Min Wan**[a,b,c], **K. Ray Chaudhuri**[a,b], **Peter Jenner**[d]

[a]King's College London, Department of Neurosciences, Institute of Psychiatry, Psychology & Neuroscience, London, United Kingdom
[b]Parkinson Foundation Centre of Excellence, King's College Hospital, London, United Kingdom
[c]Department of Psychiatry, Ng Teng Fong General Hospital, Singapore, Singapore
[d]King's College London, Neurodegenerative Diseases Research Group, Institute of Pharmaceutical Sciences, Faculty of Health Sciences and Medicine, London, United Kingdom
*Corresponding author: e-mail address: daniel.van_wamelen@kcl.ac.uk

Contents

Abstract

Stress is ubiquitous with many factors contributing to its effects, including psychological responses and associated biological factors, including cortisol related physiological responses, and inflammation. Also in Parkinson's disease there is growing evidence for the role of stress in some key symptoms, even stretching to the prodromal stage. Here we discuss the possible contributions of the range and nature of stress in PD and we aim to summarize the current knowledge about the role of stress-related responses on motor and non-motor symptoms, the underlying pathophysiology, and the potential implications for treatment.

International Review of Neurobiology, Volume 152
ISSN 0074-7742
https://doi.org/10.1016/bs.irn.2020.01.005

1. Introduction

Stress is ubiquitous and applies to PD patients in an analogous manner as to the general population. It may be psychologically driven, for example, by the stress of chronic illness, financial and interpersonal problems abut may also be biological in origin, for example, in response to physical injury and inflammation (Lederbogen et al., 2011). The latter, i.e., biologically derived stress, is an integral part of adaptive biological systems (Lucassen et al., 2014). However, stress has been proposed to be a possible mediator for the development of PD. Here we discuss the possible contributions of the range and nature of stress in PD and we aim to summarize the current knowledge about the role of stress-related responses in PD.

2. Stress as a risk factor for PD

Although the exact prevalence of (perceived) stress in Parkinson's disease (PD) is unknown, stress may underpin the expression of a range of motor and non-motor symptoms throughout all motor stages of the disease. Some have suggested that stress itself could be a contributing factor to the development of PD, although this is a controversial observation. Historically the notion of psychological stress being a possible trigger or antecedent to Parkinson's disease (PD) is not a novel idea; having been described more than 100 years ago by Sir William Richard Gowers, who wrote: "Prolonged anxiety and severe emotional shock often precede the onset of Parkinson's disease" (Gowers, 1888). While motor performance improves under acute stress, chronic anxiety and stress characteristically aggravate motor symptoms in PD, particularly rest tremor (Moore, Rose, & Grace, 2001). In the early 1990s, Sapolsky et al. suggested the phenomenon termed "neuroendangerment," when stress was indicated to increase the vulnerability of nigral neurons to an insult subsequent in life (Sapolsky, 1996). Susceptibility to psychological stress is said to be an early prodromal sign of PD, much like how major depression has been shown to be (Fang et al., 2010), with early changes in the prefrontal, parietal or limbic circuitry, leading to impairment in coping abilities. More recently, Sieurin et al. reported that a high-demand occupation is a risk factor contributing to a higher PD incidence. In this population-based cohort study of over 2 million male Swedes, higher job demands and expectations increased PD risk, especially among the higher educated subjects (Sieurin et al., 2018). However, clear and robust evidence of a link between stressful life events and risk of PD remains inconsistent (Clark, Ritz, Prescott, & Rod, 2013; Djamshidian & Lees, 2014; Hemmerle, Herman, & Seroogy, 2012; Rod, Bordelon, Thompson, Marcotte, & Ritz, 2013; Svensson, Farkas, Gradus, Lash, & Sørensen, 2016; Vlajinac et al., 2013).

3. Neurobiochemical and cortisol changes in relation to stress in PD

From a neurochemical angle, stress sensitive cortisol levels may have an effect on some PD symptoms. This observation is based on a range of studies addressing the effect of cortisol levels on motor symptoms (based on the Unified Parkinson's Disease Rating Scale), but these studies failed to show a consistent effect with some showing a worsening of motor symptoms and some an improvement (Djamshidian et al., 2011; Håglin & Bäckman, 2016; Müller & Muhlack, 2007). Higher serum cortisol levels appear to be associated with anxiety levels, risk-taking, sleep, and depressive symptoms, common and intrusive NMS of PD (Breen et al., 2014; Djamshidian et al., 2011; Müller, Welnic, & Muhlack, 2007; Růžička et al., 2015). A similar relationship between increased, and temporally disrupted, cortisol levels has been shown in Huntington disease patients where increased cortisol levels are associated with worse memory performance and depressive symptoms (Shirbin, Chua, Churchyard, Hannan, et al., 2013; Shirbin, Chua, Churchyard, Lowndes, et al., 2013). Moreover, in the transgenic R6/2 mouse model of this latter disease, expressing mutant huntingtin with about 150 CAG repeat expansions, high cortisol levels exacerbate motor performance, including worsening of chorea (Dufour & McBride, 2016). In Alzheimer's disease, increased cortisol levels have been shown to influence non-motor symptoms such as sleep (Pistollato et al., 2016), and elevated cortisol levels may predict the onset of cognitive deficits in this disease (Ennis et al., 2017). As such, cortisol appears to be a biomarker of stress-related symptoms generation in PD and also HD and AD. The implications are unclear and need systematic studies so that therapeutic strategies targeting the cortisol pathway in PD could be considered.

4. Psychological stress

Psychological stress plays a role in many disorders and the "fight and flight" catecholaminergic reactions, evoked via the sympathetic nervous system triggered by transient or acute self-limiting stressors, is thought to be an adaptive coping response (Segerstrom & Miller, 2004). Chronic psychological stress, on the other hand, either due to major life events, or accumulative minor "irritations and frustrations," provokes a prolonged activation of the hypothalamic–pituitary–adrenal (HPA) axis (McEwen, 2007), triggering an "allostatic load," which translates to a failure of adequate coping mechanisms to reduce stress (McEwen, 2003).

Psychological stress can overlap with the concept of exhaustion (Clark et al., 2013) (a state of excessive fatigue and irritability often attributed to stress) and adjustment disorder (Svensson et al., 2016) (an abnormal and excessive emotional response to the presence of an identifiable stressful event). These concepts tie in with the definition of fatigue, a key NMS of PD. Fatigue as well as adjustment disorders are considered to be possible markers of prodromal PD. In Vlajinac's case-control study on the association of the risk of PD with specific perceived stressful life events, PD was found to be significantly related to retirement (OR 18.73, 95% CI 1.9–175.4), birth of own child (OR 66.22, 95% CI 8.3–526.3) and air raids (OR 5.66, 95% CI 2.4–13.5), and the risk of PD increases with the number of perceived stressful events (Vlajinac et al., 2013). In studies on measures of stress response and its link with risk of PD, Clark et al. (2013) revealed that high vital exhaustion (a psychological response reflecting a breakdown of the adaption to stress) is associated with a higher risk of PD hospitalization in an exposure-dependent manner (HR 2.50), while Svensson et al. showed that adjustment disorder (a severe psychological reaction to stressful life events) is linked to the development of PD (standardized incidence ratio of 1.84 (95% CI 1.53–2.20)) (Svensson et al., 2016). In a cohort study (Rod et al., 2013), the number of life events was associated with risk of depression in an exposure-dependent manner in a sample of newly diagnosed PD patients, with each additional event being associated with a 56% higher risk of depression (95% CI 1.23–1.98), but revealed that social support and good coping mitigated the risk. Specific stressors associated with PD are provided in Table 1. From the literature therefore, it is surmised that future studies on the link between stress and PD need to differentiate between specific stressors as well as the response to these, and also account for the duration of exposure to each stressor.

4.1 The phenomenon of paradoxical kinesia in PD

A term first described by Souques in 1921, paradoxical kinesia (PK) is characterized by a sudden and brief period of mobility classically seen as a

Table 1 Specific stressors in Parkinson's disease.

Specific stressors	References
Hospital fire	Hammond (2010)
Earthquakes	Bonanni, Thomas, and Onofrj (2010)
Auditory stimulus and fear for safety of loved ones	Daroff (2008)
Visual motion stimuli during war	Schlesinger, Erikh, and Yarnitsky (2007)

reaction to acute stress in PD (Nathan & Scobell, 2012). Schwab and Zieper characterized cases when PD patients were able to overcome their severe akinesia under extreme circumstances (such as a house fire), but the benefit lasted only a minute or two (Schwab & Zieper, 1965). Paradoxical kinesia in subjects with advanced idiopathic PD have also been described in circumstances of extreme emotional stress such as car accidents, war or earthquakes (Bonanni, Thomas, Anzellotti, et al., 2010). The common pattern observed across the studies is that external sensory cues are pivotal in stimulating paradoxical kinesia. However, the response may be attenuated by anticipatory tight temporal constraints, which is an internal expectation that reflects knowledge that unless the movement is performed within the tight time window necessary, the intended goal will not be achieved; this internal cue producing paradoxical kinesia even in the absence of external sensory triggers (Distler, Schlachetzki, Kohl, Winkler, & Schenk, 2016). An implied sense of urgency, rather than real, is sufficient to produce paradoxical kinesia, such as seen in the study where performance in PD patients when only a visual stimulus was administered was compared to when combined with a loud auditory stimulus, and the findings revealed an increased speeding up of performance with the combined stimuli, likely due to an elicitation of a mild stress response which energized the reactions of the PD patients (Anzak et al., 2011).

Three mechanisms are proposed to explain PK which could be considered a response to acute psychological stress and include (1) noradrenergic augmentation, (2) compensatory activation of cerebellar circuitry, and (3) activation of basal ganglia reserves (Asmus, Huber, Gasser, & Schöls, 2008) (Table 2). The first is said to lead to increased vigilance and alertness (Schlesinger et al., 2007) and is exemplified by rodent model studies where administration of the dopamine blocker haloperidol renders the animal akinetic which can be overcome by stress-induced noradrenergic activation. In support of the second theory, external visual stimuli have been shown to

Table 2 Proposed mechanisms underlying paradoxical kinesia in Parkinson's disease.

Proposed mechanisms	Outcome	References
Noradrenergic augmentation	Increased vigilance as part of "fight or flight" response to stress or otherwise known as "adrenal medulla activation by environmental stress"	Yntema and Korf (1987), Marshall, Levitan, and Stricker (1976), Keefe, Salamone, Zigmond, and Stricker (1989), and Colpaert (1987)
Compensatory activation of cerebellar networks (alternative pathways)	External visual stimuli has been shown to circumvent the damaged basal ganglia, using an intact cerebellar circuit for visuomotor control	Glickstein and Stein (1991), Bostan, Dum and Strick (2010), and Wu and Hallett (2013)
Activation of basal ganglia dopamine reserves	Stress and fear being the dominant triggers in activating basal ganglia reserves	Glickstein and Stein (1991) and De La Fuente-Fernández and Stoessl (2002)

bypass damaged basal ganglia, using an intact cerebellar circuit for visuomotor control (Glickstein & Stein, 1991). Fear is the dominant trigger in activating basal ganglia reserves in the third mechanism of action, as demonstrated in Shiner et al.'s study when the speed of PD patients not on medications in a decision-making task was accelerated so as to avoid painful electric shocks (Shiner et al., 2012).

5. HPA-axis abnormalities in PD and biological stress

Normal biological stress responses appear to be formed of two different phases; the first, is a rapid, response which involves the activation of the autonomic nervous system (ANS) causing a surge in serum (nor)epinephrine, released from the adrenal medulla. This induces hypertension, tachycardia, and a subsequent increase in blood flow to the heart and skeletal muscles (Joëls, Angela Sarabdjitsingh, & Karst, 2012). The second phase in response to a stressor involves the activation of the hypothalamic–pituitary–adrenal (HPA) axis and increased serum cortisol levels as previously alluded to (Joëls et al., 2012). HPA-axis activation is triggered by corticotropin-releasing hormone (CRH) in the paraventricular nucleus (PVN), mediating the release of adrenocorticotropic hormone (ACTH)

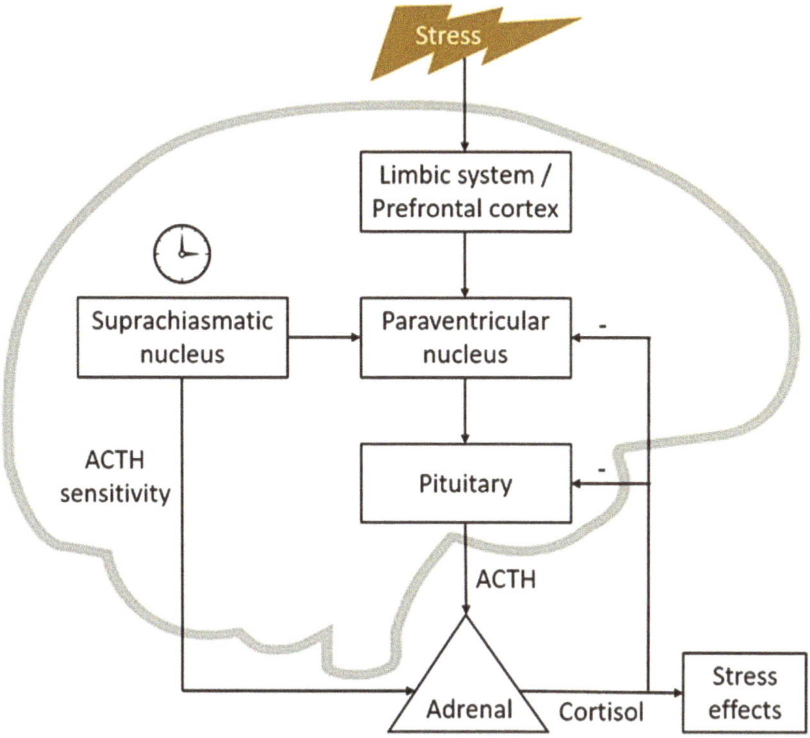

Fig. 1 Regulation of the hypothalamic–pituitary–adrenal axis. "-" indicates negative feedback.

from the pituitary, which in turn causes glucocorticoids (GC) to be released from the adrenal gland (Spencer, Chun, Hartsock, & Woodruff, 2018) (Fig. 1). This activation is responsible for down-stream emotional, cognitive, neuroendocrine and autonomic changes, determining the size and specifics of the stress response (Lucassen et al., 2014). An important role in the biological stress response is set aside for glucocorticoid (cortisol) responsive genes and for other pathways involving the gonads, adipose tissue, and the immune system (Joëls et al., 2012).

HPA-axis activation occurring during biological stress is mediated by the hypothalamus, and more specifically by the paraventricular nuclei (PVN). From here corticotropin-releasing hormone (CRH) is released into the adenohypophysis, in turn triggering adrenocorticotropic hormone (ACTH) release. ACTH mediates the release of cortisol from the adrenal gland, and the ACTH sensitivity of the adrenal cortex is regulated by the biological clock (Erkut, Pool, & Swaab, 1998) (Fig. 1). Negative feedback on this system is regulation through high-affinity mineralocorticoid (MR) and low-affinity glucocorticoid receptors (GR) (De Kloet, Joëls, & Holsboer, 2005), helping to maintain cortisol homeostasis (Erdmann, Berger, & Schütz, 2008; Kretz, Reichardt, Schütz, & Bock, 1999). Aberrant receptor expression (mainly GR) induces stress resistance, anxiety and depression (De Kloet et al., 2005; Ridder et al., 2005; Wei et al., 2007), and GR and MR expression determines cerebral stress sensitivity, modulating cognition, attention (Harris, Holmes, De Kloet, Chapman, & Seckl, 2013), and depression (Medina et al., 2013) as part of the stress response. Another important system controlling the HPA-axis and cortisol levels is the circadian system (Liston et al., 2013; Qian, Droste, Lightman, Reul, & Linthorst, 2012), exemplified by the clear diurnal pattern of cortisol secretion with highest levels approximately 30 min after awakening (Erkut et al., 1998). In addition, circannual patterns are reflected by symptom exacerbation during winter with shorter daily periods of daylight (Pierre, Schlesinger, & Androulakis, 2017).

HPA-axis abnormalities have received quite extensive attention in PD. Despite this, however, most studies have only looked at differences in cortisol levels between patients and healthy controls, without correlating these findings the clinical symptoms. In general, results related to serum cortisol levels in PD appear conflicting with serum cortisol being unchanged (four studies) (Cash & Lageman, 2015; Håglin & Bäckman, 2016; Stypuła, Kunert-Radek, Stepień, Żylińska, & Pawlikowski, 1996; Volpi et al., 1997), reduced (three studies) (Bellomo, Santambrogio, Fiacconi, Scarponi, & Ciuffetti, 1991; Müller & Muhlack, 2007; Müller et al., 2007), or higher (seven studies) in PD patients compared to healthy controls or caregivers of patients (Breen et al., 2014; Djamshidian et al., 2011; Gruden et al., 2011; Hartmann, Veldhuis, Deuschle, Standhardt, & Heuser, 1997; Ibrahimagic et al., 2016; Vardi et al., 1976) (Table 3). Most recent studies show higher cortisol levels in PD, most likely because of employing 24-h measurements of cortisol (Djamshidian et al., 2011; Hartmann et al., 1997), rather than just one-off measurements. As cortisol is under direct circadian control and its serum levels fluctuate throughout the day (Spencer et al., 2018), taking into account this circadian pattern is crucial. This is especially relevant now that is becoming increasingly clear that these circadian and circannual patterns are disrupted in PD (Logan & McClung, 2019; van Wamelen et al., 2019). The studies looking at the daily cortisol secretion patterns in PD, consistently show that not only morning concentrations are elevated compared to controls, but that cortisol levels are increased throughout the rest of the day (Breen et al., 2014;

Table 3 Cortisol levels in Parkinson's disease.

Study	Cortisol levels	Circadian pattern
Bellomo et al. (1991)	Reduced	Pattern not shifted, but levels lower throughout the day compared to healthy controls
Müller and Muhlack (2007)	Reduced after Levodopa intake; inverse correlation with UPDRS scores	N/A
Müller and Muhlack (2007)	Reduced after Levodopa intake; inverse correlation with motor scores	N/A
Stypuła et al. (1996)	Unchanged	N/A
Volpi et al. (1997)	Unchanged	N/A
Cash and Lageman (2015)	Unchanged compared to caregivers	N/A
Håglin and Bäckman (2016)	Unchanged	N/A
Vardi et al. (1976)	Raised after Levodopa intake	N/A
Hartmann et al. (1997)	Raised diurnal levels	Increase in diurnal variation of peak concentrations
Bocharov et al. (2010)	Raised	N/A
Gruden et al. (2011)	Raised compared to healthy controls	N/A
Djamshidian et al. (2011)	Raised diurnal levels	Raised diurnal levels
Breen et al. (2014)	Raised diurnal levels	Increased levels during acrophase
Ibrahimagic et al. (2016)	Raised after induction of psychological stress	N/A

Djamshidian et al., 2011). A more detailed analysis revealed that PD patients had an increased acrophase, increased amplitude, and increased area under the curve of cortisol secretion (Breen et al., 2014). In de novo untreated PD patients, 16.6% showed HPA-axis abnormalities; however, HPA-axis abnormalities in the prodromal stage of PD have been poorly researched (Håglin & Bäckman, 2016).

5.1 Studies in animal models

Experimental data from animal models of PD also support a role for stress in the neurodegenerative process (Table 4). For example, in animal models, such as the adult female Long-Evans hooded rats, chronic stress and treatment with steroids, to mimic the effects of raised cortisol levels in chronic stress, induce impairment of motor function and seems to accelerate nigral neuronal loss (Smith et al., 2008). Also, pre-treatment with corticosterone, mimicking physiological stress, induced an enhanced neuroinflammatory response to methamphetamine and increased dopaminergic neurotoxicity in rodent models of PD (Kelly et al., 2012). In a lipopolysaccharide rodent model of PD exposure to chronic stress prior to lipopolysaccharide treatment resulted severe degeneration of dopaminergic neurons and an increase in PD-like features (De Pablos et al., 2014). Unlike PD patients, only one study has looked at cortisol abnormalities in animal models of PD.

Table 4 Stress in Parkinson's disease animal models.

Animal model	Stress	Effect
Long-Evans hooded rats (Smith, Jadavji, Colwell, Katrina Perehudoff, & Metz, 2008)	Chronically raised cortisol levels	Motor impairment Nigral neuronal loss
Male C57BL/6J mice (Kelly, Miller, Bowyer, & O'Callaghan, 2012)	Corticosterone pre-treatment	Enhanced neuroinflammation Dopaminergic neurotoxicity
Lipopolysaccharide rodent model (De Pablos et al., 2014)	Chronic stress prior to Lipopolysaccharide treatments	Dopaminergic neurotoxicity Motor impairment
MPTP-treated dogs (Mizobuchi, Hineno, Kakimoto, & Hiratani, 1993)	Chronically increased cortisol and ACTH levels	Unknown

Abbreviations: ACTH, adrenocorticotrophic hormone; MPTP, 1-methyl-4-phenyl-1,2,3,6-tetrahydropyridine.

Mizobuchi et al. reported that cortisol and ACTH levels were increased in 1-methyl-4-phenyl-1,2,3,6-tetrahydropyridine (MPTP)-treated dogs (Mizobuchi et al., 1993).

5.2 The effect of HPA-axis abnormalities in PD

Cortisol levels generated from activation of the HPA-axis could have several effects (Fig. 2; Table 5). In PD, most studies correlated neuropsychiatric symptoms to cortisol levels. Seifried et al. showed that changes in the 24h cortisol secretion pattern correlated with depression scores after 3 and 6 months (Seifried et al., 2013). Similar findings were observed in a study by Müller and Muhlack (2007), but a larger study could not replicate these findings related to depression (Menza et al., 2010). Lower cortisol levels were associated with higher anxiety and weight gain (Růžička et al., 2015). In PD patients with obsessive-compulsive behavior, cortisol levels correlated with risk-taking (Djamshidian et al., 2011). In studies using other outcome measures, cortisol levels were positively correlated with UPDRS part III scores, but not with MMSE scores (Håglin & Bäckman, 2016). Breen et al. showed that PD patients (mean HY stage 1.8) had a sustained increase in cortisol levels, coinciding with a longer sleep latency, reduced sleep efficiency and decreased total sleep time. Moreover, they observed that these

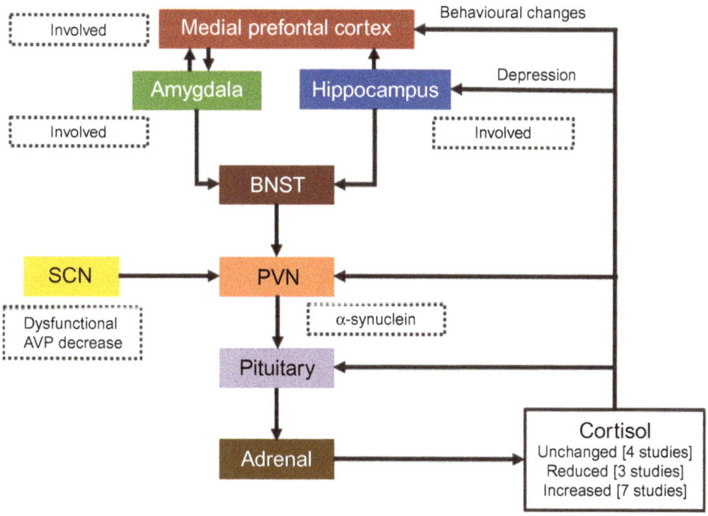

Fig. 2 Proposed mechanisms by which chronic stress in Parkinson's disease may lead to cognitive deficits, depression, and anxiety. Abbreviations: BNST, bed nucleus of stria terminalis; PVN, hypothalamic paraventricular nuclei; SCN, suprachiasmatic nuclei.

Table 5 Known correlations of serum cortisol changes and Parkinson's disease symptoms.

	Cortisol state	**Effect**
Neuropsychiatric	Increased	Higher depression scores (Müller & Muhlack, 2007; Seifried et al., 2013) Increased risk-taking (Djamshidian et al., 2011)
	Decreased	Higher anxiety (Růžička et al., 2015) Weight gain (Růžička et al., 2015)
Sleep	Increased	Longer sleep latency (Breen et al., 2014) Reduced sleep efficiency (Breen et al., 2014) Decreased total sleep time (Breen et al., 2014)
Motor	Increased	Worse motor performance (Håglin & Bäckman, 2016)

changes occurred in the presence of reduced serum levels of melatonin, and altered Bmal1 expression confirming the link between the HPA-axis and the circadian system in PD (see below) (Breen et al., 2014). This close relationship is also reflected by the observation that even partial acute disruption of night-time sleep can disrupt normal cortisol patterns, inducing an elevation in morning cortisol levels (Leproult, Copinschi, Buxton, & Van Cauter, 1997).

Although the direct effect of altered HPA-axis function has not been studied in PD, the combined clinical effect and our knowledge about the effect of cortisol and GC in general have led to several theories about the long-term effect of chronic stress. Especially the hippocampus may play a crucial role in the biological effects of chronic stress due to the extensive distribution of GC receptors in this brain area (De Kloet et al., 2005; Swaab, Bao, & Lucassen, 2005). For example, an increased expression of GC receptors has been observed in the hippocampus of patients with major depressive disorder (Wang, Joels, Swaab, & Lucassen, 2012; Wang et al., 2014). Also other brain regions, including the hypothalamus, entorhinal cortex, temporal cortex and prefrontal cortex, contain many GC receptors (Qi et al., 2013; Webster, Knable, O'Grady, Orthmann, & Weickert, 2002). Here the link with sleep, disrupted by hypercortisolemia, becomes relevant as sleep (deprivation) exerts a bi-directional control on the hippocampus and memory consolidation (Abel, Havekes, Saletin, & Walker, 2013). Smaller hippocampal volumes are associated with both cognitive decline (Manning et al., 2015; Ricci, De Caterina, & Fedorowski, 2015), and lower sleep efficiency (Elcombe et al., 2015). An explanation for this bi-directional link might be found in the recently proposed glymphatic system. This system is a glial-dependent perivascular network serving as a pseudo-lymphatic system for the brain (Plog & Nedergaard, 2018). The proposed workings of the glymphatic pathway is that cerebrospinal fluid enters the brain via periarterial spaces, passes into the interstitium via perivascular astrocytic aquaporin-4 channels, and then drives the perivenous drainage of interstitial fluid and its solute (Plog & Nedergaard, 2018). The administration of dexamethasone induces a reduction in aquaporin-4 expression in the ovine brain (Ron et al., 2005), and deletion of aquaporin-4 in rodent models results in an impairment of glymphatic transport and accumulation of cerebral β-amyloid and Tau proteins, as well as decreased working memory (Zhang et al., 2019).

5.3 Pathophysiology of HPA-axis abnormalities and links with depression

Tonic inhibition of HPA-axis activity is exerted through multiple neuronal pathways, including the bed nucleus of the stria terminalis (BST), the endocannabinoid system, and the amygdala (Herman et al., 2003; Ventura-Silva et al., 2012). The pathological involvement of these cerebral areas in PD has been shown in several studies, in both patients and animal models of PD (Bastide et al., 2017; Li, De Ren, Cao, & Huang, 2018). Moreover, direct inhibition of the HPA-axis occurs through GC-mediated negative feedback, not only at the levels of the hippocampus (Herman et al., 2003), as mentioned before, but also at the level of the pituitary and hypothalamus (Lucassen et al., 2014). Although most of the latter areas have not been extensively studied in PD, parallels can be drawn between PD and major depressive disorder (MDD), where the stress system has been a focus of more intensive research.

Indeed, a large number of MDD patients show abnormalities in HPA-axis functioning, with figures ranging between 35% and 65% (Lucassen et al., 2014). In these patients, a high number of dexamethasone non-suppressors with coinciding adrenal hypertrophy, increased plasma cortisol levels, and CRH and arginine vasopressin (AVP) neuropeptidergic expression changes in the paraventricular nucleus (PVN) of the hypothalamus (Swaab et al., 2005). Inclusions of α-synuclein, the hallmark protein in PD pathology, have been noted in the PVN of PD patients, suggesting the involvement of this structure in the PD disease process (De Pablo-Fernandez, Courtney, Holton, & Warner, 2017). When examined in more detail, MDD patients show an increase in the number of CRH immunoreactive PVN neurons, along with an increased co-expression of AVP (Raadsheer et al., 1995). These findings show a strong similarity with those in rodents injected with CRH which induces depression-like behavior (Holsboer, 2000). In PD patients, on the other hand, the only study looking at CRH expression in the hypothalamic PVN could not demonstrate differences between six patients and six controls (Hoogendijk et al., 1998).

When referring to role of AVP in depression, and its possible role in stress in PD, it is important to acknowledge the different vasopressinergic systems in the brain. To date, four different vasopressinergic systems have been identified: (1) AVP within the magnocellular neurons of the hypothalamic supraoptic and paraventricular nuclei, whose axons project to the neurohypophysis where AVP is released in the circulation (Lucassen et al., 2014); (2) AVP in the parvocellular PVN neurons secreting both CRH and AVP into the median eminence, and subsequently further into the anterior pituitary lobe potentiating ACTH release (Lucassen et al., 2014); (3) vasopressinergic fibers projecting from the hypothalamus to the hippocampus, septum, amygdala and several brainstem nuclei, where AVP takes on the role of neurotransmitter (Lucassen et al., 2014); and (4) AVP within the neurons of the biological clock (the suprachiasmatic nuclei; SCN) (Lucassen et al., 2014). All these vasopressinergic systems have been shown to be affected in MDD (Lucassen et al., 2014), but also in PD evidence is emerging that these systems are affected by the PD disease process. Specifically, it has been shown in PD that the biological clock is dysfunctional (van Wamelen et al., 2019; Wu, Li, Stavitsky Gilbert, Hu, & Cronin-Golomb, 2018), the SCN is affected in PD (De Pablo-Fernandez, Courtney, Warner, & Holton, 2018) (with a possible decrease in AVP neurons (Stopa et al., 1999)), and the PVN is affected (De Pablo-Fernandez et al., 2017). Moreover, areas in control of HPA-axis functioning have been shown to be involved in PD: the hippocampus (Das, Hwang, & Poston, 2019; Wilson, Niccolini, Pellicano, & Politis, 2019), amygdala (Kim et al., 2018; Wilson et al., 2019), and medial prefrontal cortex (Moonen et al., 2017). Fig. 2 summaries the proposed mechanisms by which chronic stress may lead to cognitive deficits, depression, and anxiety.

5.4 Genetics and epigenetics

Certain polymorphisms in the catechol-*O*-methyl-transferase (COMT) gene have been shown to be associated with glucocorticoid responsivity and associated cortisol levels. For example, Lovallo et al. showed that healthy subjects with Met/Met COMT homozygote polymorphisms are more sensitive to stressful events and have higher cortisol responses (Lovallo et al., 2017). Similar associations between COMT polymorphisms and cortisol levels have been shown in schizophrenia patients (Hirasawa-Fujitaa et al., 2018). Also other genes, relevant to PD, such as monoamine oxidase (MAO), have been shown to be related to cortisol levels. In several healthy cohorts, the effects of COMT on the release of ACTH depended on the expression levels of MAO-A, and, moreover, the Met/Met MAO-A polymorphism was associated with higher cortisol levels in males, whereas in females the Val/Val genotype had more pronounced cortisol responses to stress (Bouma, Riese, Doornbos, Ormel, & Oldehinkel, 2012; Jabbi et al., 2007). Aside from the effect certain genetic polymorphism have on cortisol levels, cortisol reciprocally influences genes that have been implicated in PD. An example of this is the observation in a rodent model of PD where the activation of Parkin by cortisol prevented dopaminergic cell death (Lee et al., 2017). Mutated Parkin, and Parkin affected by post-translational modifications, has been implicated in neuronal dopaminergic cell death in PD (Lee et al., 2017). Other rodent models have confirmed the link between stress (and elevated cortisol levels) and dopaminergic cell death. For example, Kong et al. have recently demonstrated that chronic unpredictable stress significantly aggravated behavioral dysfunction in a PD rat model and increased microglia activation and caused dopaminergic neuronal damage, possibly through an increase in several inflammatory markers, such as interleukin-1β, interleukin-6 (IL-6), and tumor necrosis factor-α (Kong et al., 2019).

Not only does the cortisol response in PD appear reliant on genetic factors, also epigenetic factors have been implicated in the stress response. These epigenetic factors are: (1) DNA methylation, which in turn regulates α-synuclein expression (Babenko, Kovalchuk, & Metz, 2012). In PD, some studies advocate that, under stressful circumstances, DJ-1 gene mutations may result in mitochondrial dysfunction and hence increased susceptibility to oxidative stress which accelerates neuronal cell death (Hemmerle et al., 2012; Kanherkar, Bhatia-Dey, Makarev, & Csoka, 2014), and (2) increased baseline circulating GC levels due to prolonged exposure to stress early in life, has also been associated with compromised neuronal survival (Cohen et al., 2012). BDNF, an active neurotrophin, primarily regulates neuronal differentiation, growth, and plasticity (Hempstead, 2015), while glial-cell-derived neurotrophic factor (GDNF) is responsible for midbrain neuroprotection, functional repair, and prevention of nigrostriatal neurodegeneration in primate models of PD (Kordower et al., 2000). Mpofana et al. demonstrated that in animal models, there is reduced response to tissue injury when under stress, as compared to when without. High levels of corticosterone augments epigenetic changes in BDNF and GDNF gene expression, suggesting that stress interferes with the expression of these, thereby decreasing the levels of both, implying compromised neuroprotection and neurogenesis, thus accelerating neuronal loss (Mpofana, Daniels, & Mabandla, 2016).

In PD, the abnormal activity of complex 1 in mitochondria directly intervenes with cellular ATP production, leading to cell death (Parker, Parks, & Swerdlow, 2008). In addition, the metabolism of brain amines such as DA by monoamine oxidase-B (MAO-B), when combined with ground state O_2, leads to the formation of ROS (Lotharius et al., 2002). PD patients have increased oxidative stress markers and related changes (including free radical damage to DNA, proteins, and fats) (Lotharius & Brundin, 2002). Increased levels of the apoptotic marker protein, "Bax" has also been observed in DA-neurons of the SNpc in MPTP-treated mice (Perier & Vila, 2012). Recently, mitochondrial DNA from PD patients were placed into neuroblastoma cells (which produced hybrid cells called "cybrid") to explore the role of mitochondria in PD (Wilkins, Carl, & Swerdlow, 2014). These cybrids were found to develop Lewy bodies. It was also thought that selective gene mutations that are involved in cell-survival mechanisms may lead to impairment of mitochondrial activity and ATP production. These results posit strong support that mitochondrial defects play a key role in the development of sporadic PD.

6. Cortisol as a target for treatment

Considering the often-cited detrimental effects of hypercortisolemia in PD, such as higher depression scores (Müller & Muhlack, 2007; Seifried et al., 2013) and sleep disorders latency (Breen et al., 2014), a reduction in stress burden by targeting elevated cortisol levels could be an interesting therapeutic strategy. Interestingly symptomatic dopamine replacement therapies such as levodopa in combination with a decarboxylase inhibitor, and apomorphine (Happe et al., 2004), has been shown to influence cortisol levels (Müller & Muhlack, 2007; Müller et al., 2007) (Fig. 3), although dopaminergic therapy appears unlikely to have long-term effects on reducing hypercortisolemia (Stypuła et al., 1996; Vogel & Ketsche, 1986). The mechanism by which dopaminergic medication influences

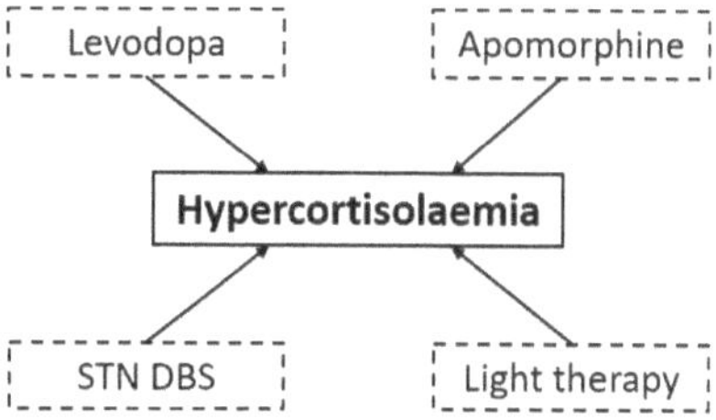

Fig. 3 Potential targets to reduce hypercortisolemia in Parkinson's disease. STN DBS, subthalamic nucleus deep brain stimulation.

cortisol levels is unclear, but it may be assumed that this is a regulatory effect of dopamine on hypothalamic mechanisms responsible for cortisol release. For example, it was shown that in the MPTP-treated non-human primate model of PD, dopamine depletion has a clear effect on circadian hormonal patterns, including that of cortisol (Fifel et al., 2014).

Non-dopaminergic therapies, such as deep brain stimulation (DBS), can also have an effect on cortisol levels. Růžička et al. showed that bilateral stimulation of the subthalamic nucleus resulted in significantly decreased cortisol levels, especially with more medial contact of the electrode with the subthalamic nucleus (Růžička et al., 2015). Similar results were also seen in a previous study by the same research group (Nováková et al., 2011). Others, however, have not been able to replicate these findings (Seifried et al., 2013).

A promising, and relatively simple, approach to improving stress and its associated symptoms in PD is bright light therapy (BLT). Rutten et al. have recently shown that after a treatment period of 3 months with 10,000 lux BLT salivary cortisol secretion was decreased in patients who received BLT, compared to placebo (Rutten et al., 2019). In addition, also physical exercise can improve dysregulated cortisol levels. Smyth et al. recently showed that 6 months of high-intensity treadmill for 30 min a day reduced cortisol secretion during the post-awakening period after nocturnal sleep (Smyth et al., 2019).

7. Conclusions

Stress is an underappreciated issue in PD and may play a role in activating neurochemical and neurohormonal responses which may increase the risk of development of PD. The centrally mediated effects of stress can also overlap with or manifest as some key NMS such as fatigue, depression and sleep dysfunction. The effect of stress has been studied and the role in the genesis or worsening of NMS such as depression and cognitive decline have been posited but not confirmed. Similarly stress responses appear to have a role on motor symptoms. So far, however, targeted treatments to prevent stress and its effects in PD are lacking, although treatments strategies such as bright light therapy may prove effective and deserve further exploration.

References

Abel, T., Havekes, R., Saletin, J. M., & Walker, M. P. (2013). Sleep, plasticity and memory from molecules to whole-brain networks. *Current Biology*, *23*, R774–R788. https://doi.org/10.1016/j.cub.2013.07.025.

Anzak, A., Tan, H., Pogosyan, A., Djamshidian, A., Ling, H., Lees, A., et al. (2011). Improvements in rate of development and magnitude of force with intense auditory stimuli in patients with Parkinson's disease. *European Journal of Neuroscience*, *34*(1), 124–132.

Asmus, F., Huber, H., Gasser, T., & Schöls, L. (2008). Kick and rush: Paradoxical kinesia in Parkinson disease. *Neurology*, *71*, 695. https://doi.org/10.1212/01.wnl.0000324618.88710.30.

Babenko, O., Kovalchuk, I., & Metz, G. A. (2012). Epigenetic programming of neurodegenerative diseases by an adverse environment. *Brain Research*, *1444*, 96–111. https://doi.org/10.1016/j.brainres.2012.01.038.

Bastide, M. F., Glangetas, C., Doudnikoff, E., Li, Q., Bourdenx, M., Fernagut, P. O., et al. (2017). Involvement of the bed nucleus of the stria terminalis in L-Dopa induced dyskinesia. *Scientific Reports*, 7(1), 2348. https://doi.org/10.1038/s41598-017-02572-9.

Bellomo, G., Santambrogio, L., Fiacconi, M., Scarponi, A. M., & Ciuffetti, G. (1991). Plasma profiles of adrenocorticotropic hormone, cortisol, growth hormone and prolactin in patients with untreated Parkinson's disease. *Journal of Neurology*, *238*(1), 19–22. https://doi.org/10.1007/BF00319704.

Bocharov, E. V., Ivanova-Smolenskaya, I. A., Poleshchuk, V. V., Kucheryanu, V. G., Il'Enko, V. A., & Bocharova, O. A. (2010). Therapeutic efficacy of the neuroprotective plant adaptogen in neurodegenerative disease (Parkinson's disease as an example). *Bulletin of Experimental Biology and Medicine*, *149*(6), 682.

Bonanni, L., Thomas, A., Anzellotti, F., Monaco, D., Ciccocioppo, F., Varanese, S., et al. (2010). Protracted benefit from paradoxical kinesia in typical and atypical parkinsonisms. *Neurological Sciences*, *31*(6), 751–756. https://doi.org/10.1007/s10072-010-0403-5.

Bonanni, L., Thomas, A., & Onofrj, M. (2010). Paradoxical kinesia in parkinsonian patients surviving earthquake. *Movement Disorders*, *25*(9), 1302–1304. https://doi.org/10.1002/mds.23075.

Bostan, A. C., Dum, R. P., & Strick, P. L. (2010). The basal ganglia communicate with the cerebellum. *Proceedings of the National Academy of Sciences of the United States of America*, *107*(18), 8452–8456. https://doi.org/10.1073/pnas.1000496107.

Bouma, E. M. C., Riese, H., Doornbos, B., Ormel, J., & Oldehinkel, A. J. (2012). Genetically based reduced MAOA and COMT functioning is associated with the cortisol stress response: A replication study. *Molecular Psychiatry*, *17*, 119–121. https://doi.org/10.1038/mp.2011.115.

Breen, D. P., Vuono, R., Nawarathna, U., Fisher, K., Shneerson, J. M., Reddy, A. B., et al. (2014). Sleep and circadian rhythm regulation in early Parkinson disease. *JAMA Neurology*, *71*(5), 589–595. https://doi.org/10.1001/jamaneurol.2014.65.

Cash, T. V., & Lageman, S. K. (2015). Randomized controlled expressive writing pilot in individuals with Parkinson's disease and their caregivers. *BMC Psychology*, *3*(1), 44. https://doi.org/10.1186/S40359-015-0101-4.

Clark, A. J., Ritz, B., Prescott, E., & Rod, N. H. (2013). Psychosocial risk factors, pre-motor symptoms and first-time hospitalization with Parkinson's disease: A prospective cohort study. *European Journal of Neurology*, *20*(8), 1113–1120. https://doi.org/10.1111/ene.12117.

Cohen, S., Janicki-Deverts, D., Doyle, W. J., Miller, G. E., Frank, E., Rabin, B. S., et al. (2012). Chronic stress, glucocorticoid receptor resistance, inflammation, and disease risk. *Proceedings of the National Academy of Sciences of the United States of America*, *109*(16), 5995–5999. https://doi.org/10.1073/pnas.1118355109.

Colpaert, F. C. (1987). Pharmacological characteristics of tremor, rigidity and hypokinesia induced by reserpine in rat. *Neuropharmacology*, *26*(9), 1431–1440. https://doi.org/10.1016/0028-3908(87)90110-9.

Daroff, R. B. (2008). Paradoxical kinesia. *Movement Disorders*, *23*(8), 1193. https://doi.org/10.1002/mds.22060.

Das, T., Hwang, J. J., & Poston, K. L. (2019). Episodic recognition memory and the hippocampus in Parkinson's disease: A review. *Cortex*, *113*, 191–209. https://doi.org/10.1016/j.cortex.2018.11.021.

De Kloet, E. R., Joëls, M., & Holsboer, F. (2005). Stress and the brain: From adaptation to disease. *Nature Reviews Neuroscience*, *6*, 463–475. https://doi.org/10.1038/nrn1683.

De La Fuente-Fernández, R., & Stoessl, A. J. (2002). The placebo effect in Parkinson's disease. *Trends in Neurosciences*, *25*, 302–306. https://doi.org/10.1016/S0166-2236(02)02181-1.

De Pablo-Fernandez, E., Courtney, R., Holton, J. L., & Warner, T. T. (2017). Hypothalamic α-synuclein and its relation to weight loss and autonomic symptoms in Parkinson's disease. *Movement Disorders*, *32*(2), 296–298. https://doi.org/10.1002/mds.26868.

De Pablo-Fernandez, E., Courtney, R., Warner, T. T., & Holton, J. L. (2018). A histologic study of the circadian system in Parkinson disease, multiple system atrophy, and progressive supranuclear palsy. *JAMA Neurology*, *75*(8), 1008–1012. https://doi.org/10.1001/jamaneurol.2018.0640.

De Pablos, R. M., Herrera, A. J., Espinosa-Oliva, A. M., Sarmiento, M., Muñoz, M. F., Machado, A., et al. (2014). Chronic stress enhances microglia activation and exacerbates death of nigral dopaminergic neurons under conditions of inflammation. *Journal of Neuroinflammation*, *11*, 34. https://doi.org/10.1186/1742-2094-11-34.

Distler, M., Schlachetzki, J. C. M., Kohl, Z., Winkler, J., & Schenk, T. (2016). Paradoxical kinesia in Parkinson's disease revisited: Anticipation of temporal constraints is critical. *Neuropsychologia*, *86*, 38–44. https://doi.org/10.1016/j.neuropsychologia.2016.04.012.

Djamshidian, A., & Lees, A. J. (2014). Can stress trigger Parkinson's disease? *Journal of Neurology, Neurosurgery and Psychiatry*, *85*, 879–882. https://doi.org/10.1136/jnnp-2013-305911.

Djamshidian, A., O'Sullivan, S. S., Papadopoulos, A., Bassett, P., Shaw, K., Averbeck, B. B., et al. (2011). Salivary cortisol levels in Parkinson's disease and its correlation to risk behaviour. *Journal of Neurology, Neurosurgery and Psychiatry*, *82*(10), 1107–1111. https://doi.org/10.1136/jnnp.2011.245746.

Dufour, B. D., & McBride, J. L. (2016). Corticosterone dysregulation exacerbates disease progression in the R6/2 transgenic mouse model of Huntington's disease. *Experimental Neurology*, *283*, 308–317. https://doi.org/10.1016/j.expneurol.2016.06.028.

Elcombe, E. L., Lagopoulos, J., Duffy, S. L., Lewis, S. J. G., Norrie, L., Hickie, I. B., et al. (2015). Hippocampal volume in older adults at risk of cognitive decline: The role of sleep, vascular risk, and depression. *Journal of Alzheimer's Disease: JAD*, *44*(4), 1279–1290. https://doi.org/10.3233/JAD-142016.

Ennis, G. E., An, Y., Resnick, S. M., Ferrucci, L., O'Brien, R. J., & Moffat, S. D. (2017). Long-term cortisol measures predict Alzheimer disease risk. *Neurology*, *88*(4), 371–378. https://doi.org/10.1212/WNL.0000000000003537.

Erdmann, G., Berger, S., & Schütz, G. (2008). Genetic dissection of glucocorticoid receptor function in the mouse brain. *Journal of Neuroendocrinology*, *20*, 655–659. https://doi.org/10.1111/j.1365-2826.2008.01717.x.

Erkut, Z. A., Pool, C., & Swaab, D. F. (1998). Glucocorticoids suppress corticotropin-releasing hormone and vasopressin expression in human hypothalamic neurons. *Journal of Clinical Endocrinology and Metabolism*, *83*(6), 2066–2073. https://doi.org/10.1210/jc.83.6.2066.

Fang, F., Xu, Q., Park, Y., Huang, X., Hollenbeck, A., Blair, A., et al. (2010). Depression and the subsequent risk of Parkinson's disease in the NIH-AARP diet and health study. *Movement Disorders*, *25*(9), 1157–1162. https://doi.org/10.1002/mds.23092.

Fifel, K., Vezoli, J., Dzahini, K., Claustrat, B., Leviel, V., Kennedy, H., et al. (2014). Alteration of daily and circadian rhythms following dopamine depletion in MPTP treated non-human primates. *PLoS One*, *9*(1), e86240. https://doi.org/10.1371/journal.pone.0086240.

Glickstein, M., & Stein, J. (1991). Paradoxical movement in Parkinson's disease. *Trends in Neurosciences*, *14*(11), 480–482. https://doi.org/10.1016/0166-2236(91)90055-Y.

Gowers, W. R. G. (1888). Review: A manual of diseases of the nervous system. *Journal of Nervous and Mental Disease*, *13*, 325–327. https://doi.org/10.2307/1411377.

Gruden, M. A., Sewell, R. D. E., Yanamandra, K., Davidova, T. V., Kucheryanu, V. G., Bocharov, E. V., et al. (2011). Immunoprotection against toxic biomarkers is retained during Parkinson's disease progression. *Journal of Neuroimmunology*, *233*(1–2), 221–227. https://doi.org/10.1016/j.jneuroim.2010.12.001.

Håglin, L., & Bäckman, L. (2016). Covariation between plasma phosphate and daytime cortisol in early Parkinson's disease. *Brain and Behavior*, *6*(12), e00556. https://doi.org/10.1002/brb3.556.

Hammond, T. C. (2010). New developments: falls, drooling & exercise in Parkinson's Disease. In *The Parkinson's Source (40).* APDA Magazine.

Happe, S., Tings, T., Helmschmeid, K., Neubert, K., Wuttke, W., Paulus, W., et al. (2004). Levodopa treatment does not affect low-dose apomorphine test in patients with Parkinson's disease. *Movement Disorders*, *19*(12), 1511–1515. https://doi.org/10.1002/mds.20244.

Harris, A. P., Holmes, M. C., De Kloet, E. R., Chapman, K. E., & Seckl, J. R. (2013). Mineralocorticoid and glucocorticoid receptor balance in control of HPA axis and behaviour. *Psychoneuroendocrinology*, *38*(5), 648–658. https://doi.org/10.1016/j.psyneuen.2012.08.007.

Hartmann, A., Veldhuis, J. D., Deuschle, M., Standhardt, H., & Heuser, I. (1997). Twenty-four hour cortisol release profiles in patients with Alzheimer's and Parkinson's disease compared to normal controls: Ultradian secretory pulsatility and diurnal variation. *Neurobiology of Aging*, *18*(3), 285–289. https://doi.org/10.1016/S0197-4580(97)80309-0.

Hemmerle, A. M., Herman, J. P., & Seroogy, K. B. (2012). Stress, depression and Parkinson's disease. *Experimental Neurology*, *233*, 79–86. https://doi.org/10.1016/j.expneurol.2011.09.035.

Hempstead, B. L. (2015). Brain-derived neurotrophic factor: Three ligands, many actions. *Transactions of the American Clinical and Climatological Association*, *126*, 9–19.

Herman, J. P., Figueiredo, H., Mueller, N. K., Ulrich-Lai, Y., Ostrander, M. M., Choi, D. C., et al. (2003). Central mechanisms of stress integration: Hierarchical circuitry controlling hypothalamo-pituitary-adrenocortical responsiveness. *Frontiers in Neuroendocrinology*, *24*(3), 151–180. https://doi.org/10.1016/j.yfrne.2003.07.001.

Hirasawa-Fujitaa, M., Tudor, L., Nikolac Pekovic, M., Kozumplik, O., Nedic Erjavec, G., Uzun, S., et al. (2018). Genotypic and haplotypic associations of catechol-O-methyltransferase (COMT) rs4680 and rs4818 with salivary cortisol in patients with schizophrenia. *Psychiatry Research, 259*, 262–264. https://doi.org/10.1016/j.psychres.2017.10.037.

Holsboer, F. (2000). The corticosteroid receptor hypothesis of depression. *Neuropsychopharmacology, 23*(5), 477–501. https://doi.org/10.1016/S0893-133X(00)00159-7.

Hoogendijk, W. J. G., Purba, J. S., Hofman, M. A., De Vos, R. A. I., Jansen, E. N. H., & Swaab, D. F. (1998). Depression in Parkinson's disease is not accompanied by more corticotropin-releasing hormone expressing neurons in the hypothalamic paraventricular nucleus. *Biological Psychiatry, 43*(12), 913–917. https://doi.org/10.1016/S0006-3223(97)00338-7.

Ibrahimagic, O. C., Jakubovic, A. C., Smajlovic, D., Dostovic, Z., Kunic, S., & Iljazovic, A. (2016). Psychological stress and changes of hypothalamic-pituitary-adrenal axis in patients with "de novo" Parkinson's disease. *Medical Archives (Sarajevo, Bosnia and Herzegovina), 70*(6), 445–448. https://doi.org/10.5455/medarh.2016.70.445-448.

Jabbi, M., Korf, J., Kema, I. P., Hartman, C., Van Der Pompe, G., Minderaa, R. B., et al. (2007). Convergent genetic modulation of the endocrine stress response involves polymorphic variations of 5-HTT, COMT and MAOA. *Molecular Psychiatry, 12*(5), 483–490. https://doi.org/10.1038/sj.mp.4001975.

Joëls, M., Angela Sarabdjitsingh, R., & Karst, H. (2012). Unraveling the time domains of corticosteroid hormone influences on brain activity: Rapid, slow, and chronic modes. *Pharmacological Reviews, 64*(4), 901–938. https://doi.org/10.1124/pr.112.005892.

Kanherkar, R. R., Bhatia-Dey, N., Makarev, E., & Csoka, A. B. (2014). Cellular reprogramming for understanding and treating human disease. *Frontiers in Cell and Development Biology, 2*, 67. https://doi.org/10.3389/fcell.2014.00067.

Keefe, K. A., Salamone, J. D., Zigmond, M. J., & Stricker, E. M. (1989). Paradoxical kinesia in parkinsonism is not caused by dopamine release: Studies in an animal model. *Archives of Neurology, 46*(10), 1070–1075. https://doi.org/10.1001/archneur.1989.00520460046012.

Kelly, K. A., Miller, D. B., Bowyer, J. F., & O'Callaghan, J. P. (2012). Chronic exposure to corticosterone enhances the neuroinflammatory and neurotoxic responses to methamphetamine. *Journal of Neurochemistry, 122*(5), 995–1009. https://doi.org/10.1111/j.1471-4159.2012.07864.x.

Kim, J., Zhang, K., Cai, W., YorkWilliams, S., Ua Cruadhlaoich, M. A. I., Llanes, S., et al. (2018). Dopamine-related dissociation of cortical and subcortical brain activations in cognitively unimpaired Parkinson's disease patients OFF and ON medications. *Neuropsychologia, 119*, 24–33. https://doi.org/10.1016/j.neuropsychologia.2018.07.024.

Kong, H., Yang, L., He, C., Zhou, J. W., Li, W. Z., Wu, W. N., et al. (2019). Chronic unpredictable mild stress accelerates lipopolysaccharide- induced microglia activation and damage of dopaminergic neurons in rats. *Pharmacology Biochemistry and Behavior, 179*, 142–149. https://doi.org/10.1016/j.pbb.2019.01.004.

Kordower, J. H., Emborg, M. E., Bloch, J., Ma, S. Y., Chu, Y., Leventhal, L., et al. (2000). Neurodegeneration prevented by lentiviral vector delivery of GDNF in primate models of Parkinson's disease. *Science, 290*(5492), 767–773. https://doi.org/10.1126/science.290.5492.767.

Kretz, O., Reichardt, H. M., Schütz, G., & Bock, R. (1999). Corticotropin-releasing hormone expression is the major target for glucocorticoid feedback-control at the hypothalamic level. *Brain Research, 818*(2), 488–491. https://doi.org/10.1016/S0006-8993(98)01277-3.

Lederbogen, F., Kirsch, P., Haddad, L., Streit, F., Tost, H., Schuch, P., et al. (2011). City living and urban upbringing affect neural social stress processing in humans. *Nature, 474*(7352), 498–501. https://doi.org/10.1038/nature10190.

Lee, Y., Ham, S., Lee, Y. I., Jo, M., Kim, H., Kang, H., et al. (2017). Hydrocortisone-induced parkin prevents dopaminergic cell death via CREB pathway in Parkinson's disease model. *Scientific Reports*, 7(1), 525. https://doi.org/10.1038/s41598-017-00614-w.

Leproult, R., Copinschi, G., Buxton, O., & Van Cauter, E. (1997). Sleep loss results in an elevation of cortisol levels the next evening. *Sleep, 20*(10), 865–870.

Li, X. R., De Ren, Y., Cao, B., & Huang, X. L. (2018). Analysis of white matter characteristics with tract-based spatial statistics according to diffusion tensor imaging in early Parkinson's disease. *Neuroscience Letters, 675*, 127–132. https://doi.org/10.1016/j.neulet.2017.11.064.

Liston, C., Cichon, J. M., Jeanneteau, F., Jia, Z., Chao, M. V., & Gan, W. B. (2013). Circadian glucocorticoid oscillations promote learning-dependent synapse formation and maintenance. *Nature Neuroscience, 16*(6), 698–705. https://doi.org/10.1038/nn.3387.

Logan, R. W., & McClung, C. A. (2019). Rhythms of life: Circadian disruption and brain disorders across the lifespan. *Nature Reviews Neuroscience, 20*, 49–65. https://doi.org/10.1038/s41583-018-0088-y.

Lotharius, J., Barg, S., Wiekop, P., Lundberg, C., Raymon, H. K., & Brundin, P. (2002). Effect of mutant α-synuclein on dopamine homeostasis in a new human mesencephalic cell line. *Journal of Biological Chemistry, 277*(41), 38884–38894. https://doi.org/10.1074/jbc.M205518200.

Lotharius, J., & Brundin, P. (2002). Pathogenesis of Parkinson's disease: Dopamine, vesicles and α-synuclein. *Nature Reviews Neuroscience, 3*(12), 932–942. https://doi.org/10.1038/nrn983.

Lovallo, W. R., Enoch, M. A., Sorocco, K. H., Vincent, A. S., Acheson, A., Cohoon, A. J., et al. (2017). Joint impact of early life adversity and COMT Val158Met (rs4680) genotypes on the adult cortisol response to psychological stress. *Psychosomatic Medicine, 79*(6), 631–637. https://doi.org/10.1097/PSY.0000000000000481.

Lucassen, P. J., Pruessner, J., Sousa, N., Almeida, O. F. X., Van Dam, A. M., Rajkowska, G., et al. (2014). Neuropathology of stress. *Acta Neuropathologica, 127*, 109–135. https://doi.org/10.1007/s00401-013-1223-5.

Manning, E. N., Macdonald, K. E., Leung, K. K., Young, J., Pepple, T., Lehmann, M., et al. (2015). Differential hippocampal shapes in posterior cortical atrophy patients: A comparison with control and typical AD subjects. *Human Brain Mapping, 36*(12), 5123–5136. https://doi.org/10.1002/hbm.22999.

Marshall, J. F., Levitan, D., & Stricker, E. M. (1976). Activation-induced restoration of sensorimotor functions in rats with dopamine-depleting brain lesions. *Journal of Comparative and Physiological Psychology, 90*(6), 536–546. https://doi.org/10.1037/h0077230.

McEwen, B. S. (2003). Mood disorders and allostatic load. *Biological Psychiatry, 54*, 200–207. https://doi.org/10.1016/S0006-3223(03)00177-X.

McEwen, B. S. (2007). Physiology and neurobiology of stress and adaptation: Central role of the brain. *Physiological Reviews, 87*, 873–904. https://doi.org/10.1152/physrev.00041.2006.

Medina, A., Seasholtz, A. F., Sharma, V., Burke, S., Bunney, W., Myers, R. M., et al. (2013). Glucocorticoid and mineralocorticoid receptor expression in the human hippocampus in major depressive disorder. *Journal of Psychiatric Research, 47*(3), 307–314. https://doi.org/10.1016/j.jpsychires.2012.11.002.

Menza, M., Dobkin, R. D. F., Marin, H., Mark, M. H., Gara, M., Bienfait, K., et al. (2010). The role of inflammatory cytokines in cognition and other non-motor symptoms of Parkinson's disease. *Psychosomatics, 51*(6), 474–479. https://doi.org/10.1176/appi.psy.51.6.474.

Mizobuchi, M., Hineno, T., Kakimoto, Y., & Hiratani, K. (1993). Increase of plasma adrenocorticotrophin and cortisol in 1-methyl-4-phenyl-1,2,3,6-tetrahydropyridine (MPTP)-treated dogs. *Brain Research*, *612*(1–2), 319–321. https://doi.org/10.1016/0006-8993(93)91678-L.

Moonen, A. J. H., Weiss, P. H., Wiesing, M., Weidner, R., Fink, G. R., Reijnders, J. S. A. M., et al. (2017). An fMRI study into emotional processing in Parkinson's disease: Does increased medial prefrontal activation compensate for striatal dysfunction? *PLoS One*, *12*(5), e0177085. https://doi.org/10.1371/journal.pone.0177085.

Moore, H., Rose, H. J., & Grace, A. A. (2001). Chronic cold stress reduces the spontaneous activity of ventral tegmental dopamine neurons. *Neuropsychopharmacology*, *24*(4), 410–419. https://doi.org/10.1016/S0893-133X(00)00188-3.

Mpofana, T., Daniels, W. M. U., & Mabandla, M. V. (2016). Exposure to early life stress results in epigenetic changes in neurotrophic factor gene expression in a parkinsonian rat model. *Parkinson's Disease, 2016*, 6438783. https://doi.org/10.1155/2016/6438783.

Müller, T., & Muhlack, S. (2007). Acute levodopa intake and associated cortisol decrease in patients with Parkinson disease. *Clinical Neuropharmacology*, *30*(2), 101–106. https://doi.org/10.1097/01.WNF.0000240954.72186.91.

Müller, T., Welnic, J., & Muhlack, S. (2007). Acute levodopa administration reduces cortisol release in patients with Parkinson's disease. *Journal of Neural Transmission*, *114*(3), 347–350. https://doi.org/10.1007/s00702-006-0552-0.

Nathan, A. J., & Scobell, A. (2012). How China sees America. *Foreign Affairs*, *91*(5), 32–47. https://doi.org/10.1017/CBO9781107415324.004.

Nováková, L., Haluzík, M., Jech, R., Urgošík, D., Růžička, F., & Růžička, E. (2011). Hormonal regulators of food intake and weight gain in Parkinson's disease after subthalamic nucleus stimulation. *Neuroendocrinology Letters*, *32*(4), 437–441.

Parker, W. D., Parks, J. K., & Swerdlow, R. H. (2008). Complex I deficiency in Parkinson's disease frontal cortex. *Brain Research*, *1189*(1), 215–218. https://doi.org/10.1016/j.brainres.2007.10.061.

Perier, C., & Vila, M. (2012). Mitochondrial biology and Parkinson's disease. *Cold Spring Harbor Perspectives in Medicine*, *2*(2), a009332. https://doi.org/10.1101/cshperspect.a009332.

Pierre, K., Schlesinger, N., & Androulakis, I. P. (2017). The hepato-hypothalamic-pituitary-adrenal-renal axis: Mathematical modeling of cortisol's production, metabolism, and seasonal variation. *Journal of Biological Rhythms*, *32*(5), 469–484. https://doi.org/10.1177/0748730417729929.

Pistollato, F., Sumalla Cano, S., Elio, I., Masias Vergara, M., Giampieri, F., & Battino, M. (2016). Associations between sleep, cortisol regulation, and diet: Possible implications for the risk of Alzheimer disease. *Advances in Nutrition: An International Review Journal*, 7(4), 679–689. https://doi.org/10.3945/an.115.011775.

Plog, B. A., & Nedergaard, M. (2018). The glymphatic system in central nervous system health and disease: Past, present, and future. *Annual Review of Pathology: Mechanisms of Disease*, *13*(1), 379–394. https://doi.org/10.1146/annurev-pathol-051217-111018.

Qi, X. R., Kamphuis, W., Wang, S., Wang, Q., Lucassen, P. J., Zhou, J. N., et al. (2013). Aberrant stress hormone receptor balance in the human prefrontal cortex and hypothalamic paraventricular nucleus of depressed patients. *Psychoneuroendocrinology*, *38*(6), 863–870. https://doi.org/10.1016/j.psyneuen.2012.09.014.

Qian, X., Droste, S. K., Lightman, S. L., Reul, J. M. H. M., & Linthorst, A. C. E. (2012). Circadian and ultradian rhythms of free glucocorticoid hormone are highly synchronized between the blood, the subcutaneous tissue, and the brain. *Endocrinology*, *153*(9), 4346–4353. https://doi.org/10.1210/en.2012-1484.

Raadsheer, F. C., Van Heerikhuize, J. J., Lucassen, P. J., Hoogendijk, W. J. G., Tilders, F. J. H., & Swaab, D. F. (1995). Corticotropin-releasing hormone mRNA levels in the paraventricular nucleus of patients with Alzheimer's disease and depression. *American Journal of Psychiatry*, *152*(9), 1372–1376. https://doi.org/10.1176/ajp.152.9.1372.

Ricci, F., De Caterina, R., & Fedorowski, A. (2015). Orthostatic hypotension: Epidemiology, prognosis, and treatment. *Journal of the American College of Cardiology*, *66*, 848–860. https://doi.org/10.1016/j.jacc.2015.06.1084.

Ridder, S., Chourbaji, S., Hellweg, R., Urani, A., Zacher, C., Schmid, W., et al. (2005). Mice with genetically altered glucocorticoid receptor expression show altered sensitivity for stress-induced depressive reactions. *Journal of Neuroscience*, *25*(26), 6243–6250. https://doi.org/10.1523/JNEUROSCI.0736-05.2005.

Rod, N. H., Bordelon, Y., Thompson, A., Marcotte, E., & Ritz, B. (2013). Major life events and development of major depression in Parkinson's disease patients. *European Journal of Neurology*, *20*(4), 663–670. https://doi.org/10.1111/ene.12019.

Ron, N. P., Kazianis, J. A., Padbury, J. F., Brown, C. M., McGonnigal, B. G., Sysyn, G. D., et al. (2005). Ontogeny and the effects of corticosteriod pretreatment on aquaporin water channels in the ovine cerebral cortex. *Reproduction, Fertility and Development*, *17*(5), 535–542. https://doi.org/10.1071/RD03044.

Rutten, S., Vriend, C., Smit, J. H., Berendse, H. W., Van Someren, E. J. W., Hoogendoorn, A. W., et al. (2019). Bright light therapy for depression in Parkinson disease: A randomized controlled trial. *Neurology*, *92*(11), E1145–E1156. https://doi.org/10.1212/WNL.0000000000007090.

Růžička, F., Jech, R., Nováková, L., Urgošík, D., Bezdíček, O., Vymazal, J., et al. (2015). Chronic stress-like syndrome as a consequence of medial site subthalamic stimulation in Parkinson's disease. *Psychoneuroendocrinology*, *52*(1), 302–310. https://doi.org/10.1016/j.psyneuen.2014.12.001.

Sapolsky, R. M. (1996). Stress, glucocorticoids, and damage to the nervous system: The current state of confusion. *Stress*, *1*(1), 1–19. https://doi.org/10.3109/10253899609001092.

Schlesinger, I., Erikh, I., & Yarnitsky, D. (2007). Paradoxical kinesia at war. *Movement Disorders*, *22*(16), 2394–2397. https://doi.org/10.1002/mds.21739.

Schwab, R. S., & Zieper, R. S. (1965). Effects of mood, motivation, stress and alertness on the performance in Parkinson's disease. *European Neurology*, *150*(6), 345–357. https://doi.org/10.1159/000127780.

Segerstrom, S. C., & Miller, G. E. (2004). Psychological stress and the human immune system: A meta-analytic study of 30 years of inquiry. *Psychological Bulletin*, *130*(4), 601–630. https://doi.org/10.1037/0033-2909.130.4.601.

Seifried, C., Boehncke, S., Heinzmann, J., Baudrexel, S., Weise, L., Gasser, T., et al. (2013). Diurnal variation of hypothalamic function and chronic subthalamic nucleus stimulation in Parkinson's disease. *Neuroendocrinology*, *97*(3), 283–290. https://doi.org/10.1159/000343808.

Shiner, T., Seymour, B., Symmonds, M., Dayan, P., Bhatia, K. P., & Dolan, R. J. (2012). The effect of motivation on movement: A study of bradykinesia in Parkinson's disease. *PLoS One*, 7(10), e47138. https://doi.org/10.1371/journal.pone.0047138.

Shirbin, C. A., Chua, P., Churchyard, A., Hannan, A. J., Lowndes, G., & Stout, J. C. (2013). The relationship between cortisol and verbal memory in the early stages of Huntington's disease. *Journal of Neurology*, *260*(3), 891–902. https://doi.org/10.1007/s00415-012-6732-y.

Shirbin, C. A., Chua, P., Churchyard, A., Lowndes, G., Hannan, A. J., Pang, T. Y., et al. (2013). Cortisol and depression in pre-diagnosed and early stage Huntington's disease. *Psychoneuroendocrinology*, *38*(11), 2439–2447. https://doi.org/10.1016/j.psyneuen.2012.10.020.

Sieurin, J., Andel, R., Tillander, A., Valdes, E. G., Pedersen, N. L., & Wirdefeldt, K. (2018). Occupational stress and risk for Parkinson's disease: A nationwide cohort study. *Movement Disorders*, *33*(9), 1456–1464. https://doi.org/10.1002/mds.27439.
Smith, L. K., Jadavji, N. M., Colwell, K. L., Katrina Perehudoff, S., & Metz, G. A. (2008). Stress accelerates neural degeneration and exaggerates motor symptoms in a rat model of Parkinson's disease. *European Journal of Neuroscience*, *27*(8), 2133–2146. https://doi.org/10.1111/j.1460-9568.2008.06177.x.
Smyth, N., Skender, D., David, F. J., Munoz, M. J., Fantuzzi, G., Clow, A., et al. (2019). Endurance exercise reduces cortisol in Parkinson's disease with mild cognitive impairment. *Movement Disorders*, *34*(8), 1238–1239.
Spencer, R. L., Chun, L. E., Hartsock, M. J., & Woodruff, E. R. (2018). Glucocorticoid hormones are both a major circadian signal and major stress signal: How this shared signal contributes to a dynamic relationship between the circadian and stress systems. *Frontiers in Neuroendocrinology*, *49*, 52–71. https://doi.org/10.1016/j.yfrne.2017.12.005.
Stopa, E. G., Volicer, L., Kuo-Leblanc, V., Harper, D., Lathi, D., Tate, B., et al. (1999). Pathologic evaluation of the human suprachiasmatic nucleus in severe dementia. *Journal of Neuropathology and Experimental Neurology*, *58*(1), 29–39. https://doi.org/10.1097/00005072-199901000-00004.
Stypuła, G., Kunert-Radek, J., Stepień, H., Żylińska, K., & Pawlikowski, M. (1996). Evaluation of interleukins, acth, cortisol and prolactin concentrations in the blood of patients with Parkinson's disease. *Neuroimmunomodulation*, *3*(2–3), 131–134. https://doi.org/10.1159/000097237.
Svensson, E., Farkas, D. K., Gradus, J. L., Lash, T. L., & Sørensen, H. T. (2016). Adjustment disorder and risk of Parkinson's disease. *European Journal of Neurology*, *23*(4), 751–756. https://doi.org/10.1111/ene.12933.
Swaab, D. F., Bao, A. M., & Lucassen, P. J. (2005). The stress system in the human brain in depression and neurodegeneration. *Ageing Research Reviews*, *4*, 141–194. https://doi.org/10.1016/j.arr.2005.03.003.
van Wamelen, D. J., Podlewska, A. M., Leta, V., Śmiłowska, K., Rizos, A., Martinez-Martin, P., et al. (2019). Slave to the rhythm: Seasonal differences in non-motor symptoms in Parkinson's disease. *Parkinsonism & Related Disorders*, *63*, 73–76. https://doi.org/10.1016/j.parkreldis.2019.02.041.
Vardi, J., Oberman, Z., Rabey, I., Streifler, M., Ayalon, D., & Herzberg, M. (1976). Weight loss in patients treated long-term with levodopa. Metabolic aspects. *Journal of the Neurological Sciences*, *30*(1), 33–40. https://doi.org/10.1016/0022-510X(76)90253-7.
Ventura-Silva, A. P., Pêgo, J. M., Sousa, J. C., Marques, A. R., Rodrigues, A. J., Marques, F., et al. (2012). Stress shifts the response of the bed nucleus of the stria terminalis to an anxiogenic mode. *European Journal of Neuroscience*, *36*(10), 3396–3406. https://doi.org/10.1111/j.1460-9568.2012.08262.x.
Vlajinac, H., Sipetic, S., Marinkovic, J., Ratkov, I., Maksimovic, J., Dzoljic, E., et al. (2013). The stressful life events and Parkinson's disease: A case-control study. *Stress and Health*, *29*(1), 50–55. https://doi.org/10.1002/smi.2424.
Vogel, H. P., & Ketsche, R. (1986). Effect of hypoglycaemia, TRH and levodopa on plasma growth hormone, prolactin, thyrotropin and cortisol in Parkinson's disease before and during therapy. *Journal of Neurology*, *233*(3), 149–152. https://doi.org/10.1007/BF00314421.
Volpi, R., Caffarra, P., Boni, S., Scaglioni, A., Malvezzi, L., Saginario, A., et al. (1997). ACTH/cortisol involvement in the serotonergic disorder affecting the parkinsonian brain. *Neuropsychobiology*, *35*(2), 73–78. https://doi.org/10.1159/000119394.
Wang, Q., Joels, M., Swaab, D. F., & Lucassen, P. J. (2012). Hippocampal GR expression is increased in elderly depressed females. *Neuropharmacology*, *62*(1), 527–533. https://doi.org/10.1016/j.neuropharm.2011.09.014.

Wang, Q., Verweij, E. W. E., Krugers, H. J., Joels, M., Swaab, D. F., & Lucassen, P. J. (2014). Distribution of the glucocorticoid receptor in the human amygdala; changes in mood disorder patients. *Brain Structure and Function*, *219*(5), 1615–1626. https://doi.org/10.1007/s00429-013-0589-4.

Webster, M. J., Knable, M. B., O'Grady, J., Orthmann, J., & Weickert, C. S. (2002). Regional specificity of brain glucocorticoid receptor mRNA alterations in subjects with schizophrenia and mood disorders. *Molecular Psychiatry*, 7(9), 985–994. https://doi.org/10.1038/sj.mp.4001139.

Wei, Q., Hebda-Bauer, E. K., Pletsch, A., Luo, J., Hoversten, M. T., Osetek, A. J., et al. (2007). Overexpressing the glucocorticoid receptor in forebrain causes an aging-like neuroendocrine phenotype and mild cognitive dysfunction. *Journal of Neuroscience*, *27*(33), 8836–8844. https://doi.org/10.1523/JNEUROSCI.0910-07.2007.

Wilkins, H. M., Carl, S. M., & Swerdlow, R. H. (2014). Cytoplasmic hybrid (cybrid) cell lines as a practical model for mitochondriopathies. *Redox Biology*, *2*, 619–631. https://doi.org/10.1016/j.redox.2014.03.006.

Wilson, H., Niccolini, F., Pellicano, C., & Politis, M. (2019). Cortical thinning across Parkinson's disease stages and clinical correlates. *Journal of the Neurological Sciences*, *398*, 31–38. https://doi.org/10.1016/j.jns.2019.01.020.

Wu, T., & Hallett, M. (2013). The cerebellum in Parkinson's disease. *Brain*, *136*, 696–709. https://doi.org/10.1093/brain/aws360.

Wu, J. Q., Li, P., Stavitsky Gilbert, K., Hu, K., & Cronin-Golomb, A. (2018). Circadian rest-activity rhythms predict cognitive function in early Parkinson's disease independently of sleep. *Movement Disorders Clinical Practice*, *5*(6), 614–619. https://doi.org/10.1002/mdc3.12692.

Yntema, O. P., & Korf, J. (1987). Transient suppression by stress of haloperidol induced catalepsy by the activation of the adrenal medulla. *Psychopharmacology*, *91*(1), 131–134. https://doi.org/10.1007/BF00690942.

Zhang, R., Liu, Y., Chen, Y., Li, Q., Marshall, C., Wu, T., et al. (2019). Aquaporin 4 deletion exacerbates brain impairments in a mouse model of chronic sleep disruption. *CNS Neuroscience & Therapeutics*, *26*, 228–239. https://doi.org/10.1111/cns.13194.

CHAPTER SEVEN

The effects of stress on cardiovascular disease and Alzheimer's disease: Physical exercise as a counteract measure

Gema Sanchis-Soler, Juan Tortosa-Martínez*, Carmen Manchado-Lopez, Juan Manuel Cortell-Tormo
University of Alicante, Alicante, Spain
*Corresponding author: e-mail address: juan.tortosa@ua.es

Contents

Abstract

AD is a complicated multi-systemic neurological disorder that involves different biological pathways. Several risk factors have been identified, including chronic stress. Chronic stress produces an alteration in the activity of the hypothalamic pituitary adrenal (HPA) system, and the autonomic nervous system (ANS), which over time increase the risk of AD and also the incidence of cardiovascular disease (CVD) and risk factors, such as hypertension, obesity and type 2 diabetes, associated with cognitive impairment and AD. Considering the multi-factorial etiology of AD, understanding the complex interrelationships between different risk factors is of potential interest for designing adequate strategies for preventing, delaying the onset or slowing down the progression of this devastating disease. Thus, in this review we will explore the general mechanisms and

International Review of Neurobiology, Volume 152
ISSN 0074-7742
https://doi.org/10.1016/bs.irn.2020.01.002

evidence linking stress, cardiovascular disease and AD, and discuss the potential benefits of physical activity for AD by counteracting the negative effects of chronic stress, CVD and risk factors.

1. Introduction

Alzheimer's disease (AD) is the most prevalent type of dementia, representing from 60% to 80% of the cases. AD is characterized by a progressive deterioration of memory function accompanied by one or more other cognitive domains including aphasia, apraxia, agnosia or disturbance of executive functioning, of sufficient severity to compromise a person's daily function. The pathological hallmarks in the brain of AD include increased levels of amyloid-β peptide, deposited in diffuse and neuritic plaques that collect between neurons and disrupt cell function, and hyper-phosphorylated forms of tau protein, that accumulates as neurofibrillary tangles inside neurons which block the neuron's transport system (Dubois et al., 2016). Mild cognitive impairment (MCI) is characterized by a cognitive deterioration greater than the expected for an individual's age and education level, but yet not severe enough for a diagnosis of dementia. The amnestic type (aMCI), in which memory loss is considered the main symptom, is more likely to progress to AD (Petersen, 2004).

Given the low effectiveness of current treatments for reversing or at least stopping the course of this neurodegenerative disease (Schneider et al., 2014), many interventional studies are focusing now in healthy individuals, without cognitive impairment, but at high risk of developing AD (Crous-Bou, Minguillón, Gramunt, & Molinero, 2017). Notwithstanding, the pathological hallmarks of AD may be present in the brain up to 20 years before the onset of clinical symptoms. This asymptomatic stage is referred to as preclinical AD (Dubois et al., 2016).

AD is a complicated multi-systemic neurological disorder that involves different biological pathways at different genetic, molecular, epigenetic, cognitive, behavioral and neurophysiological levels (Zhao, Bhattacharjee, Dua, Alexandrov, & Lukiw, 2015). Thus, other than age and genetic factors (ApoE4) (Querfurth & LaFerla, 2010) a wide vary of risk factors have been associated with the incidence and progression of the disease such as high levels of oxidative stress (Querfurth & LaFerla, 2010) and inflammatory markers (Parachikova & Cotman, 2007); depression

(Rapp, Schnaider-Beeri, Grossman, et al., 2006) and chronic stress (Tortosa-Martinez & Clow, 2012); cardiovascular disease (CVD) and risk factors (Mossello & Simoni, 2016; Wanleenuwat, Iwanowski, & Kozubski, 2019); smoking, mental inactivity, physical inactivity, and a poor diet (Barnes & Yaffe, 2011; Daviglus et al., 2010). An understanding of these risk factors, and their interactions, is thus a research priority in order to develop counter-strategies for reducing the prevalence of cognitive impairment and AD.

A risk factor that has received increased attention in the literature is stress. When a stressor occurs, the organism activates different mechanisms to respond to the stressor, increasing the activity of the hypothalamic pituitary adrenal (HPA) system, and the sympathetic nervous system (SNS), while inhibiting the parasympathetic nervous system (PNS). If there is either an inadequate stress response or an inadequate recovery to the stress response, there is a wear and tear of the body referred to as allostatic load. When this abnormal stress response is prolonged in time, it turns into an allostatic overload (McEwen, 2008), having severe health consequences including an increased risk of AD (Rothman & Mattson, 2010), cognitive deficits and the formation of amyloid-β plaques and protein Tau "tangles" (Dong et al., 2008; Green, Billings, Roozendaal, McGaugh, & LaFerla, 2006; Lee et al., 2009).

The mechanisms explaining this relationship are not fully understood but they are mostly attributed to the harmful effects on cognition of an abnormal cortisol secretion pattern seen in chronic stress (Rothman & Mattson, 2010; Tortosa-Martínez, Manchado, Chulvi-Medrano, & Cortell-Tormo, 2018). However, it is likely that the negative consequences of chronic stress on different systems produce also an indirect effect on AD. Thus, other possible pathways explaining the negative effects of stress on cognition and AD are the increased risk of CVDs and risk factors such as type II diabetes, obesity and hypertension (Rothman & Mattson, 2010; Tortosa-Martinez & Clow, 2012).

Chronic stress may play a relevant role on the development and triggering of CVD and events (Kivimäki & Steptoe, 2017; Valtorta, Kanaan, Gilbody, Ronzi, & Hanratty, 2016). In different studies, people exposed to high levels of psychosocial stress have shown a 40–60% increased risk of CVD (Steptoe & Kivimaki, 2013), and a fourfold higher risk of cardiovascular system-related death (Karasek, Baker, Marxer, Ahlbom, & Theorell, 1981). However, negative studies have also been reported (Bosma, Peter, Siegrist, & Marmot, 1998). CVD has been identified as a powerful risk factor for AD (Bendlin, 2019), and cerebrovascular disease is considered by some authors to be the second most common cause of cognitive impairment and dementia, contributing to cognitive decline in the neurodegenerative dementias (O'Brien et al., 2003). Approximately 50% of AD patients present significant cerebrovascular pathology on autopsy (Schneider, Arvanitakis, Bang, & Bennett, 2007; Sonnen et al., 2007).

CVD includes a diverse group of pathologies involving the heart and blood vessels, such as coronary artery disease, stroke, peripheral vascular disease, rheumatic heart disease, and pulmonary embolism. Common behavioral risk factors for CVDs include unhealthy dietary patterns, physical inactivity, high sodium/tobacco consumption, and alcohol misuse. In addition to these risk factors, several quantifiable physiological changes are associated with CVD, such as high blood pressure (hypertension), elevated blood glucose (diabetes mellitus), elevated cholesterol (hyperlipidemia) and obesity (Joseph et al., 2017), all considered risk factors for AD. In fact, cardiovascular-related risk factors are believed to play a key role in disease development and increased cognitive deterioration (Wanleenuwat et al., 2019).

The main theory linking CVD and AD implicates impaired cerebral perfusion resulting from vascular injury. It has been hypothesized that this hypoperfusion could lead to toxic accumulation of β-amyloid as well as to tau hyperphosphorylation and neurofibrillary tangle formation, resulting in early neuronal dysfunction (Femminella et al., 2014). Not surprisingly therefore, cerebral hypoperfusion has been found to be one of the earliest manifestations of AD (Wanleenuwat et al., 2019). Another theory suggests that increased plasma homocysteine levels related to CVD and risk factors could be responsible for an increment in the deposition of amyloid plaques (Zhuo, Wang, & Pratico', 2011). In any case, the presence of cardiovascular risk factors in midlife has been associated with amyloid deposition later in life, suggesting a direct link between cardiovascular risk factors and the pathophysiology of AD (Gottesman et al., 2017).

Thus, understanding the relationship between stress, cardiovascular disease and AD is of potential interest for developing therapeutic strategies. A possible non-pharmacological strategy that may yield positive effects on stress, cardiovascular risk factors and AD is physical activity (PA) and exercise. PA and exercise have been proved to be effective for improving cognitive function in healthy adults and older adults (Colcombe & Kramer, 2003) and reducing the risk of developing AD (Hamer & Chida, 2009). Recent meta-analysis and systematic reviews have concluded that exercise can yield modest to moderate improvements in memory and executive function in people who already have dementia and AD (Groot et al., 2016; Guitar, Connelly, Nagamatsu, Orange, & Muir-Hunter, 2018). However, not all studies have been conclusive such as the Cochrane systematic review conducted by Forbes, Forbes, Blake, Thiessen, and Forbes (2015). Furthermore, the biological mechanisms that underlie the potential benefits of exercise on cognition and AD are still to be elucidated. Interestingly enough, the main theories are closely related to stress, CVD and risk factors (Tortosa-Martinez & Clow, 2012). In this regard, exercise has been proved to be effective for reducing psychological stress responses (Lucertini et al., 2015; Rimmele et al., 2009) as well as cardiovascular risk and mortality (Mora, Cook, Buring, Ridker, & Lee, 2007; Mueller, 2007). Furthermore, O'Keefe, O'Keefe, and Lavie (2019), in a recent review study, concluded that exercise could counteract the cardiotoxicity of stress.

In this review, we will initially explore the general mechanisms and evidence linking stress, cardiovascular disease and Alzheimer's disease. Then, we will go on to analyze the existing evidence linking stress, AD, and three specific cardiovascular risk factors: hypertension, diabetes mellitus and obesity. Finally, we will discuss the potential role of exercise for counteracting the negative effects of both chronic stress and cardiovascular conditions and reducing the risk of developing AD, delay the onset or slow the rate of progression of this devastating neurodegenerative disease.

2. Stress, cardiovascular disease and Alzheimer's disease

2.1 Adrenal pituitary hypothalamic axis

In stress situations, the hypothalamus stimulates the pituitary gland by the corticotropin releasing hormone and arginine vasopressin (CRH/AVP), increasing the adrenocorticotropic hormone (ACTH) secretion and stimulating the adrenal cortex to produce cortisol. Chronic stress, alters the cortisol production mechanism causing a pathological increase in circulating cortisol levels (Stephens, Mahon, McCaul, & Wand, 2016), associated with worse cognitive performance in healthy older adults (Seeman, McEwen, Singer, Albert, & Rowe, 1997; Wolf, 2009), and in people with MCI (Arsenault-Lapierre, Chertkow, & Lupien, 2010; Wolf, Convit, Thorn, & de Leon, 2002). Not surprisingly, therefore HPA axis dysfunction is found in the early stages of patients with AD with elevated levels of cortisol being associated with higher cognitive impairment and a more rapid disease progression (Csernansky & Dong, 2006; Popp et al., 2015).

However, the cortisol secretory pattern is complex with a circadian rhythm characterized by a sharp increase from waking time to 30–45 min after, referred to as the cortisol awakening response (CAR), and a subsequent steady decline during the remainder of the day, known as the diurnal decline (DD) (Edwards, Evans, Hucklebridge, & Clow, 2001). Although is not fully clear which aspects are most closely related to cognitive function, increasing evidence indicates that a more dynamic cortisol secretion pattern, as opposed to a flatter profile, is associated with better cognitive performance (for a review, see Tortosa-Martínez et al., 2018). At a cross-sectional level, evidence with healthy older adults indicate that worse cognitive performance is associated with overall basal levels such as high morning (Beluche, Carriere, Ritchie, & Ancelin, 2010; Geerlings, Garcia, & Harris, 2015) and evening levels (Geerlings et al., 2015); high daily average (Lee et al., 2007) and area under the curve (AUC) (Lee et al., 2007; Ouanes et al., 2017); as well as flattening of the cortisol diurnal pattern (Beluche et al., 2010; Evans et al., 2011) which has been seen in people with dementia (Kovach, Woods, Logan, & Raff, 2011). The exact role played by the CAR needs to be fully clarified with evidence suggesting that a higher CAR might be associated with better cognitive performance (Evans et al., 2011; Evans, Hucklebridge, Loveday, & Clow, 2012), although some studies found just the opposite (Almela, van der Meij, Hidalgo, Villada, & Salvador, 2012; Hidalgo, Almela, Pulopulos, & Salvador, 2016), probably indicating an inverted U-shaped relationship between the CAR and cognition (Moriarty et al., 2014).

In parallel, elevated cortisol levels have been strongly associated with CVD and cardiovascular mortality risk. A recent meta-analysis of prospective studies concluded that elevated cortisol levels were associated with higher incidence of CVD (Crawford et al., 2019). Vogelzangs et al. (2010), in one prospective study, followed-up for 6 years 865 older adults finding that persons in the highest tertile of urinary cortisol levels had a fivefold increase in the risk of dying of CVD. These data are consistent with the evidence showing that in Cushing's syndrome, a condition characterized by a prolonged exposure to excessive levels of glucocorticoids such as cortisol, the rate of CVD is increased up to five times compared to the general population, driven by a frequent presence of comorbidities such as hypertension, dyslipidemia, diabetes and obesity (Nieman, 2019). The evidence linking the diurnal cycle of cortisol with CVD is more limited and deserves further research, but Kumari et al. (2010) found, in the Whitehall II study, that a flatter cortisol slope was associated with increased risk of all-cause mortality and that this excess mortality risk was mostly driven by an increased risk of cardiovascular deaths. Furthermore, a flatter cortisol profile has also been linked to the absence of nocturnal blood pressure diping (Holt-Lunstad & Steffen, 2007), diabetes (Champaneri et al., 2012) and increased levels of inflammation (De Santis et al., 2012). However, Hajat et al. (2013) found very weak or no associations between the diurnal cortisol pattern and atherosclerosis.

Abnormal cortisol secretion patterns have been related with insulin resistance, central adiposity, higher blood pressure, and immune system suppression, producing or increasing the probability of developing hypertension, diabetes type II, dyslipidemia, cholesterol or obesity (Walker, 2007; Whitworth, Williamson, Mangos, & Kelly, 2005). Additionally, the uncontrolled increase of cortisol incites the increase of phenylethanolamine *N*-methyltransferase (PNMT) (Geiger, Pitts, Feldkamp, Kirschbaum, & Wolf, 2015), an enzyme directly related with the increment of epinephrine production. As a consequence of this increment, alpha receptors are overstimulated, resulting in augmented vasoconstriction and blood pressure (Gunawardane & Grossman, 2017). In the McArthur studies of healthy aging, urinary excretion of epinephrine predicted cognitive decline in healthy older adults (Karlamangla, Singer, Greendale, & Seeman, 2005).

Excessive cortisol levels also increase inflammation markers, such as CRP, interleukin 1 and 6, which play a key in the pathogenesis of both CVD (Yao et al., 2019) and AD (Heneka et al., 2015). Inflammation increases the risk of diabetes, obesity, and hypertension (Puzianowska-Kuznicka et al., 2016; Ridker, 2018; Van Tassell, Raleigh, & Abbate, 2015), as well as atherosclerosis, atherothrombosis, myocardial infarction, stroke and sudden death (Osman, L'Allier, Elgharib, & Tardif, 2006). Neuroinflammation is present in the pathogenesis of AD (Heneka et al., 2015), and there is evidence showing that systemic illnesses with a pro-inflammatory response may contribute to neurodegeneration and worsening of dementia symptoms, although inconsistencies in the literature should be noted which could be partially explained by the complexity of interactions with life course factors such as education, excessive alcohol intake, smoking or exercise (Tsiu, Richards, & Davis, 2018). Increased levels of pro-inflammatory factors have also been also associated with decreased levels of brain-derived neurotrophic factor (BDNF), especially in the hippocampus (Cotman, Berchtold, & Christie, 2007), a key structure for memory, learning processes and spatial orientation (Rothman & Mattson, 2010). This neurotrophin is believed to play a relevant role in neurogenesis, dendritic growth, synaptogenesis and memory (Lu, Nagappan, Guan, Nathan, & Wren, 2013; Sasi et al., 2017).

High cortisol levels have also been linked to a decrease in BDNF levels (Duman & Monteggia, 2006), although evidence is not always conclusive (de Assis & Gasamov, 2019). Plasma BDNF significantly correlates with multiple risk factors for metabolic syndrome and cardiovascular dysfunction (Golden et al., 2010) and low levels of BDNF increase the incidence of cognitive impairment and AD (Querfurth & LaFerla, 2010; Ventriglia et al., 2013). Not surprisingly therefore, patients with AD show lower levels of BDNF in the hippocampus (Ventriglia et al., 2013), and atrophy in this area is found in all stages of the disease (Querfurth & LaFerla, 2010). In fact, cognitive deficits resulting from chronic adverse stress and hyperactivation of the HPA axis are likely mediated through alterations in hippocampal networks (Aisa et al., 2009), as there is a dense expression of glucocorticoid receptors in this area and is mainly responsible for the cortisol feedback mechanism system (Rothman & Mattson, 2010). Notwithstanding, corticosteroids have been found to impair neurogenesis in the dentate gyrus of the rat hippocampus (Cameron & Gould, 1994). Thus, cortisol and BDNF levels might be inversely related in AD, suggesting that cortisol/BDNF ratios could be used as a valuable biomarker for dementia and cognitive impairment (Curto et al., 2017). However, more research is needed and especially the relationship between the diurnal cycle of cortisol and BDNF levels remains to be clarified, in order to fully understand this association.

2.2 Autonomic nervous system

The ANS responds to stress by increasing the activity of the sympathetic nervous system (SNS) and decreasing the activity of the parasympathetic nervous system (PNS). Cortisol also stimulates the SNS activity, as cortisol receptors are localized on sympathetic nerves (Green, Janig, & Levine, 1997).

The increased activity of the SNS causes an overproduction of CRH/AVP, and an increase in the catecholamines production. The nuclear factor kB-dependent (NF-kB), stimulates catecholamines production. Excessive levels of norepinephrine may increase production of pro-inflammatory cytokines (Hori & Kim, 2019), oxidative stress (OS) and reactive oxygen species (ROS) (Yasunari et al., 2006). Oxidative stress, the imbalance in the production of reactive oxygen species (ROS) and antioxidative defense, plays a relevant role in CVD (Cervantes Gracia, Llanas-Cornejo, & Husi, 2017) and has been strongly linked to the pathogenesis of AD, as oxidative stress damages the mitochondria and promotes the aggregation of Aβ and the phosphorylation of tau in the brain of AD patients (Wang et al., 2014).

In parallel, the PNS action is reduced and the vagus nerve functioning is altered, inhibiting acetylcholine production which in turn inhibits the production of the nicotinic alpha7 receptor and, thus, increases the production of inflammatory factors such as tumor necrosis factor (TNF)-alpha, C-reactive protein (CRP) and other inflammatory mediators (Yao et al., 2019; Zhao et al., 2017). Furthermore, a macrophage (Finnell & Wood, 2016) and glial cells activation is observed (Leszek et al., 2016), increasing again oxidative stress and decreasing nitric oxide (NO) synthesis (de Oliveira et al., 2018; Finnell & Wood, 2016; Reus et al., 2015; Wang et al., 2018). This situation triggers the binding of glutamate to the NMDA ionotropic receptor and inhibits BDNF production, contributing to cognitive impairment. NO is important for cardiovascular homeostasis acting as a vasodilator involved in the control of vasomotor function and local blood flow (Katusic & Austin, 2014). A decrease of NO synthesis may contribute to reduced regional cerebral blood flow and increased amyloid deposition in AD (Austin, Santhanam, Hinton, Choi, & Katusic, 2013).

AD has been associated with under activity of the cholinergic nervous system (Perry, Smith, Court, & Perry, 1990) and in general a dysfunction of the autonomic nervous system, either in terms of impaired vagal parasympathetic function or as altered sympathetic response to orthostasis (Allan et al., 2007; Femminella et al., 2014). Vagal tone can be quantified via respiratory sinus arrhythmia (RSA), and high levels of RSA have been linked in older adults to better verbal episodic memory, a cognitive domain that is typically impaired in AD (Dywan, Mathewson, Choma, Rosenfeld, & Segalowitz, 2008). Furthermore, autonomic dysfunction may be present early in the preclinical stage of dementia (Braak & Braak, 1995) and people with MCI (Nicolini et al., 2014). One of the markers of autonomic dysfunction is heart rate variability (HRV), a measure of beat-to-beat temporal changes in heart rate which measures the fine balance between the SNS and the PNS. Low HRV, which reflects low parasympathetic modulation relative to sympathetic modulation, has been linked to metabolic syndrome, cardiac and cerebrovascular events, CVD and sudden death due to sympathetically mediated ventricular arrhythmias

(Licht, de Geus, & Penninx, 2013; Soares-Miranda et al., 2012), as well as to AD (de Vilhena Toledo & Junqueira Jr, 2008), although inconsistencies in the literature should be noted in this latter association (Da Silva et al., 2018).

In summary, chronic stress leads to dysfunction of the adrenal pituitary hypothalamic axis, an overstimulation of the SNS, and inhibition of the PNS and vagus nerve. As a consequence, there is a loss of internal homeostasis with cerebral hipoperfusion, overproduction of inflammatory factors and oxidative stress, autonomic dysfunction, and decreased levels of BDNF, leading to increased risk of CVD, neuronal dysfunction and AD. Next, we will go on to explore the relationship between three specific cardiovascular risk factors (hypertension, diabetes mellitus and obesity), stress and AD.

3. Hypertension

Hypertension is a highly prevalent condition which has been established as a major risk factor for CVD events, morbidity and mortality (Hansson, 1996). Higher HPA axis responses toward challenges in normotensive individuals increase the risk of hypertension (Matthews, Katholi, McCreath, et al., 2004). Recently, Chiodini et al. (2019) showed that cortisol secretion, sensitivity and activity were associated with hypertension in a sample of eucortisolemic postmenopausal women with and without type 2 diabetes. There is also evidence showing that people with higher levels of psychosocial stress, who had developed hypertension earlier, were more prone to develop angina pectoris, myocardial infarction and cerebrovascular insults compared to controls (Nedic, Filipovic, & Solak, 2001). In addition, Holt-Lunstad and Steffen (2007) found that a decreased diurnal cortisol variation significantly predicted blood pressure dipping in a cross-sectional study with 302 healthy adults. A lack in blood pressure dipping has been associated with increased cerebrovascular damage (Kario et al., 1996), as well as cardiovascular morbidity and mortality (Ohkubo et al., 1997).

In regards to the mechanisms of action, it seems that elevated levels of cortisol and catecholamines like norepinephrine (NE) present in chronic stress are responsible for elevated blood pressure levels. On one hand, it has been hypothesized that cortisol secreted by the HPA axis acts on renal mineralocorticoid receptors, resulting in salt and water retention and, thus, increasing plasma volume and blood pressure, although several aspects of this hypothesis have been refuted (Baid & Nieman, 2004). Nevertheless, hypertension prevalence has been shown to be higher in patients with hypercortisolism caused by Cushing's syndrome

(Morelli, Arosio, & Chiodini, 2018). In fact, hypertension is present in 80% of these patients, being one of the most discriminating features of this disease (Baid & Nieman, 2004). Hypertension prevalence may also be increased in people with slightly increased cortisol secretion (Chapman, Holmes, & Seckl, 2013).

On the other hand, the SNS activation incites the production of NE, which interacts with adenosine triphosphate (ATP) in the vascular smooth muscle cell (VSMC), leading to vasoconstriction, increased blood pressure, and eventually hypertension (Esch, Stefano, Fricchinoe, & Benson, 2002). This cascade of events prolonged in time affects the appearance of cardiovascular complications such as vascular injury and CVD (Curtis & O'Keefe Jr., 2002).

In different longitudinal studies, midlife hypertension has been strongly associated with cognitive impairment in late-life (Gabin, Tambs, Saltvedt, et al., 2017; Gottesman et al., 2014) and increased risk of developing AD (Mossello & Simoni, 2016). In fact, hypertension has been regarded as the most powerful vascular risk factor for AD (Wiesmann, Kiliaan, & Claassen, 2013). The strong associations found between midlife hypertension and cognitive impairment in late-life supports the idea that hypertension duration and chronicity in adulthood may be especially relevant (Walker, Power, & Gottesman, 2017), with late-life hypertension not appearing to be a risk factor for AD (Power et al., 2011). However, there are some inconsistencies in the literature and also studies finding an association between low blood pressure and increased risk of cognitive impairment and dementia (Wanleenuwat et al., 2019) indicating that, similarly to the CAR, a U-shaped relationship between blood pressure and cognition is plausible (Beeri, Ravona-Springer, Silverman, & Haroutunian, 2009). In parallel, high blood pressure has been associated with smaller hippocampus and total brain volumes (Power et al., 2016) as well as reductions in brain volume over time (Jennings et al., 2012).

Hypertension causes a thickening of vessel walls, reduced vessel elasticity, and the narrowing of the lumen, especially in small vessels, resulting in reduced cerebral blood flow, impaired cerebral blood flow auto regulation (Iadecola et al., 2016), increased oxidative stress and inflammatory processes (Pires, Dams Ramos, Matin, et al., 2013), and disruption of the blood-brain barrier (Harrison & Guzik, 2016). This cascade of events causes a faster amyloid deposition and Tau hyperphosphorylation (Wanleenuwat et al., 2019). Notwithstanding, in AD models without hypertension, a better blood flow circulation allows the body to remove amyloid plaques more efficiently compared to AD models with hypertension (Kisler, Nelson, Montagne, & Zlokovic, 2017). Angiotensin II, a peptide of the renin-angiotensin-aldosterone (RAS) cascade responsible for blood pressure regulation, is believed to be one of the major mediators in the reduction of blood flow to the brain and the negative consequences in cognition and AD (Wanleenuwat et al., 2019).

4. Diabetes type 2

In the last years, the prevalence of diabetes type 2 has increased critically, and it seems that this trend is going to continue in the coming years (Liang et al., 2018; Verdile, Fuller, & Martins, 2015). It is a disease characterized by deficient insulin production or by insulin resistance (Chicharro & Mojares, 2008) being the last one the main cause (Nolan & Prentki, 2019). There are several risk factors for developing this disease such as hypertension, obesity, a sedentary lifestyle and stress (Chicharro & Mojares, 2008; Harris et al., 2017; Siddiqui et al., 2019).

Elevated cortisol levels in the morning and at night, as well as a flatter curve during the day, have been seen in people with type 2 diabetes (Hackett, Steptoe, & Kumari, 2014; Ortiz et al., 2019; Siddiqui et al., 2019). Cortisol overproduction is related to a worse lipid metabolism and increased fat deposit in the abdomen (Hackett, Poole, Hunt, Panagi, & Steptoe, 2019; Misra et al., 2008); inhibition of insulin production by beta cells, hyperglycemia (Madhu, Siddiqui, Desai, Sharma, & Bansal, 2019); and the increase of insulin resistance (Joseph & Golden, 2017; Siddiqui et al., 2019). In addition, the uncontrolled secretion of epinephrine and norepinephrine as a result of hyperactivation of the sympathetic nervous system, alters the glucose metabolism (Liang et al., 2018; Shin, Hsieh, Kwon, Lee, & Seol, 2018; Watanabe et al., 2018) and increases oxidative stress and inflammatory cytokines (IL-1β, IL-6, C-reactive protein, TNF-α) which is directly related to the development of type II diabetes and AD (Paul, Jerrett, & Ritz, 2018).

Different studies have found a direct relationship between type 2 diabetes and the risk of developing AD (Mittal & Katare, 2016; Zhang et al., 2017), although not all evidence is conclusive (Abner et al., 2016; Dos Santos Matioli et al., 2017; Pruzin et al., 2017). The possible mechanisms involved seem to be related to the effects of hyperglycemia, insulin resistance and hyperinsulinemia present in type 2 diabetes (Pugazhenthi, Qin, & Reddy, 2017; Salas & De Strooper, 2019).

In situations of hyperglycemia or insulin resistance, AKT decreases causing the increase of GSK3β, resulting in the increase of TAU protein and β-amyloid (Arnold et al., 2018; Huang et al., 2019; Li, Song, & Leng, 2015; Pugazhenthi et al., 2017; Rosales-Corral, Tan, Manchester, & Reiter, 2015). Insulin resistance can also cause damage to different organs or systems, such as microvascular lesions, with the consequent increase in the blood-brain barrier permeability (Li et al., 2015, 2018; Salas & De Strooper, 2019), or mitochondrial dysfunction, increasing the production of ROS, oxidative stress, AGEs and reduction of the β-oxidation of ATP production (Chornenkyy, Wang, Wei, & Nelson, 2019; Pugazhenthi et al., 2017; Rosales-Corral et al., 2015).

Furthermore, hyperinsulinemia increases β-amyloid deposition (Chornenkyy et al., 2019; Li et al., 2015; Paul et al., 2018; Pugazhenthi et al., 2017; Rad et al., 2018; Rosales-Corral et al., 2015; Verdile et al., 2015) and the development of insulin resistance in the neurons, a situation that is currently beginning to be described as type 3 diabetes (Chornenkyy et al., 2019; Li et al., 2015; Mittal & Katare, 2016; Pugazhenthi et al., 2017; Rad et al., 2018), increasing the hyperactivity of CDK5 and finally causing neuronal death (Chow et al., 2019).

In summary, chronic stress causes systemic dysfunction with the consequent alteration of glucose metabolism (hyperglycemia, hyperinsulinemia and insulin resistance) and high risk of diabetes type II, resulting in an increase in the production of inflammatory factors and ROS as well as the phosphorylation of TAU and deposits of β-amyloid, whose final result may be AD development (Paul et al., 2018; Rad et al., 2018; Verdile, Keane, et al., 2015).

5. Obesity

Obesity is defined as an excess of body fat or adipose tissue (Alford, Patel, Perakakis, & Mantzoros, 2018; Lloret, Monllor, Esteve, Cervera-Ferri, & Lloret, 2019). The latest data provided by the World Health Organization (WHO) indicate that 39% and 13% of adults over 18 are overweight and obese, respectively. Specially alarming is the prevalence in children and adolescents, since it already account for 340 million cases worldwide (World Health Organization, 2018). Obesity increases the risk of suffering from both cardiovascular and neurodegenerative diseases (Bluher, 2019).

Stress has been associated with obesity although it does not affect all people equally, causing either hyperphagia or hypophagia, and leading to weight loss or gain, respectively (Hausl, Balsevich, Gassen, & Schmidt, 2019). The type of stress is also relevant. In situations of acute stress caused, for example, by low temperatures, there is an activation of sympathetically activated b-adrenergic receptors, resulting in a negative energy balance, with the consequent weight loss (Hausl et al., 2019; Razzoli & Bartolomucci, 2016; Robertson et al., 2017; Wei et al., 2019). However, the stress derived from daily situations such as sedentary behaviors, improper diet, nervousness, or depression may cause a positive energy balance, with a higher caloric consumption (Jung, Bae, Kratzsch, Riedel-Heller, & Luck-Sikorski, 2019; Mason, Schleicher, Coccia, Epel, & Aschbacher, 2018).

In the latter case, the mechanisms involved are quite complex. On the one hand, chronic stress can cause hyperactivation of the HPA axis, causing an uncontrolled increase in cortisol levels (Jung et al., 2019; Ouakinin, Barreira, & Gois, 2018; Wei et al., 2019), leptin resistance and neuropeptide Y (NPY) stimulation (Ouakinin et al., 2018; Papargyri et al., 2018; Razzoli & Bartolomucci, 2016). These changes in internal homeostasis are directly responsible for the alteration of the metabolism and distribution of body fat, increased fat deposition at the abdominal level (Jung et al., 2019; Mazgelyte et al., 2019; Wei et al., 2019) and triglycerides in the plasma (Lopez Lopez, Escobar Villanueva, Brianza Padilla, Bonilla Jaime, & Alarcon Aguilar, 2018), hyperglycemia, increased insulin secretion (Razzoli & Bartolomucci, 2016), disappearance of the feeling of satiety, decreased energy expenditure and altered caloric intake. Elevated levels of cortisol and insulin may also have as a consequence an increased need to consume carbohydrates and fats (Choudhary et al., 2017; Mason et al., 2018; Ouakinin et al., 2018; Wei et al., 2019), with a higher risk of developing visceral obesity (Razzoli & Bartolomucci, 2016). Ghrelin, one of the hormones responsible for regulating appetite by estimating the NPY, among others, increases in situations of fasting, triggering appetite. In situations of chronic stress, ghrelin increases, affecting the normal functioning of different systems, such as the hypothalamic and mesolimbic, causing the increase in glucose, insulin levels and the feeling of hunger (Abizaid, 2019; Mani & Zigman, 2017; Tomiyama, 2019; Yousufzai, Harmatz, Shah, Malik, & Goosens, 2018).

Obesity together with the presence of the APOE4 allele increases the risk of Alzheimer's disease (Jones & Rebeck, 2019; Moser & Pike, 2017) being a modifiable risk factor (Alford et al., 2018; Bluher, 2019;

Christensen & Pike, 2017; Lizarbe, Soares, Larsson, & Duarte, 2018). The production of pro-inflammatory cytokines (IL-6, IL-1β, TNF-α, MCP-1) by macrophages of adipose tissue as well as the decrease in adiponectin production and increase in leptin levels, develop an environment of oxidative stress as well as the activation of NF-kB. This environment of inflammation and oxidative stress (ROS/RNS), in which DNA and mitochondrial damage is observed, together with the situation of hyperlipidemia, triggers insulin resistance causing hyperglycemia and deregulation in insulin production (Alford et al., 2018; Lizarbe et al., 2018; Lloret et al., 2019; Maslov et al., 2019; Verdile, Keane, et al., 2015). Insulin resistance and high levels of inflammation, ROS, glucose, lipids, as well as the activation of glycogen synthase kinase 3 (GSK3), at the brain level, promote the accumulation of beta amyloid (Aβ) and hyperphosphorylation of Tau at the brain level. Furthermore, BDNF levels are also decreased in the hypothalamus in obese mice (Agrimi et al., 2019; Wosiski-Kuhn, Erion, Gomez-Sanchez, Gomez-Sanchez, & Stranahan, 2014).

6. Exercise for stress, cardiovascular health, cognition and AD

Despite the evidence supporting the benefits of exercise on cognition and AD, the mechanisms through which physical activity and exercise exert their neuroprotective effects in humans remain to be elucidated, as most evidence in this regard comes from animal studies (Paillard et al., 2015). One of the most recent theories suggests that these benefits could be mediated by the effects of exercise on the stress neuroendocrine system (Tortosa-Martinez & Clow, 2012).

In animal models, exercise has been shown to improve HPA axis regulation with an adequate glucocorticoid response to stress (Droste, Chandramohan, Hill, Linthorst, & Reul, 2007; Droste et al., 2003; Droste, Schweizer, Ulbricht, & Reul, 2006; Hill, Droste, Nutt, Linthorst, & Reul, 2010; Kannangara et al., 2011), contributing to a reversal of the effects of uncontrolled stress (Greenwood, Strong, Dorey, & Fleshner, 2007). In human studies, higher levels of physical activity and fitness seem to improve the ability for coping with psychological stress, lowering cortisol, heart rate and reactivity of the autonomic nervous system (Lucertini et al., 2015; Rimmele et al., 2007, 2009; O'Keefe et al., 2019), although some inconsistencies in the literature should be noted (Moyna et al., 1999) probably due to the complexity of different stressors, measurement instruments, individual responses to psychological stress, and the relative lack of evidence looking at the diurnal pattern of cortisol secretion and not just a one-point measurement. In this regard, cross-sectional studies in healthy older adults have shown a positive relationship between a more dynamic cortisol secretion pattern and physical performance but not physical activity levels (for a review, see Tortosa-Martínez et al., 2018). Research linking the effects of exercise on cortisol and cognition is scarce. At a cross-sectional level, Dijckmans et al. (2017) showed that a more dynamic cortisol secretion pattern was correlated positively with better fitness and cognition in a sample of healthy older adults and people with mild cognitive impairment. Drogos et al. (2019) concluded that 6 months of aerobic training increased the magnitude of the CAR in a population of healthy older adults, although these changes were not correlated with cognitive improvements. In patients with MCI, Tortosa-Martinez et al. (2015) found that a 3-month aerobic exercise program resulted in better executive function, a tendency for an increased peak of cortisol at 30 min after awakening and an increase in the magnitude of the CAR, as well as a significant increase in the drop between the peak of cortisol at 30 min and the cortisol values at 12:00 p.m. Although these results are promising, there is a clear need for more research in this area.

It is also possible that exercise may benefit cognition and AD through reducing the autonomic dysfunction that might be present in both chronic stress and CVD. Low-intensity aerobic training has been shown to improve vagal function and reduce SNS activation, measured by HRV, in healthy adults (Myslivecek, Brown, & Wolfe, 2002), patients with peripheral artery disease (Brenner et al., 2020), patients following coronary artery bypass graft surgery (Brown, Wolfe, Hains, Pym, & Parker, 1994), and hypertensive men (Hua, Brown, Hains, Godwin, & Parlow, 2009) and women (Goldie, Brown, Hains, Parlow, & Birtwhistle, 2013). Furthermore, autonomic dysfunction and elevated levels of cortisol increase the production of epinephrine and norepinephrine, both associated with cognitive impairment and AD. Although acute exercise increases catecholamines production, Qi et al. (2019) found that 8 weeks of aerobic training reduced epinephrine and norepinephrine plasma levels in rats. However, to the best of our knowledge, there are no data available on AD patients.

Considering the multi-systemic etiology of AD, it is likely that exercise can impact cognition and AD through different pathways at the same time. Notwithstanding, exercise has been shown to counteract most of the consequences of chronic stress, considered AD risk factors, as for example cardiovascular disease and risk factors (Mora et al., 2007; Mueller, 2007).

Acute exercise increases cardiac output, leading to increased cerebral blood flow, which triggers various neurobiological mechanisms in the brain tissues (Paillard, Rolland, & de Barreto, 2015). Repeated increases in brain vascularization, and hence blood flow, may induce angiogenesis, neurogenesis, synaptogenesis, and neurotransmitter synthesis in the different cerebral areas involved in cognition (Paillard et al., 2015; Radak et al., 2010) such as the dentate gyrus of the hippocampus (Pereira et al., 2017), minimizing or delaying the cognitive decline associated with aging (Isaacs, Anderson, Alcantara, Black, & Greenough, 1992). Gegentonglaga et al. (2013) also proposed that exercise-induced hippocampal hyper/hypo oxygen condition may participate in the beneficial effects of exercise on brain function. In parallel, Lange-Asschenfeldt and Kojda (2008) proposed that the molecular mechanisms of the beneficial effects of exercise on CVD and AD, could be mediated by the activation of the vascular nitric oxide (NO), building a vascular reserve, as well as maintaining neuronal plasticity. In this line, Hasegawa et al. (2018) found that high intensity aerobic training may reduce central arterial stiffness via the increase in aortic nitric oxide bioavailability.

Exercise may also benefit cognition through reducing amyloid and TAU (Adlard, Perreau, Pop, & Cotman, 2005), inflammation (Cotman et al., 2007), oxidative stress (Radak et al., 2010); while increasing levels of BDNF (Kramer & Erickson, 2007) and hippocampal volume (Erickson et al., 2011; Erickson, Gildengers, & Butters, 2013), all factors being related to stress and CVD.

The most direct link of exercise with AD is the reduction in amyloid and Tau deposition seen in animal studies. In mice with AD, exercise has been shown to reduce β-amyloid levels in the prefrontal cortex, the cortex near the hippocampus and the hippocampus (Adlard et al., 2005; Koo, Kang, Oh, Yang, & Cho, 2017; Um et al., 2008), ameliorating Aβ-induced spatial learning and memory impairment (Azimi, Gharakhanlou, Naghdi, Khodadadi, & Heysieat-talab, 2018; Khodadadi et al., 2018). Aerobic exercise has also been shown to decrease tau protein concentrations in the hippocampus (Belarbi et al., 2011; Belviranli & Okudan, 2019). In addition, exercise may decrease the neuroinflammatory responses consequent of the increase in tau proteins (Leem, Lee, Son, & Lee, 2011). Notwithstanding, exercise has consistently been shown to reduce inflammatory markers in patients with chronic heart failure (Adamopoulos et al., 2002), coronary heart disease (Goldhammer et al., 2005), sedentary lean and obese men with or without diabetes type 2 (Dekker et al., 2007), and healthy adults (Mattusch, Dufaux, Heine, Mertens, & Rost, 2000).

Ryan and Kelly (2016), in a review article summarizing evidence from animal studies, concluded that the benefits of exercise on cognition in AD were partially mediated by the alteration of the expression of inflammatory molecules involved in cognitive restoration and hippocampal neurogenesis.

In parallel, exercise has been proved to increase the production of antioxidant enzymes and reduce the production of ROS by increasing the activity of the proteasome, uncoupling protein 2 (UCP2) and PGC-1α among others, resulting in a decrease in amyloid plaques, especially in the hippocampus (Azimi et al., 2018; Lange-Asschenfeldt & Kojda, 2008; Radak et al., 2010).

Aerobic training seems to increase brain plasticity and regional gray matter, especially in the hippocampus and the prefrontal areas, in healthy older adults (Erickson et al., 2011, 2013; Voelcker-Rehage & Niemann, 2013). In a study conducted with healthy older adults, those physically active presented higher volumes of lateral prefrontal cortex and hippocampus irrespective of their levels of pro-inflammatory markers (Papenberg et al., 2016). In a mouse model of AD, 16 weeks of aerobic exercise also resulted in higher hippocampal volume (Yuede et al., 2009). These structural changes are associated with vascular factors (Voelcker-Rehage & Niemann, 2013) as well as to compelling evidence showing that physical exercise can increase BDNF levels regardless of age and gender (de Assis & de Almondes, 2017). A single session of aerobic exercise has been shown to significantly increase BDNF plasma levels in older adults with depression (Laske et al., 2010) and AD patients (Coelho et al., 2014). Long-term exercise programs, based on either strength or aerobic training, have been proved to increase BDNF levels and improve cognition in healthy older adults (Coelho et al., 2011; Erickson et al., 2011; Ruscheweyh et al., 2011) and MCI patients (Baker et al., 2010). Jahangiri, Gholamnezhad, and Hosseini (2019) reviewed the benefits of exercise on AD mice models, with studies showing increases in BDNF levels, and reductions in inflammatory markers and oxidative stress.

6.1 Exercise and hypertension

Several studies have shown that medication treatment to lower blood pressure in midlife, could help to retain or improve cognitive function by reducing the risk of AD (Santos et al., 2017). However, current guidelines advise being cautious when starting antihypertensive pharmacological treatment in people with AD (Qaseem et al., 2017), due to the limited evidence available about the safety of these treatments in this population (van der Wardt, Logan, Conroy, Harwood, & Gladman, 2014). Exercise may be a non-pharmachological strategy with no side effects, as different types of exercise such as aerobic, dynamic and isometric strength training, have been proved effective for lowering blood pressure in healthy adults and older adults (Cornelissen & Fagard, 2005). In a study using an AD animal model, long term exercise reduced blood pressure, promoted endothelial nitric oxide synthase expression, ameliorated Alzheimer's pathology, and improved cognitive function (Zhang et al., 2018).

6.2 Exercise and diabetes

Different studies have demonstrated the benefits of physical exercise for reducing the risk of developing type 2 diabetes (Arena et al., 2017; Kolb & Martin, 2017) by improving the risk factors of the disease such as the decrease in body fat mass, the increase in muscle mass and HDL cholesterol, as well as the decrease in blood pressure and LDL cholesterol among others (Pan, Wu, Liu, Li, & Chen, 2019; Zanuso, Jimenez, Pugliese, Corigliano, & Balducci, 2010). Regular physical activity may also improve insulin sensitivity and resistance, glycosylated hemoglobin, regulate blood glucose levels and improve physical fitness in people already affected (Colberg et al., 2016; Karstoft & Pedersen, 2016; Way, Hackett, Baker, & Johnson, 2016).

An exercise intervention found a significant effect on cognitive function in sedentary older adults with, but not without diabetes, indicating that the mechanism of the cognitive benefits of exercise could be different for those with diabetes mellitus (Espeland et al., 2017), and supports the importance of physical exercise in people with type 2 diabetes as a preventive method to reduce the biomarkers of AD (Bertram, Brixius, & Brinkmann, 2016). In line with this, Shih, Paul, Haan, Yu, and Ritz (2018) demonstrated the protective role of regular physical exercise against the development of dementia in diabetic patients carrying the ApoE-4 allele. In AD patients with diabetes type 2, insulinic therapy has been shown to be effective in slowing cognitive decline (Plastino et al., 2010) but, to the best of our knowledge, there are no studies replicating these results with exercise.

6.3 Exercise and obesity

There is compelling evidence about the benefits of physical exercise and the maintenance of an active lifestyle for preventing and treating people with obesity for weight control, improvement of body composition and intestinal microbiota (Budui et al., 2019; Cerdá et al., 2016; Rocha-Rodrigues, Goncalves, Beleza, Ascensao, & Magalhaes, 2018). Exercise also seems to reduce the inflammatory situation and hormonal imbalance caused by obesity (low levels of adiponectin and high levels of leptin, IL-1, IL-6, TNF-α) (Batatinha, Rosa, & Krüger, 2019; Figueiredo, Nunes, Marmett, de Sá, & Arbex, 2017; Goldfield et al., 2018; Rocha-Rodrigues et al., 2018; Sirico et al., 2018). It seems that the best strategy for obtaining improvements is one that combines multi-component programs: exercise, nutrition, psychology, etc. (Trouwborst et al., 2018). However, the best type, frequency and intensity of physical exercise to obtain greater benefits remains to be elucidated.

Do et al. (2018), in a study conducted with mice, observed improvements after a physical training program in regards to gene expressions of neuropeptides, inflammation and apoptosis related to appetite. More studies, conducted with animals with high-fat diets, support these findings, demonstrating that physical exercise can reduce insulin resistance, improve mitochondrial function, suppress apoptosis, oxidative stress and increase BDNF levels (Graham et al., 2019; Park, Park, Kim, Shin, & Kim, 2019; Shi et al., 2018).

7. Conclusions

Chronic stress alters the functioning of the hypothalamic pituitary adrenal (HPA) system, over activating the sympathetic nervous system (SNS), while inhibiting the parasympathetic nervous system (PNS). As a consequence, there is an abnormal cortisol secretion pattern, elevated levels of epinephrine and norepinephrine, and autonomic dysfunction with reduced HRV. This cascade of events for prolonged periods of time increases the risk of CVD and risk factors, leading to higher blood pressure, reduced blood flow and hypoperfusion in the brain. Both stress and CVD lead to increased levels of inflammation markers, oxidative stress, and insulin resistance, decreased NO synthesis, BDNF levels and hippocampus volume, and elevated accumulations of amyloid and Tau in the brain, all key factors for cognitive impairment and AD. Thus, chronic stress, CVD and risk factors seem to play a key role in the complex etiology of AD, although the interrelations between these variables deserve further research as some inconsistencies in the literature should be noted and the exact mechanisms still need to be elucidated. In parallel, exercise seems to reduce the risk of developing MCI and AD, and improve memory in patients with MCI and AD. The exact mechanisms and pathways of these benefits are not fully clear but accumulating evidence links them to the role of exercise for reducing the risk and counteracting the consequences of stress, CVD and risk factors such as hypertension, diabetes type 2 and obesity. Exercise seems to lower overall cortisol levels and increase the dynamics of the diurnal cortisol secretion pattern, decrease excessive levels of norepinephrine and epinephrine, and improve autonomic dysfunction with an increase in HRV. In parallel, exercise increases blood flow, brain vascularization, NO availability and BDNF levels, while reduces insulin resistance, oxidative stress and inflammatory markers. As a consequence, there is an increase in angiogenesis, neurogenesis, synaptogenesis, and neurotransmitter synthesis, increasing brain volume and improving cognition. Furthermore, amyloid deposition and TAU deposition seem to be partially decreased. While most available evidence is based on animal studies and further research is needed, there is enough evidence to promote the use of exercise in mid- and late-life as a key intervention for reducing the risk of and consequences of stress, CVD, and AD.

References

Abizaid, A. (2019). Stress and obesity: The ghrelin connection. *Journal of Neuroendocrinology, 31*(7), 1–8.

Abner, E. L., Nelson, P. T., Kryscio, R. J., Schmitt, F. A., Fardo, D. W., Woltjer, R. L., et al. (2016). Diabetes is associated with cerebrovascular but not Alzheimer's disease neuropathology. *Alzheimer's & Dementia, 12*(8), 882–889.

Adamopoulos, S., Parissis, J., Karatzas, D., Kroupis, C., Georgiadis, M., Karavolias, G., et al. (2002). Physical training modulates proinflammatory cytokines and the soluble Fas/soluble Fas ligand system in patients with chronic heart failure. *Journal of the American College of Cardiology, 39*, 653–663.

Adlard, P. A., Perreau, V. M., Pop, V., & Cotman, C. W. (2005). Voluntary exercise decreases amyloid load in a transgenic model of Alzheimer's disease. *The Journal of Neuroscience, 25*, 4217–4221.

Agrimi, J., Spalletti, C., Baroni, C., Keceli, G., Zhu, G., Caragnano, A., et al. (2019). Obese mice exposed to psychosocial stress display cardiac and hippocampal dysfunction associated with local brain-derived neurotrophic factor depletion. *eBioMedicine, 47*, 384–401.

Aisa, B., Elizalde, N., Tordera, R., Lasheras, B., Del Río, J., & Ramírez, M. J. (2009). Effects of neonatal stress on markers of synaptic plasticity in the hippocampus: Implications for spatial memory. *Hippocampus, 19*(12), 1222–1231.

Alford, S., Patel, D., Perakakis, N., & Mantzoros, C. S. (2018). Obesity as a risk factor for Alzheimer's disease: Weighing the evidence. *Obesity Reviews, 19*(2), 269–280.

Allan, L. M., Ballard, C. G., Allen, J., Murray, A., Davidson, A. W., McKeith, I. G., et al. (2007). Autonomic dysfunction in dementia. *Journal of Neurology, Neurosurgery, and Psychiatry, 78*, 671–677.

Almela, M., van der Meij, L., Hidalgo, V., Villada, C., & Salvador, A. (2012). The cortisol awakening response and memory performance in older men and women. *Psychoneuroendocrinology, 37*, 1929–1940. https://doi.org/10.1016/j.psyneuen.2012.04.009.

Arena, R., Sagner, M., Byrne, N. M., Williams, A. D., McNeil, A., Street, S. J., et al. (2017). Novel approaches for the promotion of physical activity and exercise for prevention and management of type 2 diabetes. *European Journal of Clinical Nutrition*, *71*(7), 858–864.

Arnold, S. E., Arvanitakis, Z., Macauley-Rambach, S. L., Koenig, A. M., Wang, H. Y., Ahima, R. S., et al. (2018). Brain insulin resistance in type 2 diabetes and Alzheimer disease: Concepts and conundrums. *Nature Reviews Neurology*, *14*(3), 168.

Arsenault-Lapierre, G., Chertkow, H., & Lupien, S. (2010). Seasonal effects on cortisol secretion in normal aging, mild cognitive impairment and Alzheimer's disease. *Neurobiology of Aging*, *31*, 1051–1054.

Austin, S. A., Santhanam, A. V., Hinton, D. J., Choi, D. S., & Katusic, Z. S. (2013). Endothelial nitric oxide deficiency promotes Alzheimer's disease pathology. *Journal of Neurochemistry*, *127*, 691–700.

Azimi, M., Gharakhanlou, R., Naghdi, N., Khodadadi, D., & Heysieat-talab, S. (2018). Moderate treadmill exercise ameliorates amyloid-β-induced learning and memory impairment, possibly via increasing AMPK activity and up-regulation of the PGC-1α/FNDC5/BDNF pathway. *Peptides*, *102*, 78–88.

Baid, S., & Nieman, L. K. (2004). Glucocorticoid excess and hypertension. *Current Hypertension Reports*, *6*, 493–499.

Baker, L. D., Frank, L. L., Foster-Schubert, K., Green, P. S., Wilkinson, C. W., McTiernan, A., et al. (2010). Effects of aerobic exercise on mild cognitive impairment: A controlled trial. *Archives of Neurology*, *67*, 71–79.

Barnes, D. E., & Yaffe, K. (2011). The projected effect of risk factor reduction on Alzheimer's disease prevalence. *Lancet Neurology*, *10*(9), 819–828.

Batatinha, H., Rosa, N., & Krüger, K. (2019). Inflammatory features of obesity and smoke exposure and the immunologic effects of exercise. *Exercise Immunology Review*, *25*, 96–111.

Beeri, M. S., Ravona-Springer, R., Silverman, J. M., & Haroutunian, V. (2009). The effects of cardio-vascular risk factors on cognitive compromise. *Dialogues in Clinical Neuroscience*, *11*, 201–212.

Belarbi, K., Burnouf, S., Fernandez-Gomez, F. J., Laurent, C., Lestavel, S., Figeac, M., et al. (2011). Beneficial effects of exercise in a transgenic mouse model of Alzheimer's disease-like Tau pathology. *Neurobiology of Disease*, *43*, 486–494.

Beluche, I., Carriere, I., Ritchie, K., & Ancelin, M. L. (2010). A prospective study of diurnal cortisol and cognitive function in community-dwelling elderly people. *Psychological Medicine*, *40*, 1039–1049.

Belviranli, M., & Okudan, N. (2019). Voluntary, involuntary and forced exercises almost equally reverse behavioral impairment by regulating hippocampal neurotrophic factors and oxidative stress in experimental Alzheimer's disease model. *Behavioural Brain Research*, *368*, 245–255.

Bendlin, B. B. (2019). Antidiabetic therapies and Alzheimer disease. *Dialogues in Clinical Neuroscience*, *21*(1), 83–91.

Bertram, S., Brixius, K., & Brinkmann, C. (2016). Exercise for the diabetic brain: How physical training may help prevent dementia and Alzheimer's disease in T2DM patients. *Endocrine*, *53*(2), 350–363.

Bluher, M. (2019). Obesity: Global epidemiology and pathogenesis. *Nature Reviews Endocrinology*, *15*(5), 288–298.

Bosma, H., Peter, R., Siegrist, J., & Marmot, M. (1998). Two alternative job stress models and the risk of coronary heart disease. *American Journal of Public Health*, *88*, 68–74.

Braak, H., & Braak, E. (1995). Staging of Alzheimer's disease-related neurofibrillary changes. *Neurobiology of Aging*, *16*, 271–278.

Brenner, I. K. M., Brown, C. A., Hains, S. J. M., Tranmer, J., Zelt, D. T., & Brown, P. M. (2020). Low-intensity exercise training increases heart rate variability in patients with peripheral artery disease. *Biological Research for Nursing*, *22*(1), 24–33.

Brown, C. A., Wolfe, L. A., Hains, S., Pym, J., & Parker, J. O. (1994). Early low intensity home exercise after coronary artery bypass graft surgery. *Journal of Cardiopulmonary Rehabilitation, 14*, 238–245.

Budui, S., Bigolin, F., Giordano, F., Leoni, S., Berteotti, M., Sartori, E., et al. (2019). Effects of an intensive inpatient rehabilitation program in elderly patients with obesity. *Obesity Facts, 12*(2), 199–210.

Cameron, H. A., & Gould, E. (1994). Adult neurogenesis is regulated by adrenal steroids in the dentate gyrus. *Neuroscience, 61*, 203–209.

Cerdá, B., Pérez, M., Pérez-Santiago, J. D., Tornero-Aguilera, J. F., González-Soltero, R., & Larrosa, M. (2016). Gut microbiota modification: Another piece in the puzzle of the benefits of physical exercise in health? *Frontiers in Physiology, 7*, 51.

Cervantes Gracia, K., Llanas-Cornejo, D., & Husi, H. (2017). CVD and oxidative stress. *Journal of Clinical Medicine, 6*, 22.

Champaneri, S., Xu, X., Carnethon, M. R., Bertoni, A. G., Seeman, T., Roux, A. D., et al. (2012). Diurnal salivary cortisol and urinary catecholamines are associated with diabetes mellitus: The Multi-Ethnic Study of Atherosclerosis. *Metabolism, 61*, 986–995.

Chapman, K., Holmes, M., & Seckl, J. (2013). 11b-hydroxysteroid dehydrogenases: Intracellular gate-keepers of tissue glucocorticoid action. *Physiological Reviews, 93*(3), 1139–1206.

Chicharro, J. L., & Mojares, L. M. L. (2008). *Fisiología clínica del ejercicio*. Madrid (España): Ed. Médica Panamericana.

Chiodini, I., Gaudio, A., Eller-Vainicher, C., Morelli, V., Aresta, C., Zhukouskaya, V. V., et al. (2019). Cortisol secretion, sensitivity, and activity are associated with hypertension in postmenopausal eucortisolemic women. *The Journal of Clinical Endocrinology and Metabolism, 104*(10), 4441–4448. https://doi.org/10.1210/jc.2019-00037.

Chornenkyy, Y., Wang, W. X., Wei, A., & Nelson, P. T. (2019). Alzheimer's disease and type 2 diabetes mellitus are distinct diseases with potential overlapping metabolic dysfunction upstream of observed cognitive decline. *Brain Pathology, 29*(1), 3–17.

Chow, H. M., Shi, M., Cheng, A., Gao, Y., Chen, G., Song, X., et al. (2019). Age-related hyperinsulinemia leads to insulin resistance in neurons and cell-cycle-induced senescence. *Nature Neuroscience, 22*(11), 1806–1819.

Christensen, A., & Pike, C. J. (2017). Age-dependent regulation of obesity and Alzheimer-related outcomes by hormone therapy in female 3xTg-AD mice. *PLoS One, 12*(6) e0178490.

Coelho, F. M., Pereira, D. S., Lustosa, L. P., Silva, J. P., Dias, J. M. D., Dias, R. C. D., et al. (2011). Physical therapy intervention (PTI) increases plasma brain-derived neurotrophic factor (BDNF) levels in non-frail and pre-frail elderly women. *Archives of Gerontology and Geriatrics, 54*(3), 415–420.

Coelho, F. G., Vital, T. M., Stein, A. M., Arantes, F. J., Rueda, A. V., Camarini, R., et al. (2014). Acute aerobic exercise increases brain-derived neurotrophic factor levels in elderly with Alzheimer's disease. *Journal of Alzheimer's Disease, 39*, 401–408.

Colberg, S. R., Sigal, R. J., Yardley, J. E., Riddell, M. C., Dunstan, D. W., Dempsey, P. C., et al. (2016). Physical activity/exercise and diabetes: A position statement of the American Diabetes Association. *Diabetes Care, 39*(11), 2065–2079.

Colcombe, S., & Kramer, A. (2003). Fitness effects on the cognitive function of older adults: A meta-analytic study. *Psychological Science, 14*(2), 125–130.

Cornelissen, V. A., & Fagard, R. H. (2005). Effect of resistance training on resting blood pressure: A meta-analysis of randomized controlled trials. *Journal of Hypertension, 23*, 251–259.

Cotman, C. W., Berchtold, N. C., & Christie, L. A. (2007). Exercise builds brain health: Key roles of growth factor cascades and inflammation. *Trends in Neurosciences, 30*(9), 464–472.

Crawford, A. A., Söderberg, S., Kirschbaum, C., Murphy, L., Eliasson, M., Ebrahim, S., et al. (2019). Morning plasma cortisol as a cardiovascular risk factor: Findings from prospective cohort and Mendelian randomization studies. *European Journal of Endocrinology, 181*(4), 429–438. https://doi.org/10.1530/EJE-19-0161.

Crous-Bou, M., Minguillón, C., Gramunt, N., & Molinero, J. L. (2017). Alzheimer's disease prevention: From risk factors to early intervention. *Alzheimer's Research & Therapy, 9* 71 https://doi.org/10.1186/s13195-017-0297-z.

Csernansky, J. G., & Dong, H. (2006). Plasma cortisol and progression of dementia in subjects with Alzheimer-type dementia. *The American Journal of Psychiatry, 163*, 2164–2169.

Curtis, B. M., & O'Keefe, J. H., Jr. (2002). Autonomic tone as a cardiovascular risk factor: The dangers of chronic fight or flight. *Mayo Clinic Proceedings*, 77(1), 45–54.

Curto, M., Martocchia, A., Ferracuti, S., Comite, F., Scaccianoce, S., Girardi, P., et al. (2017). Increased total urinary cortisol (tUC) and serum brain-derived neurotrophic factor (BDNF) ratio in Alzheimer disease (AD)-affected patients. *Alzheimer Disease and Associated Disorders, 31*, 173–176.

Da Silva, V. P., Oliveira, B. R. R., Mello, R. G. T., Moraes, H., Deslandes, A. C., & Laks, J. (2018). Heart rate variability indexes in dementia: A systematic review with a quantitative analysis. *Current Alzheimer Research*, 1*5*(1), 80–88.

Daviglus, M. L., et al. (2010). National Institutes of Health State-of-the-Science Conference statement: Preventing Alzheimer disease and cognitive decline. *Annals of Internal Medicine, 153*(3), 176–181.

de Assis, G. G., & de Almondes, K. M. (2017). Exercise-dependent BDNF as a modulatory factor for the executive processing of individuals in course of cognitive decline. A systematic review. *Frontiers in Psychology, 8*, 1–8. https://doi.org/10.3389/fpsyg.2017.00584.

de Assis, G. G., & Gasamov, E. V. (2019). BDNF and cortisol integrative system—Plasticity vs. degeneration: Implications of the Val66Met polymorphism. *Frontiers in Neuroendocrinology, 55*, 1–10. https://doi.org/10.1016/j.yfrne.2019.100784.

de Oliveira, J. F., Wiener, C. D., Jansen, K., Portela, L. V., Lara, D. R., Souza, L. D. M., et al. (2018). Serum levels of interleukins IL-6 and IL-10 in individuals with posttraumatic stress disorder in a population-based sample. *Psychiatry Research, 260*, 111–115.

De Santis, A. S., Diez Roux, A. V., Hajat, A., Aiello, A. E., Golden, S. H., Jenny, N. S., et al. (2012). Associations of salivary cortisol levels with inflammatory markers: The Multi Ethnic Study of Atherosclerosis. *Psychoneuroendocrinology, 37*, 1009–1018.

de Vilhena Toledo, M. A., & Junqueira, L. F., Jr. (2008). Cardiac sympathovagal modulation evaluated by short-term heart interval variability is subtly impaired in Alzheimer's disease. *Geriatrics & Gerontology International, 8*, 109–118.

Dekker, M. J., Lee, S., Hudson, R., Kilpatrick, K., Graham, T. E., Ross, R., et al. (2007). An exercise intervention without weight loss decreases circulating interleukin-6 in lean and obese men with and without type 2 diabetes mellitus. *Metabolism, 56*, 332–338.

Dijckmans, B., Tortosa-Martínez, J., Caus, N., González-Caballero, G., MartínezPelegrin, B., Manchado-Lopez, C., et al. (2017). Does the diurnal cycle of cortisol explain the relationship between physical performance and cognitive function in older adults? *European Review of Aging and Physical Activity, 14*(1), 6, https://doi.org/10.1186/s11556-017-0175-5.

Do, K., Laing, B. T., Landry, T., Bunner, W., Mersaud, N., Matsubara, T., et al. (2018). The effects of exercise on hypothalamic neurodegeneration of Alzheimer's disease mouse model. *PLoS One, 13*(1) e0190205.

Dong, H., Yuede, C. M., Yoo, H. S., Martin, M. V., Deal, C., Mace, A. G., et al. (2008). Corticosterone and related receptor expression are associated with increased b-amyloid plaques in isolated Tg2576 mice. *Neuroscience, 155*, 154–163.

Dos Santos Matioli, M. N. P., Suemoto, C. K., Rodriguez, R. D., Farias, D. S., da Silva, M. M., Leite, R. E. P., et al. (2017). Diabetes is not associated with Alzheimer's disease neuropathology. *Journal of Alzheimer's Disease*, *60*(3), 1035–1043.

Droste, S. K., Chandramohan, Y., Hill, L. E., Linthorst, A. C., & Reul, J. M. H. M. (2007). Voluntary exercise impacts on the rat hypothalamic pituitary-adrenocortical axis mainly at the adrenal level. *Neuroendocrinology*, *86*, 26–37.

Droste, S. K., Gesing, A., Ulbricht, S., Müller, M. B., Linthorst, A. C., & Reul, J. M. (2003). Effects of long-term voluntary exercise on the mouse hypothalamic-pituitary-adrenocortical axis. *Endocrinology*, *144*(7), 3012–3023.

Droste, S. K., Schweizer, M. C., Ulbricht, S., & Reul, J. M. H. M. (2006). Longterm voluntary exercise and the mouse hypothalamic-pituitary adrenocortical axis: Impact of concurrent treatment with theantidepressant drug tianeptine. *Journal of Neuroendocrinology*, *18*, 915–925.

Drogos, L. L., Wynne-Edwards, K., Zhou, R., Hall, S. E., Tyndall, A. V., Longman, R. S., et al. (2019). Aerobic exercise increases cortisol awakening response in older adults. *Psychoneuroendocrinology*, *103*, 241–248.

Dubois, B., et al. (2016). Preclinical Alzheimer's disease: Definition, natural history, and diagnostic criteria. *Alzheimer's & Dementia*, *12*(3), 292–323.

Duman, R. S., & Monteggia, L. M. (2006). A neurotrophic model for stress related mood disorders. *Biological Psychiatry*, *59*, 1116–1127.

Dywan, J., Mathewson, K. J., Choma, B. L., Rosenfeld, B., & Segalowitz, S. J. (2008). Autonomic and electrophysiological correlates of emotional intensity in older and younger adults. *Psychophysiology*, *45*, 389–397.

Edwards, S., Evans, P., Hucklebridge, F., & Clow, A. (2001). Association between time of awakening and diurnal cortisol secretory activity. *Psychoneuroendocrinology*, *26*, 613–622.

Erickson, K. I., Gildengers, A. G., & Butters, M. A. (2013). Physical activity and brain plasticity in late adulthood. *Dialogues in Clinical Neuroscience*, *15*, 99–108.

Erickson, K. I., Voss, M. W., Prakash, R. S., Basak, C., Szabo, A., Chaddock, L., et al. (2011). Exercise training increases size of hippocampus and improves memory. *Proceedings of the National Academy of Sciences of the United States of America*, *108*, 3017–3022.

Esch, T., Stefano, G. B., Fricchinoe, G. L., & Benson, H. (2002). Stress in cardiovascular disease. *Medical Science Monitor*, *8*, 93–101.

Espeland, M. A., Lipska, K., Miller, M. E., Rushing, J., Cohen, R. A., & Verghese, J. (2017). Effects of physical activity intervention on physical and cognitive function in sedentary adults with and without diabetes. *The Journals of Gerontology Series A, Biological Sciences and Medical Sciences*, 72(6), 861–866.

Evans, P. D., Fredhoi, C., Loveday, C., Hucklebridge, F., Aitchison, E., Forte, D., et al. (2011). The diurnal cortisol cycle and cognitive performance in the healthy old. *International Journal of Psychophysiology*, *79*, 371–377. https://doi.org/10.1016/j.ijpsycho.2010.12.006.

Evans, P., Hucklebridge, F., Loveday, C., & Clow, A. (2012). The cortisol awakening response is related to executive function in older age. *International Journal of Psychophysiology*, *84*, 201–204. https://doi.org/10.1016/j.ijpsycho.2012.02.008.

Femminella, G. D., Rengo, G., Komici, K., Iacotucci, P., Petraglia, L., Pagano, G., et al. (2014). Autonomic dysfunction in Alzheimer's disease: Tools for assessment and review of the literature. *Journal of Alzheimer's Disease*, *42*, 369–377.

Figueiredo, L., Nunes, R. B., Marmett, B., de Sá, L. B. P. C., & Arbex, A. K. (2017). Anti-inflammatory effects of physical exercise on obesity. *Open Journal of Endocrine and Metabolic Diseases*, *07*(01), 44–51.

Finnell, J. E., & Wood, S. K. (2016). Neuroinflammation at the interface of depression and cardiovascular disease: Evidence from rodent models of social stress. *Neurobiology of Stress, 4,* 1–14.
Forbes, D., Forbes, S. C., Blake, C. M., Thiessen, E. J., & Forbes, S. (2015). Exercise programs for people with dementia. *The Cochrane Database of Systematic Reviews. 4,* 1–75. https://doi.org/10.1002/14651858.CD006489.pub4.
Gabin, J. M., Tambs, K., Saltvedt, I., et al. (2017). Association between blood pressure and Alzheimer disease measured up to 27 years prior to diagnosis: The HUNT study. *Alzheimer's Research & Therapy, 9,* 37.
Geerlings, M. I., Garcia, M. E., & Harris, T. B. (2015). Salivary cortisol, brain volumes, and cognition in community-dwelling elderly without dementia. *Neurology, 85*(11), 976–983.
Gegentonglaga, Yoshizato, H., Higuchi, Y., Toyota, Y., Hanai, Y., Ando, Y., et al. (2013). Variable alteration of regional tissue oxygen pressure in rat hippocampus by acute swimming exercise. *Life Sciences, 93,* 773–777.
Geiger, A. M., Pitts, K. P., Feldkamp, J., Kirschbaum, C., & Wolf, J. M. (2015). Cortisol-dependent stress effects on cell distribution in healthy individuals and individuals suffering from chronic adrenal insufficiency. *Brain, Behavior, and Immunity, 50,* 241–248.
Golden, E., Emiliano, A., Maudsley, S., Windham, B. G., Carlson, O. D., Egan, J. M., et al. (2010). Circulating brain-derived neurotrophic factor and indices of metabolic and cardiovascular health: Data from the Baltimore Longitudinal Study of Aging. *PLoS ONE, 5*(4), 1–24.
Goldfield, G. S., Kenny, G. P., Prud'homme, D., Holcik, M., Alberga, A. S., Fahnestock, M., et al. (2018). Effects of aerobic training, resistance training, or both on brain-derived neurotrophic factor in adolescents with obesity: The hearty randomized controlled trial. *Physiology & Behavior, 191,* 138–145.
Goldhammer, E., Tanchilevitch, A., Maor, I., Beniamini, Y., Rosen-schein, U., & Sagiv, M. (2005). Exercise training modulates cytokines activity in coronary heart disease patients. *International Journal of Cardiology, 100,* 93–99.
Goldie, C. L., Brown, C. A., Hains, S. M., Parlow, J. L., & Birtwhistle, R. (2013). Synergistic effects of low-intensity exercise conditioning and beta-blockade on cardiovascular and autonomic adaptation in pre- and postmenopausal women with hypertension. *Biological Research for Nursing, 15,* 433–442.
Gottesman, R. F., Schneider, A. L., Albert, M., Alonso, A., Bandeen-Roche, K., Coker, L., et al. (2014). Midlife hypertension and 20-year cognitive change: The atherosclerosis risk in communities neurocognitive study. *JAMA Neurology, 71,* 1218–1227.
Gottesman, R. F., et al. (2017). Association between midlife vascular risk factors and estimated brain amyloid deposition. *JAMA, 317*(14), 1443–1450.
Graham, L. C., Grabowska, W. A., Chun, Y., Risacher, S. L., Philip, V. M., Saykin, A. J., et al. (2019). Exercise prevents obesity-induced cognitive decline and white matter damage in mice. *Neurobiology of Aging, 80,* 154–172.
Green, K. N., Billings, L. M., Roozendaal, B., McGaugh, J. L., & LaFerla, F. M. (2006). Glucocorticoids increase amyloid-b and tau pathology in a mouse model of Alzheimer's disease. *The Journal of Neuroscience, 26,* 9047–9056.
Green, P. G., Janig, W., & Levine, J. D. (1997). Negative feedback neuroendocrine control of inflammatory response in the rat is dependent on the sympathetic postganglionic neuron. *The Journal of Neuroscience, 17*(9), 3234–3238.
Greenwood, B. N., Strong, P. V., Dorey, A. A., & Fleshner, M. (2007). Therapeutic effects of exercise: Wheel running reverses stress induced interference with shuttle box escape. *Behavioural Neuroscience, 21*(5), 992–1000.

Groot, C., Hooghiemstra, A. M., Raijmakers, P. G. H. M., Van Berckel, B. N. M., Scheltens, P., Scherder, E. J. A., et al. (2016). The effect of physical activity on cognitive function in patients with dementia: A meta analysis of randomized control trials. *Ageing Research Reviews*, *25*, 13–23.

Guitar, N. A., Connelly, D. M., Nagamatsu, L. S., Orange, J. B., & Muir-Hunter, S. W. (2018). The effects of physical exercise on executive function in community-dwelling older adults living with Alzheimer's-type dementia: A systematic review. *Ageing Research Reviews*, *47*, 159–167.

Gunawardane, P. T. K., & Grossman, A. (2017). Phaeochromocytoma and paraganglioma. *Advances in Experimental Medicine and Biology*, *956*, 239–259.

Hackett, R. A., Poole, L., Hunt, E., Panagi, L., & Steptoe, A. (2019). Loneliness and biological responses to acute stress in people with Type 2 diabetes. *Psychophysiology*, *56*(6) e13341.

Hackett, R. A., Steptoe, A., & Kumari, M. (2014). Association of diurnal patterns in salivary cortisol with type 2 diabetes in the Whitehall II study. *The Journal of Clinical Endocrinology and Metabolism*, *99*(12), 4625–4631.

Hajat, A., Diez-Roux, A. V., Sánchez, B. N., Holvoet, P., Lima, J. A., Merkin, S. S., et al. (2013). Examining the association between salivary cortisol levels and subclinical measures of atherosclerosis: The Multi-Ethnic Study of Atherosclerosis. *Psychoneuroendocrinology*, *38*, 1036–1046.

Hamer, M., & Chida, Y. (2009). Physical activity and risk of neurodegenerative disease: A systematic review of prospective evidence. *Psychological Medicine*, *39*(1), 3–11.

Hansson, L. (1996). The benefits of lowering elevated blood pressure: A critical review of studies of cardiovascular morbidity and mortality in hypertension. *Journal of Hypertension*, *14*, 537–544.

Harris, M. L., Oldmeadow, C., Hure, A., Luu, J., Loxton, D., & Attia, J. (2017). Stress increases the risk of type 2 diabetes onset in women: A 12-year longitudinal study using causal modelling. *PLoS One*, *12*(2) e0172126.

Harrison, D. G., & Guzik, T. J. (2016). Macrophages come to mind as keys to cognitive decline. *The Journal of Clinical Investigation*, *126*, 4393–4395.

Hasegawa, N., Fujie, S., Horii, N., Miyamoto-Mikami, E., Tsuji, K., Uchida, M., et al. (2018). Effects of different exercise models on arterial stiffness and nitric oxide synthesis. *Medicine and Science in Sports and Exercise*, *50*(6), 1177–1185.

Hausl, A. S., Balsevich, G., Gassen, N. C., & Schmidt, M. V. (2019). Focus on FKBP51: A molecular link between stress and metabolic disorders. *Molecular Metabolism*, *29*, 170–181.

Heneka, M. T., Carson, M. J., Khoury, J. E., Landreth, G. E., Brosseron, F., Feinstein, D. L., et al. (2015). Neuroinflammation in Alzheimer's disease. *Lancet Neurology*, *14*, 388–405.

Hidalgo, V., Almela, M., Pulopulos, M. M., & Salvador, A. (2016). Memory performance is related to the cortisol awakening response in older people, but not to the diurnal cortisol slope. *Psychoneuroendocrinology*, *71*, 136–146.

Hill, L. E., Droste, S. K., Nutt, D. J., Linthorst, A. C. E., & Reul, J. M. H. M. (2010). Voluntary exercise alters GABAA receptor subunit and glutamic acid decarboxylase-67 gene expression in the rat forebrain. *Journal of Psychopharmacology*, *24*, 745–756.

Holt-Lunstad, J., & Steffen, P. R. (2007). Diurnal cortisol variation is associated with nocturnal blood pressure dipping. *Psychosomatic Medicine*, *69*, 339–343.

Hori, H., & Kim, Y. (2019). Inflammation and post-traumatic stress disorder. *Psychiatry and Clinical Neurosciences*, *73*(4), 143–153.

Hua, L. P., Brown, C. A., Hains, S. J., Godwin, M., & Parlow, J. L. (2009). Effects of low-intensity exercise conditioning on blood pressure, heart rate, and autonomic modulation of heart rate in men and women with hypertension. *Biological Research for Nursing*, *11*, 129–143. https://doi.org/10.1177/1099800408324853.

Huang, H. J., Chen, S. L., Huang, H. Y., Sun, Y. C., Lee, G. C., Lee-Chen, G. J., et al. (2019). Chronic low dose of AM404 ameliorates the cognitive impairment and pathological features in hyperglycemic 3xTg-AD mice. *Psychopharmacology*, *236*(2), 763–773.

Iadecola, C., Yaffe, K., Biller, J., Bratzke, L. C., Faraci, F. M., Gorelick, P. B., et al. (2016). Impact of hypertension on cognitive function: A scientific statement from the American Heart Association. *Hypertension*, *68*, e67–e94. https://doi.org/10.1161/HYP.0000000000000053.

Isaacs, K. R., Anderson, B. J., Alcantara, A. A., Black, J. E., & Greenough, W. T. (1992). Exercise and the brain: angiogenesis in the adult rat cerebellum after vigorous physical activity and motor skill learning. *Journal of Cerebral Blood Flow & Metabolism*, *12*(1), 110–119.

Jahangiri, Z., Gholamnezhad, Z., & Hosseini, M. (2019). Neuroprotective effects of exercise in rodent models ofmemory deficit and Alzheimer's. *Metabolic Brain Disease*, *34*, 21–37.

Jennings, J. R., Mendelson, D. N., Muldoon, M. F., Ryan, C. M., Gianaros, P. J., Raz, N., et al. (2012). Regional grey matter shrinks in hypertensive individuals despite successful lowering of blood pressure. *Journal of Human Hypertension*, *26*, 295–305.

Jones, N. S., & Rebeck, G. W. (2019). The synergistic effects of APOE genotype and obesity on Alzheimer's disease risk. *International Journal of Molecular Sciences*, *20*(1), pii: E63.

Joseph, J. J., & Golden, S. H. (2017). Cortisol dysregulation: The bidirectional link between stress, depression, and type 2 diabetes mellitus. *Annals of the New York Academy of Sciences*, *1391*(1), 20–34.

Joseph, P., Leong, D., McKee, M., Anand, S. S., Schwalm, J. D., Teo, K., et al. (2017). Reducing the global burden of cardiovascular disease, part 1: The epidemiology and risk factors. *Circulation Research*, *121*, 677–694.

Jung, F. U., Bae, Y. J., Kratzsch, J., Riedel-Heller, S. G., & Luck-Sikorski, C. (2019). Internalized weight bias and cortisol reactivity to social stress. *Cognitive, Affective, & Behavioral Neuroscience*, 1–10. https://doi.org/10.3758/s13415-019-00750-y.

Kannangara, T. S., Lucero, M. J., Gil-Mohapel, J., Drapala, R. J., Simpson, J. M., Christie, B. R., et al. (2011). Running reduces stress and enhances cell genesis in aged mice. *Neurobiological Aging*, *32*(12), 2279–2286.

Karasek, R. A., Baker, D., Marxer, F., Ahlbom, A., & Theorell, T. (1981). Job decision latitude, job demands, and cardiovascular disease: A prospective study of Swedish men. *American Journal of Public Health*, *71*, 694–705.

Kario, K., Matsuo, T., Kobayashi, H., Imiya, M., Matsuo, M., & Shimada, K. (1996). Nocturnal fall of blood pressure and silent cerebrovascular damage in elderly hypertensive patients: Advanced silent cerebrovascular damage in extreme dippers. *Hypertension*, *27*, 130–135.

Karlamangla, A. S., Singer, B. H., Greendale, G. A., & Seeman, T. E. (2005). Increase in epinephrine excretion is associated with cognitive decline in elderly men: MacArthur studies of successful aging. *Psychoneuroendocrinology*, *30*, 453–460.

Karstoft, K., & Pedersen, B. K. (2016). Exercise and type 2 diabetes: Focus on metabolism and inflammation. *Immunology and Cell Biology*, *94*(2), 146–150.

Katusic, Z. S., & Austin, S. A. (2014). Endothelial nitric oxide: Protector of a healthy mind. *European Heart Journal*, *35*, 888–894.

Khodadadi, D., Gharakhanlou, R., Naghdi, N., Salimi, M., Azimi, M., Shahed, A., et al. (2018). Treadmill exercise ameliorates spatial learning and memory deficits through improving the clearance of peripheral and central amyloid-beta levels. *Neurochemical Research*, *43*(8), 1561–1574.

Kisler, K., Nelson, A. R., Montagne, A., & Zlokovic, B. V. (2017). Cerebral blood flow regulation and neurovascular dysfunction in Alzheimer disease. *Nature Reviews. Neuroscience*, *18*, 419–434.

Kivimäki, M., & Steptoe, A. (2017). Effects of stress on the development and progression of cardiovascular disease. *Nature Reviews. Cardiology*, *15*, 215–229.

Kolb, H., & Martin, S. (2017). Environmental/lifestyle factors in the pathogenesis and prevention of type 2 diabetes. *BMC Medicine*, *15*(1), 131.

Koo, J.-H., Kang, E.-B., Oh, Y.-S., Yang, D.-S., & Cho, J.-Y. (2017). Treadmill exercise decreases amyloid-β burden possibly via activation of SIRT-1 signaling in a mouse model of Alzheimer's disease. *Experimental Neurology*, *288*, 142–152.

Kovach, C. R., Woods, D. L., Logan, B. R., & Raff, H. (2011). Diurnal variation of cortisol in people with dementia: Relationship to cognition and illness burden. *American Journal of Alzheimer's Disease and Other Dementias*, *26*, 145–150. https://doi.org/10.1177/1533317510397329.

Kramer, A. F., & Erickson, K. I. (2007). Capitalizing on cortical plasticity: Influence of physical activity on cognition and brain function. *Trends in Cognitive Sciences*, *11*, 342–348.

Kumari, M., Badrick, E., Sacker, A., Kirschbaum, C., Marmot, M., & Chandola, T. (2010). Identifying patterns in cortisol secretion in an older population. Findings from the Whitehall II study. *Psychoneuroendocrinology*, *35*, 1091–1099.

Lange-Asschenfeldt, C., & Kojda, G. (2008). Alzheimer's disease, cerebrovascular dysfunction and the benefits of exercise: From vessels to neurons. *Experimental Gerontology*, *43*, 499–504.

Laske, C., Banschbach, S., Stranskya, E., Bosch, S., Straten, G., Machann, J., et al. (2010). Exercise-induced normalization of decreased BDNF serum concentration in elderly women with remitted major depression. *International Journal of Neuropsychopharmacology*, *13*(5), 595–602.

Lee, B. K., Glass, T. a., McAtee, M. J., Wand, G. S., Bandeen-Roche, K., Bolla, K. I., et al. (2007). Associations of salivary cortisol with cognitive function in the Baltimore memory study. *Archives of General Psychiatry*, *64*, 810–818. https://doi.org/10.1001/archpsyc.64.7.810.

Lee, K. W., Kim, J. B., Seo, J. S., Kim, T. K., Im, J. Y., Baek, I. S., et al. (2009). Behavioral stress accelerates plaque pathogenesis in the brain of Tg2576 mice via generation of metabolic oxidative stress. *Journal of Neurochemistry*, *108*, 165–175.

Leem, Y. H., Lee, Y. I., Son, H. J., & Lee, S. H. (2011). Chronic exercise ameliorates the neuroinflammation in mice carrying NSE/htau23. *Biochemical and Biophysical Research Communications*, *406*, 359–365.

Leszek, J., Barreto, G. E., Gasiorowski, K., Koutsouraki, E., Avila-Rodrigues, M., & Aliev, G. (2016). Inflammatory mechanisms and oxidative stress as key factors responsible for progression of neurodegeneration: Role of brain innate immune system. *CNS & Neurological Disorders Drug Targets*, *15*(3), 329–336.

Li, W., Risacher, S. L., Gao, S., Boehm, S. L., II, Elmendorf, J. S., Saykin, A. J., et al. (2018). Type 2 diabetes mellitus and cerebrospinal fluid Alzheimer's disease biomarker amyloid β1-42 in Alzheimer's disease neuroimaging initiative participants. *Alzheimer's & Dementia*, *10*, 94–98.

Li, X., Song, D., & Leng, S. X. (2015). Link between type 2 diabetes and Alzheimer's disease: From epidemiology to mechanism and treatment. *Clinical Interventions in Aging*, *10*, 549–560.

Liang, Y. Z., Dong, J., Zhang, J., Wang, S., He, Y., & Yan, Y. X. (2018). Identification of neuroendocrine stress response-related circulating microRNAs as biomarkers for type 2 diabetes mellitus and insulin resistance. *Frontiers in Endocrinology*, *9*, 132.

Licht, C. M., de Geus, E. J., & Penninx, B. W. (2013). Dysregulation of the autonomic nervous system predicts the development of the metabolic syndrome. *The Journal of Clinical Endocrinology and Metabolism*, *98*(6), 2484–2493.

Lizarbe, B., Soares, A. F., Larsson, S., & Duarte, J. M. N. (2018). Neurochemical modifications in the hippocampus, cortex and hypothalamus of mice exposed to long-term high-fat diet. *Frontiers in Neuroscience*, *12*, 985.

Lopez Lopez, A. L., Escobar Villanueva, M. C., Brianza, Padilla M., Bonilla, Jaime H., & Alarcon Aguilar, F. J. (2018). Chronic unpredictable mild stress progressively disturbs glucose metabolism and appetite hormones in rats. *Acta Endocrinology*, *14*(1), 16–23.

Lloret, A., Monllor, P., Esteve, D., Cervera-Ferri, A., & Lloret, M. A. (2019). Obesity as a risk factor for Alzheimer's disease: Implication of leptin and glutamate. *Frontiers in Neuroscience*, *13*(May), 1–8.

Lu, B., Nagappan, G., Guan, X., Nathan, P. J., & Wren, P. (2013). BDNF-based synaptic repair as a disease-modifying strategy for neurodegenerative diseases. *Nature Reviews. Neuroscience*, *14*, 401–416. https://doi.org/10.1038/nrn3505.

Lucertini, F., Ponzio, E., Di Palma, M., Galati, C., Federici, A., Barbadoro, P., et al. (2015). High cardiorespiratory fitness is negatively associated with daily cortisol output in healthy aging men. *PLoS One*, *10*, e0141970. https://doi.org/10.1371/journal.pone.0141970.

Madhu, S. V., Siddiqui, A., Desai, N. G., Sharma, S. B., & Bansal, A. K. (2019). Chronic stress, sense of coherence and risk of type 2 diabetes mellitus. *Diabetes and Metabolic Syndrome: Clinical Research and Reviews*, *13*(1), 18–23.

Mani, B. K., & Zigman, J. M. (2017). Ghrelin as a survival hormone. *Trends in Endocrinology and Metabolism*, *28*(12), 843–854.

Maslov, L. N., Naryzhnaya, N. V., Boshchenko, A. A., Popov, S. V., Ivanov, V. V., & Oeltgen, P. R. (2019). Is oxidative stress of adipocytes a cause or a consequence of the metabolic syndrome? *Journal of Clinical & Translational Endocrinology*, *15*, 1–5.

Mason, A. E., Schleicher, S., Coccia, M., Epel, E. S., & Aschbacher, K. (2018). Chronic stress and impulsive risk-taking predict increases in visceral fat over 18 months. *Obesity*, *26*(5), 869–876.

Matthews, K. A., Katholi, C. R., McCreath, H., et al. (2004). Blood pressure reactivity to psychological stress predicts hypertension in the CARDIA study. *Circulation*, *110*(1), 74–78.

Mattusch, F., Dufaux, B., Heine, O., Mertens, I., & Rost, R. (2000). Reduction of the plasma concentration of C-reactive protein following nine months of endurance training. *International Journal of Sports Medicine*, *21*, 21–24.

Mazgelyte, E., Karciauskaite, D., Linkeviciute, A., Mazeikiene, A., Burokiene, N., Matuzeviciene, R., et al. (2019). Association of hair cortisol concentration with prevalence of major cardiovascular risk factors and allostatic load. *Medical Science Monitor*, *25*, 3573–3582.

McEwen, B. S. (2008). Central effects of stress hormones in health and disease: Understanding the protective and damaging effects of stress and stress mediators. *European Journal of Pharmacology*, *583*, 174–185.

Misra, M., Bredella, M. A., Tsai, P., Mendes, N., Miller, K. K., & Klibanski, A. (2008). Lower growth hormone and higher cortisol are associated with greater visceral adiposity, intramyocellular lipids, and insulin resistance in overweight girls. *American Journal of Physiology. Endocrinology and Metabolism*, *295*(2), e385–e392.

Mittal, K., & Katare, D. P. (2016). Shared links between type 2 diabetes mellitus and Alzheimer's disease: A review. *Diabetes and Metabolic Syndrome: Clinical Research and Reviews*, *10*(2), S144–S149.

Mora, S., Cook, N., Buring, J. E., Ridker, P. M., & Lee, I. M. (2007). Physical activity and reduced risk of cardiovascular events: Potential mediating mechanisms. *Circulation*, *116*, 2110–2118.

Morelli, V., Arosio, M., & Chiodini, I. (2018). Cardiovascular mortality in patients with subclinical Cushing. *Annales d'Endocrinologie*, *79*(3), 149–152.

Moriarty, A. S., Bradley, A. J., Anderson, K. N., Watson, S., Gallagher, P., & McAllisterWilliams, R. H. (2014). Cortisol awakening response and spatial working memory in man: A U-shaped relationship. *Human Psychopharmacology*, *29*, 295–298. https://doi.org/10.1002/hup.2399.

Moser, V. A., & Pike, C. J. (2017). Obesity accelerates Alzheimer-related pathology in APOE4 but not APOE3 mice. *eNeuro*, *4*(3), 1–18.

Mossello, E., & Simoni, D. (2016). High blood pressure in older subjects with cognitive impairment. *Monaldi Archives for Chest Disease*, *84*, 730–735.

Moyna, N. M., Bodnar, J. D., Goldberg, H. R., Shurin, M. S., Robertson, R. J., & Rabin, B. S. (1999). Relation between aerobic fitness level and stress induced alterations in neuroendocrine and immune function. *International Journal of Sports Medicine*, *20*, 136–141.

Mueller, P. J. (2007). Exercise training and sympathetic nervous system activity: Evidence for physical activity dependent neural plasticity. *Clinical and Experimental Pharmacology & Physiology*, *34*, 377–384.

Myslivecek, P. R., Brown, C. A., & Wolfe, L. A. (2002). Effects of physical conditioning on cardiac autonomic function in healthy middle-aged women. *Canadian Journal of Applied Physiology*, *27*, 1–18.

Nedic, O., Filipovic, D., & Solak, Z. (2001). Job stress and cardiovascular diseases with health workers. *Medicinski Pregled*, *54*, 423–431.

Nicolini, P., Ciulla, M. M., Malfatto, G., Abbate, C., Mari, D., Rossi, P. D., et al. (2014). Autonomic dysfunction in mild cognitive impairment: Evidence from power spectral analysis of heart rate variability in a cross-sectional case-control study. *PLoS One*, *9*(5), e96656.

Nieman, L. K. (2019). Hypertension and cardiovascular mortality in patients with Cushing syndrome. *Endocrinology and Metabolism Clinics of North America*, *48*(4), 717–725.

Nolan, C. J., & Prentki, M. (2019). Insulin resistance and insulin hypersecretion in the metabolic syndrome and type 2 diabetes: Time for a conceptual framework shift. *Diabetes & Vascular Disease Research*, *16*(2), 118–127.

O'Brien, J. T., Erkinjuntti, T., Reisberg, B., Roman, G., Sawada, T., Pantoni, L., et al. (2003). Vascular cognitive impairment. *Lancet Neurology*, *2*, 89–98.

Ohkubo, T., Imai, Y., Tsuji, I., Nagai, K., Watanabe, N., Minami, N., et al. (1997). Relation between nocturnal decline in blood pressure and mortality. *American Journal of Hypertension*, *10*, 1201–1207.

O'Keefe, E. L., O'Keefe, J. H., & Lavie, C. J. (2019). Exercise counteracts the cardiotoxicity of psychosocial stress. *Mayo Clinic Proceedings*, *94*(9), 1852–1864.

Ortiz, R., Kluwe, B., Odei, J. B., Echouffo Tcheugui, J. B., Sims, M., Kalyani, R. R., et al. (2019). The association of morning serum cortisol with glucose metabolism and diabetes: The Jackson Heart Study. *Psychoneuroendocrinology*, *103*, 25–32. 8.

Osman, R., L'Allier, P., Elgharib, N., & Tardif, J. C. (2006). Critical appraisal of C-reactive protein throughout the spectrum of cardiovascular disease. *Vascular Health and Risk Management*, *2*(3), 221–237. https://doi.org/10.2147/vhrm.2006.2.3.221.

Ouakinin, S. R. S., Barreira, D. P., & Gois, C. J. (2018). Depression and obesity: Integrating the role of stress, neuroendocrine dysfunction and inflammatory pathways. *Frontiers in Endocrinology*, *9*, 431.

Ouanes, S., Castelao, E., von Gunten, A., Vidal, P. M., Preisig, M., & Popp, J. (2017). Personality, cortisol, and cognition in non-demented elderly subjects: Results from a population-based study. *Frontiers in Aging Neuroscience*, *9*, 1–9. https://doi.org/10.3389/fnagi.2017.00063.

Paillard, T., Rolland, Y., & de Barreto, P. S. (2015). Protective effects of physical exercise in Alzheimer's disease and Parkinson's disease: A narrative review. *Journal of Clinical Neurology*, *11*(3), 212–219.

Pan, Q., Wu, J., Liu, Y., Li, X., & Chen, J. (2019). Involvement of hepatic SHIP2 and PI3K/Akt signalling in the regulation of plasma insulin by Xiaoyaosan in chronic immobilization-stressed rats. *Molecules*, *24*(3), 480–497.

Papargyri, P., Zapanti, E., Salakos, N., Papargyris, L., Bargiota, A., & Mastorakos, G. (2018). Links between HPA axis and adipokines: Clinical implications in paradigms of stress-related disorders. *Expert Review of Endocrinology and Metabolism*, *13*(6), 317–332.

Papenberg, G., Ferencz, B., Mangialasche, F., Mecocci, P., Cecchetti, R., Kalpouzos, et al. (2016). Physical activity and inflammation: Effects on gray-matter volume and cognitive decline in aging. *Human Brain Mapping*, *37*, 3462–3473.

Parachikova, A., & Cotman, C. W. (2007). Reduced CXCL12/CXCR4 results in impaired learning and is downregulated in a mouse model of Alzheimer disease. *Neurobiology of Disease*, *28*, 143–153.

Park, H. S., Park, S. S., Kim, C. J., Shin, M. S., & Kim, T. W. (2019). Exercise alleviates cognitive functions by enhancing hippocampal insulin signaling and neuroplasticity in high-fat diet-induced obesity. *Nutrients*, *11*(7), 1603–1616.

Paul, K. C., Jerrett, M., & Ritz, B. (2018). Type 2 diabetes mellitus and Alzheimer's disease: Overlapping biologic mechanisms and environmental risk factors. *Current Environmental Health Reports*, *5*(1), 44–58.

Pereira, R. M., Botezelli, J. D., da Cruz Rodrigues, K. C., Mekary, R. A., Cintra, D. E., Pauli, J. R., et al. (2017). Fructose consumption in the development of obesity and the effects of different protocols of physical exercise on the hepatic metabolism. *Nutrients*, *9*(4), 405–426.

Perry, E. K., Smith, C. J., Court, J. A., & Perry, R. H. (1990). Cholinergic nicotinic and muscarinic receptors in dementia of Alzheimer, Parkinson and Lewy body types. *Journal of Neural Transmission Parkinson's Disease and Dementia Section*, *2*, 149–158.

Petersen, R. C. (2004). Mild cognitive impairment as a diagnostic entity. *Journal of Internal Medicine*, *256*(3), 183–194.

Pires, P. W., Dams Ramos, C. M., Matin, N., et al. (2013). The effects of hypertension on the cerebral circulation. *American Journal of Physiology. Heart and Circulatory Physiology*, *304*, 1598–1614.

Plastino, M., Fava, A., Pirritano, D., Cotronei, P., Sacco, N., Sperlı, T., et al. (2010). Effects of insulinic therapy on cognitive impairment in patients with Alzheimer disease and diabetes mellitus type-2. *Journal of the Neurological Sciences*, *288*, 112–116.

Popp, J., Wolfsgruber, S., Heuser, I., Peters, O., Hüll, M., Schröder, J., et al. (2015). Cerebrospinal fluid cortisol and clinical disease progression in MCI and dementia of Alzheimer's type. *Neurobiology of Aging*, *36*(2), 601–607.

Power, M. C., Schneider, A. L., Wruck, L., Griswold, M., Coker, L. H., & Alonso, A. (2016). Life-course blood pressure in relation to brain volumes. *Alzheimer's & Dementia*, *16*, 890–899.

Power, M. C., Weuve, J., Gagne, J. J., McQueen, M. B., Viswanathan, A., & Blacker, D. (2011). The association between blood pressure and incident Alzheimer disease: A systematic review and meta-analysis. *Epidemiology*, *22*, 646–659.

Pruzin, J. J., Schneider, J. A., Capuano, A. W., Leurgans, S. E., Barnes, L. L., Ahima, R. S., et al. (2017). Diabetes, hemoglobin A1C, and regional Alzheimer disease and infarct pathology. *Alzheimer Disease and Associated Disorders*, *31*(1), 41–47.

Pugazhenthi, S., Qin, L., & Reddy, P. H. (2017). Common neurodegenerative pathways in obesity, diabetes, and Alzheimer's disease. *Biochimica et Biophysica Acta—Molecular Basis of Disease*, *1863*(5), 1037–1045.

Puzianowska-Kuznicka, M., Owczarz, M., Wieczorowska-Tobis, K., Nadrowski, P., Chudek, J., Slusarczyk, P., et al. (2016). Interleukin-6 and C-reactive protein, successful aging, and mortality: The PolSenior study. *Immunity & Ageing*, *13*, 21.

Qaseem, A., Wilt, T. J., Rich, R., Humphrey, L. L., Frost, J., Forciea, M. A., et al. (2017). Pharmacologic treatment of hypertension in adults aged 60 years or older to higher versus lower blood pressure targets: a clinical practice guideline from the American College of Physicians and the American Academy of Family Physicians. *Annals of the Intern Medicine, 166*, 430–437.

Qi, J., Yu, X. J., Fu, L. Y., Liu, K. L., Gao, T. T., Tu, J. W., et al. (2019). Exercise training attenuates hypertension through TLR4/MyD88/NF-κB signaling in the hypothalamic paraventricular nucleus. *Frontiers in Neuroscience, 24*(13), 1138.

Querfurth, H. W., & LaFerla, F. M. (2010). Mechanisms of disease. Alzheimer's disease. *The New England Journal of Medicine, 362*, 329–344.

Rad, S. K., Arya, A., Karimian, H., Madhavan, P., Rizwan, F., Koshy, S., et al. (2018). Mechanism involved in insulin resistance via accumulation of β-amyloid and neurofibrillary tangles: Link between type 2 diabetes and Alzheimer's disease. *Drug Design, Development and Therapy, 12*, 3999–4021.

Radak, Z., Hart, N., Sarga, L., Koltai, E., Atalay, M., Ohno, H., et al. (2010). Exercise plays a preventive role against Alzheimer's disease. *Journal of Alzheimer's Disease, 20*, 777–783.

Rapp, M. A., Schnaider-Beeri, M., Grossman, H. T., et al. (2006). Increased hippocampal plaques and tangles in patients with Alzheimer disease with a lifetime history of major depression. *Archives of General Psychiatry, 63*(2), 161–167. https://doi.org/10.1001/archpsyc.63.2.161.

Razzoli, M., & Bartolomucci, A. (2016). The dichotomous effect of chronic stress on obesity. *Trends in Endocrinology and Metabolism, 27*(7), 504–515.

Reus, G. Z., Fries, G. R., Stertz, L., Badawy, M., Passos, I. C., Barichello, T., et al. (2015). The role of inflammation and microglial activation in the pathophysiology of psychiatric disorders. *Neuroscience, 300*, 141–154.

Ridker, P. M. (2018). Interleukin-1 inhibition and ischaemic stroke: Has the time for a major outcomes trial arrived? *European Heart Journal, 39*(38), 3518–3520.

Rimmele, U., Zellweger, B. C., Marti, B., Seiler, R., Mohiyeddini, C., Ehlert, U., et al. (2007). Trained men show lower cortisol, heart rate and psychological responses to psychosocial stress compared with untrained men. *Psychoneuroendocrinology, 32*, 627–635.

Rimmele, U., Seiler, R., Marti, B., Wirtz, P. H., Ehlert, U., & Heinrichs, M. (2009). The level of physical activity affects adrenal and cardiovascular reactivity to psychosocial stress. *Psychoneuroendocrinology, 34*(2), 190–198.

Robertson, B. A., Rathbone, L., Cirillo, G., D'Eath, R. B., Bateson, M., Boswell, T., et al. (2017). Food restriction reduces neurogenesis in the avian hippocampal formation. *PLoS One, 12*(12), e0189158.

Rocha-Rodrigues, S., Goncalves, I. O., Beleza, J., Ascensao, A., & Magalhaes, J. (2018). Physical exercise mitigates high-fat diet-induced adiposopathy and related endocrine alterations in an animal model of obesity. *Journal of Physiology and Biochemistry, 74*(2), 235–246.

Rosales-Corral, S., Tan, D. X., Manchester, L., & Reiter, R. J (2015). Diabetes and Alzheimer disease, two overlapping pathologies with the same background: Oxidative stress. *Oxidative Medicine and Cellular Longevity, 2015* 985845.

Rothman, S. M., & Mattson, M. P. (2010). Adverse stress, hippocampal networks, and Alzheimer's disease. *Neuromolecular Medicine, 12*(1), 56–70.

Ruscheweyh, R., Willemer, C., Kruger, K., Duning, T., Warnecke, T., Sommer, J., et al. (2011). Physical activity and memory functions: An interventional study. *Neurobiology of Aging, 30*(7), 1114–1124.

Ryan, S. M., & Kelly, A. M. (2016). Exercise as a pro-cognitive, pro-neurogenic and anti-inflammatory intervention in transgenic mouse models of Alzheimer's disease. *Ageing Research Reviews, 27*, 77–92.

Salas, I. H., & De Strooper, B. (2019). Diabetes and Alzheimer's disease: A link not as simple as it seems. *Neurochemical Research*, *44*(6), 1271–1278.

Santos, C. Y., Snyder, P. J., Wu, W.-C., Zhang, M., Echeverria, A., & Alber, J. (2017). Pathophysiologic relationship between Alzheimer's disease, cerebrovascular disease, and cardiovascular risk: A review and synthesis. *Alzheimer's Dementia*, 7, 69–87.

Sasi, M., Vignoli, B., Canossa, M., Blum, R., Manju, S., Vignoli, B., et al. (2017). Neurobiology of local and intercellular BDNF signaling. *Pflügers Archiv/European Journal of Physiology*, *469*, 1–18. https://doi.org/10.1007/s00424-017-1964-4.

Schneider, J. A., Arvanitakis, Z., Bang, W., & Bennett, D. A. (2007). Mixed brain pathologies account for most dementia cases in community-dwelling older persons. *Neurology*, *69*, 2197–2204.

Schneider, L. S., et al. (2014). Clinical trials and late-stage drug development for Alzheimer's disease: An appraisal from 1984 to 2014. *Journal of Internal Medicine*, *275*(3), 251–283.

Seeman, T. E., McEwen, B. S., Singer, B. K., Albert, M. S., & Rowe, J. W. (1997). Increase in urinary cortisol excretion and memory declines: MacArthur studies of successful aging. *The Journal of Clinical Endocrinology and Metabolism*, *82*(8), 2458–2465.

Shi, Z., Li, C., Yin, Y., Yang, Z., Xue, H., Mu, N., et al. (2018). Aerobic interval training regulated SIRT3 attenuates high-fat-diet-associated cognitive dysfunction. *BioMed Research International*, *2018*, 2708491.

Shih, I. F., Paul, K., Haan, M., Yu, Y., & Ritz, B. (2018). Physical activity modifies the influence of apolipoprotein E epsilon 4 allele and type 2 diabetes on dementia and cognitive impairment among older Mexican Americans. *Alzheimers & Dementia*, *14*(1), 1–9.

Shin, Y. K., Hsieh, Y. S., Kwon, S., Lee, H. S., & Seol, G. H. (2018). Linalyl acetate restores endothelial dysfunction and hemodynamic alterations in diabetic rats exposed to chronic immobilization stress. *Journal of Applied Physiology (Bethesda, MD: 1985)*, *124*(5), 1274–1283.

Siddiqui, A., Desai, N. G., Sharma, S. B., Aslam, M., Sinha, U. K., & Madhu, S. V. (2019). Association of oxidative stress and inflammatory markers with chronic stress in patients with newly diagnosed type 2 diabetes. *Diabetes/Metabolism Research and Reviews*, *35*(5) e3147.

Sirico, F., Bianco, A., D'Alicandro, G., Castaldo, C., Montagnani, S., Spera, R., et al. (2018). Effects of physical exercise on adiponectin, leptin, and inflammatory markers in childhood obesity: Systematic review and meta-analysis. *Childhood Obesity*, *14*(4), 207–217.

Soares-Miranda, L., Sandercock, G., Vale, S., Santos, R., Abreu, S., Moreira, C., et al. (2012). Metabolic syndrome, physical activity and cardiac autonomic function. *Diabetes/Metabolism Research and Reviews*, *28*, 363–369.

Sonnen, J. A., Larson, E. B., Crane, P. K., Haneuse, S., Li, G., Schellenberg, G. D., et al. (2007). Pathological correlates of dementia in a longitudinal, population-based sample of aging. *Annals of Neurology*, *62*, 406–413.

Stephens, M. A., Mahon, P. B., McCaul, M. E., & Wand, G. S. (2016). Hypothalamic-pituitary-adrenal axis response to acute psychosocial stress: Effects of biological sex and circulating sex hormones. *Psychoneuroendocrinology*, *66*, 47–55.

Steptoe, A., & Kivimaki, M. (2013). Stress and cardiovascular disease: An update on current knowledge. *Annual Review of Public Health*, *34*, 337–354.

Tomiyama, A. J. (2019). Stress and obesity. *Annual Review of Psychology*, *70*, 703–718.

Tortosa-Martinez, J., & Clow, A. (2012). Does physical activity reduce risk for Alzheimer's disease through interaction with the stress neuroendocrine system? *Stress*, *15*(3), 243–261.

Tortosa-Martinez, J., Clow, A., Caus-Pertegaz, N., Gonzalez-Caballero, G., Abellan-Miralles, I., & Saenz, M. J. (2015). Exercise increases the dynamics of diurnal cortisol secretion and executive function in people with amnestic mild cognitive impairment. *Journal of Aging and Physical Activity*, *23*(4), 550–558.

Tortosa-Martínez, J., Manchado, C., Chulvi-Medrano, I., & Cortell-Tormo, J. M. (2018). Exercise, the diurnal cycle of cortisol and cognitive impairment in older adults. *Neurobiology of Stress*, *9*, 40–47.

Trouwborst, I., Verreijen, A., Memelink, R., Massanet, P., Boirie, Y., Weijs, P., et al. (2018). Exercise and nutrition strategies to counteract sarcopenic obesity. *Nutrients*, *10*(5), 605.

Tsiu, A., Richards, M., & Davis, D. (2018). Systemic inflammation and modifiable risk factors for cognitive impairment in older persons: Findings from a British birth cohort. *Aging Medicine*, *1*(3), 243–248. https://doi.org/10.1002/agm2.12044.

Um, H. S., Kang, E. B., Leem, Y. H., Cho, I. H., Yang, C. H., Chae, K. R., et al. (2008). Exercise training acts as a therapeutic strategy for reduction of the pathogenic phenotypes for Alzheimer's disease in an NSE/APPsw-transgenic model. *International Journal of Molecular Medicine*, *22*, 529–539.

Valtorta, N. K., Kanaan, M., Gilbody, S., Ronzi, S., & Hanratty, B. (2016). Loneliness and social isolation as risk factors for coronary heart disease and stroke: Systematic review and meta-analysis of longitudinal observational studies. *Heart*, *102*, 1009–1016.

van der Wardt, V., Logan, P., Conroy, S., Harwood, R., & Gladman, J. (2014). Antihypertensive treatment in people with dementia. *Journal of the American Medical Directors Association*, *15*(9), 620–629.

Van Tassell, B. W., Raleigh, J. M., & Abbate, A. (2015). Targeting interleukin-1 in heart failure and inflammatory heart disease. *Current Heart Failure Reports*, *12*(1), 33–41.

Ventriglia, M., Zanardini, R., Bonomini, C., Zanetti, O., Volpe, D., Pasqualetti, P., et al. (2013). Serum brain-derived neurotrophic factor levels in different neurological diseases. *BioMed Research International*, *2013*, 901082,. https://doi.org/10.1155/2013/901082.

Verdile, G., Fuller, S. J., & Martins, R. N. (2015). The role of type 2 diabetes in neurodegeneration. *Neurobiology of Disease*, *84*, 22–38.

Verdile, G., Keane, K. N., Cruzat, V. F., Medic, S., Sabale, M., Rowles, J., et al. (2015). Inflammation and oxidative stress: The molecular connectivity between insulin resistance, obesity, and Alzheimer's disease. *Mediators of Inflammation*, *2015* 105828.

Voelcker-Rehage, C., & Niemann, C. (2013). Structural and functional brain changes related to different types of physical activity across the life span. *Neuroscience and Biobehavioral Reviews*, *37*, 2268–2295.

Vogelzangs, N., Beekman, A. T., Milaneschi, Y., Bandinelli, S., Ferrucci, L., & Penninx, B. W. (2010). Urinary cortisol and six-year risk of all-cause and cardiovascular mortality. *Journal of Clinical Endocrinology and Metabolism*, *95*, 4959–4964.

Walker, B. R. (2007). Glucocorticoids and cardiovascular disease. *European Journal of Endocrinology*, *157*(5), 545–559.

Walker, K. A., Power, M. C., & Gottesman, R. F. (2017). Defining the relationship between hypertension, cognitive decline, and dementia: A review. *Current Hypertension Reports*, *19*(3), 24.

Wang, Y. L., Han, Q. Q., Gong, W. Q., Pan, D. H., Wang, L. Z., Hu, W., et al. (2018). Microglial activation mediates chronic mild stress-induced depressive- and anxiety-like behavior in adult rats. *Journal of Neuroinflammation*, *15*(1), 21.

Wang, X., Wang, W., Li, L., Perry, G., Lee, H., & Zhu, X. (2014). Oxidative stress and mitochondrial dysfunction in Alzheimer's disease. *Biochimica et Biophysica Acta—Molecular Basis of Disease*, *1842*, 1240–1247.

Wanleenuwat, P., Iwanowski, P., & Kozubski, W. (2019). Alzheimer's dementia: Pathogenesis and impact of cardiovascular risk factors on cognitive decline. *Postgraduate Medicine*, *131*(7), 415–422.

Watanabe, K., Sakuraya, A., Kawakami, N., Imamura, K., Ando, E., Asai, Y., et al. (2018). Work-related psychosocial factors and metabolic syndrome onset among workers: A systematic review and meta-analysis. *Obesity Reviews*, *19*(11), 1557–1568.

Way, K. L., Hackett, D. A., Baker, M. K., & Johnson, N. A. (2016). The effect of regular exercise on insulin sensitivity in type 2 diabetes mellitus: A systematic review and meta-analysis. *Diabetes and Metabolism Journal*, *40*(4), 253–271.

Wei, N. L., Quan, Z. F., Zhao, T., Yu, X. D., Xie, Q., Zeng, J., et al. (2019). Chronic stress increases susceptibility to food addiction by increasing the levels of DR2 and MOR in the nucleus accumbens. *Neuropsychiatric Disease and Treatment*, *15*, 1211–1229.

Whitworth, J. A., Williamson, P. M., Mangos, G., & Kelly, J. J. (2005). Cardiovascular consequences of cortisol excess. *Vascular Health and Risk Management*, *1*(4), 291.

Wiesmann, M., Kiliaan, A. J., & Claassen, J. A. H. R. (2013). Vascular aspects of cognitive impairment and dementia. *Journal of Cerebral Blood Flow and Metabolism*, *33*, 1696–1706.

Wolf, O. T. (2009). Stress and memory in humans: Twelve years of progress? *Brain Research*, *1293*, 142–154.

Wolf, O. T., Convit, A., Thorn, E., & de Leon, M. J. (2002). Salivary cortisol day profiles in elderly with mild cognitive impairment. *Psychoneuroendocrinology*, *27*, 777–789.

World Health Organization. (2018). *Obesity and overweight*. Retrieved from https://www.who.int/news-room/fact-sheets/detail/obesity-and-overweight.

Wosiski-Kuhn, M., Erion, J. R., Gomez-Sanchez, E. P., Gomez-Sanchez, C. E., & Stranahan, A. M. (2014). Glucocorticoid receptor activation impairs hippocampal plasticity by suppressing BDNF expression in obese mice. *Psychoneuroendocrinology*, *42*, 165–177.

Yao, B. C., Meng, L. B., Hao, M. L., Zhang, Y. M., Gong, T., & Guo, Z. G. (2019). Chronic stress: A critical risk factor for atherosclerosis. *The Journal of International Medical Research*, *47*(4), 1429–1440.

Yasunari, K., Matsui, T., Maeda, K., Nakamura, M., Watanabe, T., & Kiriike, N. (2006). Anxiety-induced plasma norepinephrine augmentation increases reactive oxygen species formation by monocytes in essential hypertension. *American Journal of Hypertension*, *19*(6), 573–578.

Yousufzai, M., Harmatz, E. S., Shah, M., Malik, M. O., & Goosens, K. A. (2018). Ghrelin is a persistent biomarker for chronic stress exposure in adolescent rats and humans. *Translational Psychiatry*, *8*(1), 74.

Yuede, C. M., Zimmerman, S. D., Dong, H., Kling, M. J., Bero, A. W., Holtzman, D. M., et al. (2009). Effects of voluntary and forced exercise on plaque deposition, hippocampal volume, and behavior in the Tg2576 mouse model of Alzheimer's disease. *Neurobiology of Disease*, *35*, 426–432. https://doi.org/10.1016/j.nbd.2009.06.002.

Zanuso, S., Jimenez, A., Pugliese, G., Corigliano, G., & Balducci, S. (2010). Exercise for the management of type 2 diabetes: A review of the evidence. *Acta Diabetologica*, *47*(1), 15–22.

Zhang, S., Chai, R., Yang, Y. Y., Guo, S. Q., Wang, S., Guo, T., et al. (2017). Chronic diabetic states worsen Alzheimer neuropathology and cognitive deficits accompanying disruption of calcium signaling in leptin-deficient APP/PS1 mice. *Oncotarget*, *8*(27), 43617.

Zhang, L., Zheng, H., Luo, J., Li, L., Pan, X., & Jiang, T. (2018). Inhibition of endothelial nitric oxide synthase reverses the effect of exercise on improving cognitive function in hypertensive rats. *Hypertension Research*, *41*(6), 414–425.

Zhao, Y., Bhattacharjee, S., Dua, P., Alexandrov, P. N., & Lukiw, W. J. (2015). microRNA-based biomarkers and the diagnosis of Alzheimer's disease. *Frontiers in Neurology*, *6*, 162.

Zhao, D., Xu, X., Pan, L., Zhu, W., Fu, X., Guo, L., et al. (2017). Pharmacologic activation of cholinergic alpha7 nicotinic receptors mitigates depressive-like behavior in a mouse model of chronic stress. *Journal of Neuroinflammation*, *14*(1), 234.

Zhuo, J. M., Wang, H., & Pratico', D. (2011). Is hyperhomocysteinemia an Alzheimer's disease (AD) risk factor, an AD marker, or neither? *Trends in Pharmacological Sciences*, *32*(9), 562–571.

Further reading

Brasure, M., Desai, P., Davila, H., Nelson, V. A., Calvert, C., Jutkowitz, E., et al. (2018). Physical activity interventions in preventing cognitive decline and Alzheimer-type dementia: A systematic review. *Annals of Internal Medicine*, *168*(1), 30–38.

Frederiksen, K. S., Gjerum, L., Waldemar, G., & Hasselbalch, S. G. (2018). Effects of physical exercise on Alzheimer's disease biomarkers: A systematic review of intervention studies. *Journal of Alzheimer's Disease*, *61*(1), 359–372.

Kjeldsen, S. E. (2018). Hypertension and cardiovascular risk: General aspects. *Pharmacological Research*, *129*, 95–99.

Köhler, S., Baars, M. A., Spauwen, P., Schievink, S., Verhey, F. R., & van Boxtel, M. J. (2014). Temporal evolution of cognitive changes in incident hypertension: Prospective cohort study across the adult age span. *Hypertension*, *63*, 245–251.

Macedo, I. C., Rozisky, J. R., Oliveira, C., Oliveira, C. M., Laste, G., Nonose, Y., et al. (2015). Chronic stress associated with hypercaloric diet changes the hippocampal BDNF levels in male Wistar rats. *Neuropeptides*, *51*, 75–81.

Meissner, A. (2016). Hypertension and the brain: A risk factor for more than heart disease. *Cerebrovascular Diseases*, *42*(3–4), 255–262.

Panza, G. A., Taylor, B. A., MacDonald, H. V., Johnson, B. T., Zaleski, A. L., Livingston, J., et al. (2018). Can exercise improve cognitive symptoms of Alzheimer's disease? *Journal of the American Geriatrics Society*, *66*(3), 487–495.

Sampaio, A., Marques, E. A., Mota, J., & Carvalho, J. (2019). Effects of a multicomponent exercise program in institutionalized elders with Alzheimer's disease. *Dementia (London)*, *18*(2), 417–431.

Skoog, I., Lernfelt, B., Landahl, S., Palmertz, B., Andreasson, L. A., & Nilsson, L. (1996). 15-year longitudinal study of blood pressure and dementia. *Lancet*, *347*, 1141–1145.

Svensson, M., Lexell, J., & Deierborg, T. (2015). Effects of physical exercise on neuroinflammation, neuroplasticity, neurodegeneration, and behavior: What we can learn from animal models in clinical settings. *Neurorehabilitation and Neural Repair*, *29*(6), 577–589.

CHAPTER EIGHT

The intersections of stress, anxiety and epilepsy

Jay A. Salpekar[a,*], Trina Basu[b], Swami Thangaraj[a], Jamie Maguire[b]
[a]Kennedy Krieger Institute, Johns Hopkins University Medical School, Baltimore, MD, United States
[b]Department of Neuroscience, Tufts University School of Medicine, Boston, MA, United States
*Corresponding author: e-mail address: salpekar@kennedykrieger.org

Contents

Abstract

Stress is ubiquitous in chronic medical conditions; however, the connections to psychiatric and neurologic conditions are not always clearly established. Epilepsy is a unique illness that is intimately intertwined with stress and anxiety not only as a result of the disease process but also as a cause of disease exacerbation. Anxiety and depression also involve stress management and often overlap with epilepsy. Anxiety symptoms themselves may be present as intrinsic aspects of seizure phenomena, either during the events or closely related to them. The pathways of stress and anxiety involve the hypothalamic pituitary adrenal (HPA) axis and explain at least in part how stress may lead to worsening seizure control. Ultimately, the study of stress, anxiety, and epilepsy offers insight into mind and body connections, and furthers understanding of neuropsychiatric illness.

International Review of Neurobiology, Volume 152
ISSN 0074-7742
https://doi.org/10.1016/bs.irn.2020.02.001

1. Introduction

Psychiatric conditions and epilepsy have been known to overlap for millennia. However, the relationship of anxiety and stress to seizure events has been difficult to reliably assess. Anecdotally, many caregivers and patients themselves are aware that anxiety or fear often occurs in the context of seizures. Some patients have even noticed being more prone to having seizures if they are stressed or nervous about certain situations. Children may have a seizure in the context of increased stimuli or fearfulness. The same may be true for adults. Ultimately, emotional stress may be associated with epilepsy in ways that we are only beginning to understand.

One challenge of connecting the two conditions is the dichotomy of clinical conventions in modern psychiatry and neurology. Few psychiatrists are well versed in epilepsy and few neurologists are experienced in recognizing anxiety or stress in the context of neurologic conditions. Despite evidence to the contrary, many neurologists still consider anxiety or depressed mood to be expected in the course of chronic illness (Salpekar, Mishra, & Hauptman, 2015). However, the reality is that such conditions are not typical nor should they be considered acceptable in patients. Relegating mental illness to the realm of what should be expected is a superficial approach. A comprehensive and expanded approach to epilepsy management is necessary if stress is a marked component. Aggressive treatment of comorbid conditions is essential for an individual's neuropsychiatric well-being and to enhance of quality of life (Salpekar, 2016).

The good news is that the gulf between the fields is changing, and treatment studies assessing both anxiety and epilepsy are actively being undertaken. The phenomenology of stress in a variety of neurologic conditions is beginning to be better understood. Stress management and diagnosis of anxiety disorder have become key components of many comprehensive treatment plans both in neurology and in psychiatry.

In this chapter, a broad overview of the overlap of anxiety and epilepsy will be explored. Anxiety is variable and may manifest as a full psychiatric disorder or as a temporary acute stress reaction. The brain reacts to stress in different ways depending upon whether the stressors are acute or chronic and may impact seizure thresholds. Because anxiety and depression frequently co-occur, significant emphasis will be given to both conditions. Finally, treatment aspects will be briefly explored to the extent that reliable evidence is available.

2. Classification of anxiety

Anxiety is the most common psychiatric illness in the general population of the United States and likely worldwide. However, like epilepsy, anxiety itself is heterogeneous, and presentations vary in different age groups. Details of anxiety disorders are described in Table 1.

Anxiety disorders are defined categorically by the Diagnostic Statistical Manual (DSM) of Mental Disorders published by the American Psychiatric Association (APA, 2013). Generalized anxiety refers to symptoms that are globally present, chronically, and in multiple situations, not restricted to the circumstances of social exposure or potential separation. In social anxiety, patients experience marked fear or anxiety in one or more social situations in which the individual is exposed to social situations that are potentially embarrassing. Common triggers include talking in classrooms, performing in front of an audience or meeting new people. However, symptoms may also extend to any social exposure such as eating in a cafeteria or using public restrooms.

Separation anxiety is primarily defined as an age-inappropriate fear or worry regarding separation from home or primary attachment figures. Children can worry about separation from parent due to catastrophic

Table 1 Spectrum of anxiety disorders.

	Spectrum of anxiety disorders			
	Generalized anxiety disorder	**Separation anxiety disorder**	**Social anxiety disorder**	**Panic disorder**
Core symptoms	Excessive anxiety and worry; out-of-proportion to the situation; co-occurrence of somatic symptoms; time frame of at least 6 months	Age-inappropriate fear or worry regarding separation from home or primary attachment figures; time frame of at least 4 weeks	Marked fear or anxiety about one or more social situations; can be influenced by temperament and environment	Recurrent, unexpected abrupt surges of intense fear or discomfort [panic attacks]; may be associated with agoraphobia
Common age range	Any childhood age	Early childhood or middle school age	Any childhood age	Typically adolescence and early adulthood

circumstances such as abduction or death. This can lead to school avoidance in order to prevent separation. Somatic symptoms such as stomach aches and headaches may exacerbate when separation is anticipated. Younger children often express their fears through irritability, crying spells and tantrums.

Both social anxiety and separation anxiety are common in the context of epilepsy, since an individual may be nervous about having a public seizure or being away from a trusted caregiver who may provide acute care. Such symptoms may lead to marked resistance to leaving home or being in novel social situations. At times parents may unwittingly reinforce anxiety symptoms in their children, and vice versa, for fear of a medical emergency occurring without sufficient support.

Panic disorder is more common in the adolescent or young adult population. Typical symptoms include sudden onset of shortness of breath, chest pain, palpitations, dizziness and sweating. Patients with panic disorder are more likely to have somatic sensitivities. Panic symptoms may seem very similar to early stages of a focal seizure emerging from the temporal lobe. In both cases, a sense of doom may suddenly emerge and be accompanied by physiologic signs of hypersympathetic states. Adrenergic reactions such as increased heart rate, palpitations, sweating, and hyperventilation may occur in both conditions. Spikes in heart rate are very common in acute seizures and have been the basis for some retail epilepsy monitoring devices such as watches.

3. Overlap of depression and anxiety

Not only is anxiety heterogeneous, but it commonly overlaps with depression in patients with or without epilepsy. Adolescents or young adults often have symptoms of agitation or restlessness in response to anxiety. In this age group, anxiety may be prominent even in the context of a primary depressive disorder. Depression may be secondary in some cases, but the overlap between the conditions is so common that similar brain regions and even pathophysiology are relevant.

The medial temporal lobe is often implicated in the neural circuitry of mood and anxiety disorders (Drevets, 2003). The amygdala particularly is a key weigh station in the brain that is involved in emotional processing. The amygdala is connected with the processing of sensory information and is viewed as applying emotional significance to incoming stimuli. The amygdala is particularly active in states of fear. Classically, the 19th Century neurologist, Hughlings Jackson, connected symptoms of fear and changes in consciousness to seizure foci in the amygdala (Salpekar & Dadson, 2012). In the classic condition of Kluver-Bucy syndrome, bilateral amygdala lesions are associated with a marked lack of anxiety (Masia & Devinsky, 2000). Apparently, without the influence of the amygdalae, fear is markedly reduced.

The pathophysiology of depression also involves the amygdala. Early positron emission tomography imaging noted increased activity in the amygdala in familial pure depressive disorder (Drevets et al., 1992). Other subcortical brain regions are also involved in depression, such as the thalamus and caudate nucleus. A reverberating circuit has been proposed that involves underactive inhibitory networks, presumably GABAergic from the region of the caudate nucleus, which insufficiently prevents overactivation of the amygdala (Drevets, Price, & Furey, 2008). The overactivation leads to anxiety symptoms that are excessive.

Treatments overlap between the two conditions as well. Antidepressants are the mainstays of treatment for both depression and anxiety disorders. Although benzodiazepines are effective in cases of acute anxiety or panic symptoms, treatment with an antidepressant is often viewed as more effective over the long term. A medicine such as alprazolam is an atypical benzodiazepine considered to be particularly effective for panic disorder, yet also has some antidepressant effect as well.

4. Anxiety as a peri-ictal phenomenon

Anxiety and panic symptoms have been noted as related to seizure events. Sometimes, a seizure prodrome involves an unsettled or restless sensation, or even a foreboding feeling. Seizure auras may involve a sensation of fear or panic, especially if a seizure focus is in the temporal lobe. A classic study reported observations during invasive EEG monitoring for medically refractory epilepsy (Gloor, Olivier, Quesney, Andermann, & Horowitz, 1982). During active stimulation of the amygdala in patients with temporal lobe seizures, an overwhelming sensation of fear was common. Thus it may not be surprising to expect that anxiety may be a characteristic part of certain seizure types and localization.

However, it may also be that seizure propagation overlaps with neural pathways of anxiety. Seizure propagation and spread is considered to depend upon involvement of both cortical and subcortical regions. Mood and anxiety disorders also seem to depend upon resonance or possibly reverberating activity in multiple brain regions (Timofeev, Grenier, & Steriade, 1998).

Although involvement of the amygdala appears to be key in anxiety phenomena, with or without epilepsy, it may be that other regions are unable to counter or inhibit over-activity of the amygdala or other brain regions that are involved in anxiety (Adamec & Shallow, 2000).

5. Psychogenic nonepileptic events

Non-epileptic events are phenomena that appear like seizures, with visible convulsions or tremulousness, but are not associated with epileptiform findings on an EEG. The events usually occur independently but also happen in some patients with electrical seizures. The condition is considered to be either a conversion reaction or a part of a larger somatization disorder. Often at the core of these events is anxiety, sometimes resulting from trauma and taking the form of atypical panic attacks (Salpekar, 2019). While the physiologic overlap of nonepileptic events does not appear to correlate with epilepsy pathophysiology, similar brain regions may be involved. Treatment of anxiety or mood disorder, sometimes even with anticonvulsant medicines, may mitigate nonepileptic events.

6. Stress and epilepsy

6.1 Stress exacerbating epilepsy

Psychogenic nonepileptic events are appropriately considered to be stress conditions. However, stress may also overlap with electrical seizures. Efforts to monitor stress have led to interesting insights into potential triggers for seizure events. An early, blinded study had 12 patients with epilepsy journal their seizures and record stressful occurrences in their lives for 3 months. In 58% of the patients, there was a significant increase in seizure frequency on self-reported "high stress" days compared to days that were identified as "low stress" (Temkin & Davis, 1984). Notably, measures of stress are highly subjective, especially in studies relying upon self-reports. Although anecdotally correlated, it is nevertheless difficult to measure stress since similar experiences between individuals evoke different responses based on numerous factors including psychological triggers and prior life experiences.

Several investigators have addressed the problem of subjectivity, by studying whether stress provokes quantifiable changes in electroencephalogram (EEG) recordings. An early study conducted in the 1950s was the first to report subtle, but highly distinct changes in EEG bandwidth and regional frequency in patients with epilepsy that underwent individualized verbal stress stimulation compared to non-epileptic controls (Stevens, 1959). Though the control subjects did not exhibit "pathological" changes in EEG, there were still transient changes in alpha and theta waves in the presence of verbal stressors (Stevens, 1959). The finding of EEG changes in patients with epilepsy who undergo stress was validated in another study (Berkhout, Walter, & Adey, 1969), suggesting that acute stressors can alter neural network activity and potentially contribute to lowering seizure thresholds in patients with established epilepsy.

Outside of laboratory settings, uncontrolled environments that are inherently high stress, such as war affected regions or natural disaster zones, provide additional insight to the relationship between stress and seizure susceptibility in patients with epilepsy. During a flood evacuation in the Netherlands in 1995, 27% of individuals with epilepsy who were forced to evacuate as a result of residing in flood affected regions reported an increase in seizure frequency compared to patients with epilepsy living outside of the evacuation zone (Swinkels et al., 1998). Approximately 92% of the patients with epilepsy living in the flood affected region who experienced worsening seizures reported stress as the triggering factor for the increased seizure occurrence.

However, some studies are counter to the above findings. Retrospective studies of the frequency of seizures during the Persian Gulf War found that though 82 of the 100 participants that were diagnosed with epilepsy prior to the war reported that they experienced stress during the conflict, only 8 recounted an increase in seizure frequency during the war. This study concluded a weak association between stress and epilepsy (Neufeld, Sadeh, Cohn, & Korczyn, 1994). This is in contrast to a retrospective study conducted on children entangled in the Croatian War. Children that were diagnosed with epilepsy prior to the start of the war reported having greater seizure events compared to children who were not in war affected regions (Bosnjak, Vukovic-Bobic, & Mejaski-Bosnjak, 2002).

Aside from the different age groups investigated in these two studies, the disparate findings could be due to the subjective nature of recalling and reporting stress. The study on the Persian Gulf War was uncontrolled, and solely looked at patients with epilepsy, basing conclusions on self-reported seizure events made years after the fact (Neufeld et al., 1994). The study looking into seizure frequency in children caught in the Croatian War used defined stressors, e.g., witnessing bombing or shelling first hand, witnessing a family member being abducted and presumably taken away to prison camp. Medical reports were conducted more closely in real time with routine doctor visits before and after the war assessing seizure stability (Bosnjak et al., 2002). Such methodology in high stress environments more convincingly shows that there is a population of individuals with epilepsy exhibiting stress-sensitive seizures.

6.2 New onset epilepsy resulting from stress

The fact that stress seems to powerfully influence seizure thresholds is significant and implies that stress possibly alters brain function even in people without diagnosed epilepsy. The idea of de novo epilepsy resulting from stress represents a huge paradigm change in ways to think about psychiatric and neurologic illness. The phenomenon of bidirectionality, where either of the two illnesses may promote the development of the other, has been frequently discussed in the literature (Hesdorffer et al., 2012). It is known from epidemiologic studies that a diagnosis of epilepsy makes it more likely that a diagnosis of depression or anxiety may develop. However, the converse is also true. Patients with depressive disorders have a higher risk for developing epilepsy compared to those that do not suffer from mood or anxiety disorders (Forsgren & Nyström, 1990; Hesdorffer et al., 2012; Kanner, 2011; Salpekar, 2017).

The phenomenon of bidirectionality may hold true for stress and seizure disorders as well. Bosnjak et al. (2002) reported that approximately 28% of the children who were not diagnosed with epilepsy prior to the start of the Croatian war exhibited their first seizure during the war (Bosnjak et al., 2002). The de novo appearance of epilepsy as a result of a psychological stressor has been documented in several other instances. In a retrospective study conducted in Denmark, there was a 50% increased risk for a parent who experienced the death of their child in being diagnosed with epilepsy compared to a control group of parents who did not have a deceased child (Christensen, Li, Vestergaard, & Olsen, 2007). The increased risk for developing epilepsy remained high in fathers during the first 3 years following the death of their child, after which the risk was statistically equivalent to the control group. However, the risk for developing epilepsy in mothers following their child's death remained high for the duration of the study, which continued until 18 years after the child's death (Christensen, Li, et al., 2007). This suggests that stress may have sexually dimorphic mechanisms in provoking and maintaining seizure susceptibility.

Another study examined whether occupational stress increased the incidence of de novo epilepsy. The participants of this particular study were recent Israeli military recruits (Moshe et al., 2008). Military recruits that did not have a history of seizures prior to joining the army and who were assigned to higher stress positions (combat units) had a much higher seizure incidence compared to those assigned to low stress positions (administrative or maintenance units) (Moshe et al., 2008). Those assigned to combat units described the work as mentally strenuous and physically challenging in addition to experiencing sleep deprivation, a known trigger for reducing seizure thresholds (Ferlisi & Shorvon, 2014; Moshe et al., 2008; Nakken et al., 2005). However, the assignment designation did not alter seizure frequency for military recruits that had a history of seizures prior to joining the military, suggesting that the increased occupation induced stress was a risk factor for only those without a history of seizures (Moshe et al., 2008).

In a recent study, investigators studied whether an acute, but serious, emotional life stressor could trigger the onset of epilepsy in patients with no history of seizures. Though uncommon, out of over 4600 patients with epilepsy evaluated at an epilepsy center in Montpellier, France, 22 patients reported severe emotional trauma as the onset of their epilepsy. Death of a loved one preceding an epileptic seizure accounted for 54% of these cases, while others included divorce, dissolution of a partnership, domestic abuse without head trauma, and job loss (Gélisse, Genton, Coubes, Tang, & Crespel, 2015). These data suggest that though uncommon, extreme stress induced by an acute life event or chronic stress brought on by a psychologically taxing profession can contribute to the de novo appearance of epilepsy in previously healthy individuals.

7. Anxiety and depression comorbidity in epilepsy

Depression and anxiety are the two most common psychiatric comorbidities in epilepsy, with incidence rates that far exceed both that of the general population and those suffering from other types of chronic illness (Brandt et al., 2010; Kanner, 2011a; Kwon & Park, 2014; Lambert & Robertson, 1999; Moore, Elliott, Lu, Klatte, & Charyton, 2009). Within the population of patients with epilepsy, the rates for comorbid psychiatric disorders are even higher among patients with medically refractory epilepsy (Boylan et al., 2004; Kwon & Park, 2014). Psychiatric comorbidities associated with epilepsy can severely impact a patient's quality of life. Suicide rates among those with epilepsy are at least threefold and up to five times higher than the general population (Christensen, Vestergaard, Mortensen, Sidenius, & Agerbo, 2007; Harris & Barraclough, 1997). Among patients with epilepsy, those that exhibit an affective disorder have a 32-fold higher risk for suicide compared to patients with epilepsy that do not have a comorbid affective disorder (Christensen, Vestergaard, et al., 2007). Though it is known that affective states can severely worsen the quality of life for a person with epilepsy, the literature is underdeveloped regarding the mechanisms underlying this negative relationship.

However, it is well established that seizure severity heavily correlates with anxiety or depressive disorders in patients with epilepsy (Smith, Baker, Dewey, Jacoby, & Chadwick, 1991), supporting a direct relationship between seizures and affective states. Furthermore, both depressive and anxiety episodes share a temporal relationship with seizure occurrences.

Symptoms associated with an anxiety or depressive episode can occur either before (preictal), during (ictal), or after (postictal) the onset of a seizure. Preictal depressive episodes can present in the hours or days preceding a seizure, and manifest as periods of dysphoric mood or increased anxiety (Kanner, 2016). Ictal symptoms most commonly manifest as an aura of fear, panic, or anguish in the moment immediately leading up to the seizure (Beyenburg et al., 2001; Goldstein & Mellers, 2006; Kanner, 2016; Vazquez & Devinsky, 2003). Depression is the second most reported ictal symptom (Kanner, 2016). Finally, postictal anxiety or depressive episodes occur in the hours and up to 7 days following a seizure and affect more than 50% of patients with epilepsy, with depression and anxiety affecting 43% and 45% of these patients, respectively (Kanner, 2003, 2016). In addition to the onset of psychiatric symptoms related to ictal activity, some patients also have state anxiety or depression that is not temporally linked to an ictal event. If left untreated, the presence of psychiatric comorbidities in epilepsy can worsen seizure severity (Hitiris, Mohanraj, Norrie, Sills, & Brodie, 2007), negatively impact efficacy of antiseizure drugs (ASDs) (Kanner, Barry, Gilliam, Hermann, & Meador, 2012) or surgical resection (Kanner, Byrne, Chicharro, Wuu, & Frey, 2009), and significantly increase a patient's risk of suicide.

7.1 Anxiety in epilepsy

Recently referred to as the "forgotten psychiatric comorbidity" in epilepsy (Gandy et al., 2015; Kanner, 2011a), anxiety disorders are reported in up to 25% of patients with epilepsy (Brandt et al., 2010), though epidemiological estimates range based on the population studied. Of the nearly one dozen different types of anxiety disorders described in the DSM, patients with epilepsy predominantly suffer from general anxiety disorder, panic attacks and panic disorder, posttraumatic stress disorder, social anxiety, and phobia (Beyenburg, Mitchell, Schmidt, Elger, & Reuber, 2005; Kwon & Park, 2014; Mula, 2013). Anxiety disorders are especially prevalent among patients with refractory TLE compared to patients with generalized epilepsy or patients with other chronic illnesses (Beyenburg et al., 2005; Brandt et al., 2010; Gaitatzis, Carroll, Majeed, & Sander, 2004). Like depression, anxiety is highly prevalent among patients with epilepsy, but it is underdiagnosed and therefore may go untreated in many patients (Gandy et al., 2015).

Like depression, stress is a trigger for both anxiety and epilepsy, and neuropathology underlying both disorders have some commonality (Jones et al., 2015). The hippocampus and amygdala are critical in the pathophysiology of anxiety (Martin et al., 2009) and in generation of epileptic discharges in patients with epilepsy (Goddard, 1983; Lothman, 1994; Pitkänen, Tuunanen, Kälviäinen, Partanen, & Salmenperä, 1998). Chronic stress in rats provokes irreversible pathological changes in neuronal morphology in the amygdala, enlarged amygdala volume, and increased anxiety-like behavior (Vyas, Pillai, & Chattarji, 2004). Also, MRI scans detect enlarged amygdala volumes in patients with temporal lobe epilepsy (TLE) (Mitsueda-Ono et al., 2011), suggesting a common structural abnormality that may contribute to the comorbid appearance of anxiety in epilepsy.

7.2 Depression in epilepsy

According to epidemiological estimates, approximately 55% of patients with temporal lobe epilepsy exhibit comorbid depression (Jackson & Turkington, 2005; Lambert & Robertson, 1999). Population studies have shown that depression is six times more frequent in newly diagnosed patients with epilepsy compared to controls without epilepsy (Forsgren & Nyström, 1990). Recent studies have reinforced the nature of a bidirectional relationship between depression and epilepsy in patients (Hesdorffer et al., 2012; Kanner, 2006). For example, older patients with depression have a fourfold higher risk for having an unprovoked seizure (Hesdorffer et al., 2012). Unfortunately, certain anticonvulsants prescribed to control seizures may have adverse side effects that include increased risk for depression. Recent reports suggest that there is an increased incidence of adverse events in patients with epilepsy that take anticonvulsants compared to asymptomatic patients with epilepsy (Kanner et al., 2012). Depression in epilepsy is also markedly underdiagnosed (Kanner, 2009).

The comorbid appearance of depression in epilepsy and the overlapping mechanisms underlying each disorder is emphasized by similarities in structural damage associated with each disorder. The temporal lobe regions are vulnerable to disease associated damage in both epilepsy and depression, with significant atrophy and volumetric decreases in the hippocampus (Bernasconi et al., 2003; Garcia, 2012; Mazarati & Sankar, 2016) and increased amygdalar volume (van Elst, Woermann, Lemieux, & Trimble, 1999). Binding to serotonin receptors is also reduced in the hippocampi in patients with both depression and epilepsy (Theodore et al., 2007).

Furthermore, stress is a trigger for both epilepsy and depression (Kanner, 2003) and chronic exposure to stress hormones has been shown to elicit pathological alterations in neuronal structure and functionality in limbic structures, similar to those seen in separate depression and epilepsy studies. These changes include atrophy of dendrites in hippocampal CA3 pyramidal neurons, and reorganization and decrease in dendritic branching and spine loss in the prefrontal cortex (Arnsten, 2009; Seib & Wellman, 2003; Wellman, 2001; Woolley, Gould, & McEwen, 1990). The changes brought on by prolonged glucocorticoid exposure worsen disease progression for both epilepsy and depression.

8. How the HPA axis contributes to comorbidity

The bidirectional relationship of psychiatric comorbidities and epilepsy suggests a common pathophysiological mechanism that underlies their neurological manifestations. In addition to being upregulated in epilepsy, HPA axis hyperactivity is a hallmark feature of mood disorders and has been implicated in anxiety states as well. The shared pathological feature of HPA axis hyperactivity may be a contributing factor in the comorbid appearance of affective states in epilepsy (Maguire & Salpekar, 2013).

Increased HPA axis function is also a common trait between anxiety and epilepsy (Arborelius, Owens, Plotsky, & Nemeroff, 1999), though this overlapping feature has not been extensively explored. Acute exogenous corticosterone exposure in rodents provokes morphological alterations in amygdala neurons and elicits anxiety-like behavior in the elevated plus maze paradigm in rats (Mitra & Sapolsky, 2008), suggesting that HPA axis hyperactivity alone can lead to anxiety-like states. As discussed earlier, both rodent models of epilepsy and human studies exhibit increased HPA axis function. Additionally, rodent models of TLE exhibit anxiety-like behavior (Müller, Bankstahl, Gröticke, & Löscher, 2009; Muller, Groticke, Bankstahl, & Loscher, 2009; Zeidler et al., 2018). These studies suggest a potential overlap in HPA axis hyperexcitability in contributing to pathology in epilepsy and anxiety. In support of this, mice that undergo pilocarpine induced status epilepticus exhibit elevated corticosterone levels at baseline and during stress exposure, reduced FOS expression in the amygdala and hippocampus, and display anxiety-like behavior in the elevated plus maze and open field tests (Wulsin et al., 2018).

Numerous studies have demonstrated that the HPA axis is pathologically overactive in depression (de Kloet, Joels, & Holsboer, 2005; Rubin, Phillips, McCracken, & Sadow, 1996; Rubin, Poland, Lesser, & Martin, 1987; Swaab, Bao, & Lucassen, 2005; Wassink, Piven, & Patil, 2001). Patients with major depressive disorder exhibit an increase in plasma, urine and cerebrospinal fluid concentration of corticosterone equivalents (Gold, Goodwin, & Chrousos, 1988; Holsboer & Barden, 1996; Pariante & Miller, 2001). Postmortem studies of suicide victims, presumably thought to be depressed, show an increase in adrenal gland volume and weight, which is indicative of adrenal gland hypertrophy (Dorovini-Zis & Zis, 1987). A more precise assessment using magnetic resonance imaging supports this finding in clinically depressed, unmedicated patients (Rubin, Mandell, & Crandall, 1966). Patients suffering from depression also exhibit elevated concentration of CRH in the cerebrospinal fluid and an increase in CRH mRNA expression and protein levels in the paraventricular nucleus (PVN) of the hypothalamus (Gold et al., 1988; Raadsheer et al., 1995). The hypersecretion of CRH in these studies suggests that the HPA axis is overactive in clinically depressed patients.

While rodent models of TLE have been shown to exhibit depression-like behavior (Koh, Magid, Chung, Stine, & Wilson, 2007; Mazarati et al., 2008; Tchekalarova et al., 2011; Vrinda et al., 2017), only a few recent rodent studies have implicated a hyperactive HPA axis in contributing to the comorbid appearance of epilepsy and depression-like states. Elevations in HPA axis activity positively correlate with increased immobility time, an indication of depression-like behavior, in the forced swim test in chronically epileptic mice (Hooper, Paracha, & Maguire, 2018; Mazarati et al., 2009). However, there are studies that show that status epilepticus induced by either pilocarpine (Groticke, Hoffmann, & Loscher, 2007; Müller et al., 2009; Muller et al., 2009) or dorsal intrahippocampal kainic acid (IHKA) (Gröticke, Hoffmann, & Löscher, 2008) injections elicits increased mobility in the forced swim test. These contradictory results may be due to a range of factors including age, rodent strain, type of status epilepticus induction paradigm used, and lab to lab variability in conducting the assays. Interestingly, a recent study found that ventral intrahippocampal kainic acid (IHKA) injections elicit stronger affective behavioral states in mice compared to the widely used dorsal IHKA injections (Zeidler et al., 2018). In contrast to the dorsal region, the ventral hippocampus projects to the amygdala and contributes to emotional processing (Cenquizca & Swanson, 2007; Fanselow & Dong, 2010). Moreover, the ventral hippocampus is critical in negatively regulating the HPA axis by sending indirect inhibitory projections to the PVN of the hypothalamus (Jankord & Herman, 2008). Lesions to the ventral hippocampal region disinhibit the HPA axis (Herman, Cullinan, Morano, Akil, & Watson, 1995; Jankord & Herman, 2008). These data suggest that the location of seizure generation and seizure-induced damage may influence depression-like states in experimental epilepsy models.

Pharmacologically blocking glucocorticoid production prior to (Smith-Swintosky et al., 1995) or inhibiting steroidogenesis simultaneously with (Stein & Sapolsky, 1988) kainic acid seizure induction reduces seizure-induced hippocampal damage, suggesting that corticosterone production during seizures also contributes to epileptogenic neuronal injury. Furthermore, following pilocarpine induced status epilepticus, pharmacologically blocking glucocorticoid receptors with mifeprestone (RU486) reduces status epilepticus induced hippocampal damage (Wulsin, Herman, & Danzer, 2016). The proconvulsant actions of corticosterone are thought to be mediated predominantly by mineralocorticoids (Roberts & Keith, 1994). In line with this idea, pharmacologically blocking mineralocorticoid (MR) activity in vitro following exogenous corticosterone application eliminates excitation in the CA1 region whereas blocking glucocorticoids does not (Joëls & de Kloet, 1992). Collectively, these data support the direct impact of stress hormones on neuronal excitability and in turn, seizure susceptibility.

Chronic stress has also been shown to decrease GABAergic neurosteroid synthesis (Serra, Pisu, Mostallino, Sanna, & Biggio, 2008). Because neurosteroids are potent positive modulators of $GABA_A$ receptors, the

reduced neurosteroid concentration supports the idea that chronic stress impairs neuronal inhibition. Consistent with this, brain concentration of the neurosteroids allopregnanolone, pregnenolone, progesterone, and allotetrahydrodeoxycorticoster
one are all reduced following chronic social isolation (Serra et al., 2000). The reduced neurosteroid concentration in the brain could be explained by the findings in another study demonstrating a decrease in the expression of 5α-reductase, an enzyme that is important in the synthesis of the GABAergic neurosteroids (Dong et al., 2001). A reduction in neurosteroidogenesis following chronic stress may compromise GABAergic inhibition and create a neuronal environment that is prone to hyperexcitability and seizures. The effects may be long lasting as well. Recent studies demonstrate long term elevations in the levels of corticosterone in chronically epileptic mice, lasting 10 weeks poststatus epilepticus (Wulsin et al., 2018).

9. Treatment strategies

Reduction in stress seems reasonable as an initial treatment strategy for epilepsy. Given the strong circumstantial evidence suggesting a relationship of stress and seizure susceptibility, it is intuitive to consider stress reduction as a valid treatment. Stress cannot be eliminated of course, but may be managed. Psychotherapeutic approaches may be very helpful to reduce stress and offer coping strategies for anxiety management. However, given the complex physiologic changes in response to acute and chronic stressors, pharmacologic agents for treatment are often considered.

The intuitive notion that stress reduction may prevent seizures reinforces that idea that medication for anxiety may also lead to improvement in epilepsy. However, this notion is not well accepted given the lack of clear evidence. The idea of using a psychotropic medication to improve epilepsy, may be unconventional, but still may be valid nevertheless, especially when considering physiologic stress responses and overlaps of epilepsy and mood and anxiety disorder pathophysiology. Anticonvulsants themselves often have psychiatric benefit, and have been observed in an epilepsy population to simultaneously improve seizures as well as behavior (Salpekar et al., 2006). The idea that there are common therapeutic options for treating affective disorders and epilepsy is becoming better accepted.

9.1 Antidepressants

Antidepressants have become the mainstay of treatment for both anxiety and depressive disorders. Antidepressants, especially newer generation compounds, avoid the negative aspects of benzodiazepines including sedation, and propensity to be habit forming. Antidepressants have been widely studied and are among the most commonly prescribed medicines worldwide.

Antidepressants are still widely used despite the fact that they are uncommonly studied in an epilepsy population. However, some evidence exists suggesting that newer antidepressants at least do not worsen epilepsy. An open label study of 100 adults with epilepsy and depression taking sertraline found only a few that seemed to have seizure exacerbation (Kanner, Kozak, & Frey, 2000). Many had improved depressive symptoms.

However, first-generation antidepressants, most notably tricyclic antidepressants, were found to potentiate neuronal excitability and exacerbate preexisting epilepsy. In contrast, second-generation antidepressants have been found to have a decreased impact on neuronal excitability and accordingly have been somewhat more successful in simultaneously alleviating symptoms of mood disorders and epilepsy. Selective serotonin reuptake inhibitors (SSRIs) and serotonin-norepinephrine reuptake inhibitors (SNRIs) are the two most common classes of drugs prescribed to patients with depression and anxiety (Cardamone, Salzberg, O'Brien, & Jones, 2013). Both SSRIs and SNRIs have been shown to exhibit anticonvulsant properties (Cardamone et al., 2013). Analogously, anticonvulsants like pregabalin, valproate, carbamazepine and lamotrigine are also successful mood stabilizers. Antidepressants, particularly SSRIs, are thought to suppress HPA axis function in both human (Himmerich et al., 2007; Pariante, 2006) and animal studies (Jensen et al., 1999; Jongsma, Bosker, Cremers, Westerink, & Boer, 2005), potentially through modulating expression of mineralocorticoid and glucocorticoid receptors (Barden, 1999; Pepin, Beaulieu, & Barden, 1989; Reul, 1993; Reul, Labeur, Grigoriadis, Souza, & Holsboer, 1994). Furthermore, SSRIs have been found to contribute to increasing GABAergic signaling (Krystal et al., 2002), in part through production of neurosteroids (Longone et al., 2008; Matsumoto, Puia, Dong, & Pinna, 2009; Nemeroff & Owens, 2004), which is normally compromised in epilepsy, depression and HPA axis hyperactivity. Several studies have demonstrated that antidepressants may also be involved in inhibiting sodium, potassium, or calcium channels to exert their anticonvulsant actions (Cardamone et al., 2013). These drug studies not only emphasize overlapping mechanisms underlying affective states and epilepsy, but they also highlight a promising therapeutic avenue in simultaneously treating comorbid psychiatric disorders in epilepsy with fewer drugs and without compromising the state of either disorder.

9.2 Benzodiazepines

Benzodiazepines are essential parts of the therapeutic repertoire for epilepsy. Rescue medicines include lorazepam and diazepam, and are often used for status epilepticus when other treatments are unsuccessful. Benzodiazepines are also excellent medicines for acute anxiety or panic attacks. Alprazolam is often used for that purpose, and may even be considered a hybrid medicine that has the structure of a benzodiazepine but more antidepressant effect than others in its class. However, the utility of this class of medicine is markedly limited by its propensity to cause sedation and cognitive slowing or dulling. Furthermore, the medicines may be habit forming, with tolerance preventing consistent responses from routine usage.

Another option includes use of a benzodiazepine variant. Clobazam in particular, is intriguing as it is an atypical benzodiazepine that targets GABA subunits that are different from other typical 1,4 benzodiazepines. This may allow better tolerability in terms of sedation and cognitive side effects. Clobazam is widely used in Europe and Asia for anxiety yet only has regulatory approval in the United States for Lennox Gastaut syndrome. It is structured slightly differently, in that the nitrogen atoms are in different locations on the aromatic ring. For clobazam the nitrogen atoms are in the 1,5 positions as shown in Fig. 1. Successful use of clobazam would strongly suggest that an anticonvulsant approach to treating both anxiety and epilepsy simultaneously may be valid. Preliminary evidence suggests that clobazam may be effective in treating both conditions (Brodie et al., 2016).

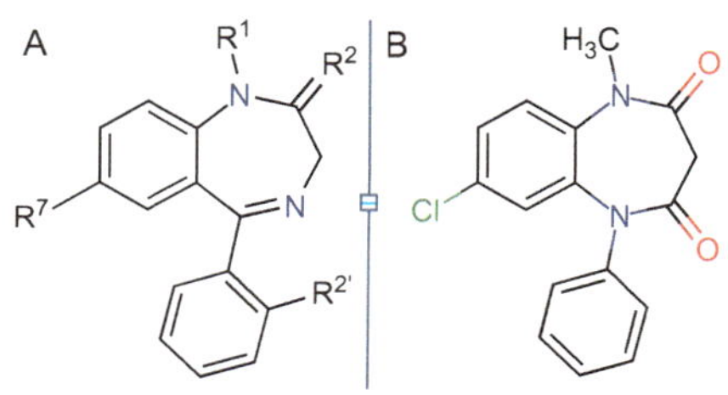

Fig. 1 Typical benzodiazepine structure (A) and clobazam structure (B).

10. Conclusion

Stress, anxiety, and epilepsy appear to be intimately intertwined. Stress itself influences the likelihood for seizures to occur and may prove to have long term effects upon the development of recurrent seizures. The effects upon the regulation and stability of the hypothalamic pituitary adrenal axis may be profound. Anxiety and depression often overlap symptomatically but also in terms of treatment strategies. Antidepressants and benzodiazepines are the mainstay for treatment of anxiety with or without epilepsy and are commonly used in this comorbidity. However, evidence guiding treatment is uncommon. Anticonvulsants themselves may also prove to be useful for direct anxiety control. Atypical benzodiazepines like alprazolam or clobazam may prove to be especially efficacious in addressing anxiety and epilepsy comorbidity. Future studies will be enlightening in clarifying methods for stress reduction and anxiety management, and will undoubtedly make use of a variety of treatment modalities.

References

Adamec, R., & Shallow, T. (2000). Effects of baseline anxiety on response to kindling of the right medial amygdala. *Physiology & Behavior, 70*(1–2), 67–80.

American Psychiatric Association. (2013). *Diagnostic and statistical manual of mental disorders* (5th ed.). Arlington, VA: American Psychiatric Association.

Arborelius, L., Owens, M. J., Plotsky, P. M., & Nemeroff, C. B. (1999). The role of corticotropin-releasing factor in depression and anxiety disorders. *Journal of Endocrinology, 160*(1), 1–12. https://doi.org/10.1677/joe.0.1600001.

Arnsten, A. F. T. (2009). Stress signalling pathways that impair prefrontal cortex structure and function. *Nature Reviews. Neuroscience, 10*(6), 410–422. https://doi.org/10.1038/nrn2648.

Barden, N. (1999). Regulation of corticosteroid receptor gene expression in depression and antidepressant action. *Journal of Psychiatry & Neuroscience, 24*(1), 25–39.

Berkhout, J., Walter, D. O., & Adey, W. R. (1969). Alterations of the human electroencephalogram induced by stressful verbal activity. *Electroencephalography and Clinical Neurophysiology, 27*(5), 457–469. https://doi.org/10.1016/0013-4694(69)90186-2.

Bernasconi, N., Bernasconi, A., Caramanos, Z., Antel, S. B., Andermann, F., & Arnold, D. L. (2003). Mesial temporal damage in temporal lobe epilepsy: A volumetric MRI study of the hippocampus, amygdala and parahippocampal region. *Brain, 126*(2), 462–469. https://doi.org/10.1093/brain/awg034.

Beyenburg, S., Mitchell, A. J., Schmidt, D., Elger, C. E., & Reuber, M. (2005). Anxiety in patients with epilepsy: Systematic review and suggestions for clinical management. *Epilepsy & Behavior,* 7(2), 161–171. https://doi.org/10.1016/j.yebeh.2005.05.014.

Beyenburg, S., Stoffel-Wagner, B., Bauer, J., Watzka, M., Blümcke, I., Bidlingmaier, F., et al. (2001). Neuroactive steroids and seizure susceptibility. *Epilepsy Research, 44*(2–3), 141–153. https://doi.org/10.1016/s0920-1211(01)00194-2.

Bosnjak, J., Vukovic-Bobic, M., & Mejaski-Bosnjak, V. (2002). Effect of war on the occurrence of epileptic seizures in children. *Epilepsy & Behavior, 3*(6), 502–509. https://doi.org/10.1016/s1525-5050(02)00602-9.

Boylan, L. S., Flint, L. A., Labovitz, D. L., Jackson, S. C., Starner, K., & Devinsky, O. (2004). Depression but not seizure frequency predicts quality of life in treatment-resistant epilepsy. *Neurology, 62*(2), 258–261. https://doi.org/10.1212/01.wnl.0000103282.62353.85.

Brandt, C., Schoendienst, M., Trentowska, M., May, T. W., Pohlmann-Eden, B., Tuschen-Caffier, B., et al. (2010). Prevalence of anxiety disorders in patients with refractory focal epilepsy—A prospective clinic based survey. *Epilepsy & Behavior, 17*(2), 259–263. https://doi.org/10.1016/j.yebeh.2009.12.009.

Brodie, M. J., Chung, S., Wade, A., Quelen, C., Guiraud-Diawara, A., Francois, C., et al. (2016). Clobazam and clonazepam use in epilepsy: Results from a UK database incident user cohort study. *Epilepsy Research, 123*, 68–74. https://doi.org/10.1016/j.eplepsyres.2016.04.003.

Cardamone, L., Salzberg, M. R., O'Brien, T. J., & Jones, N. C. (2013). Antidepressant therapy in epilepsy: Can treating the comorbidities affect the underlying disorder? *British Journal of Pharmacology, 168*(7), 1531–1554. https://doi.org/10.1111/bph.12052.

Cenquizca, L. A., & Swanson, L. W. (2007). Spatial organization of direct hippocampal field CA1 axonal projections to the rest of the cerebral cortex. *Brain Research Reviews, 56*(1), 1–26. https://doi.org/10.1016/j.brainresrev.2007.05.002.

Christensen, J., Li, J., Vestergaard, M., & Olsen, J. (2007). Stress and epilepsy: A population-based cohort study of epilepsy in parents who lost a child. *Epilepsy & Behavior, 11*(3), 324–328. https://doi.org/10.1016/j.yebeh.2007.06.003.

Christensen, J., Vestergaard, M., Mortensen, P. B., Sidenius, P., & Agerbo, E. (2007). Epilepsy and risk of suicide: A population-based case–control study. *The Lancet Neurology, 6*(8), 693–698. https://doi.org/10.1016/s1474-4422(07)70175-8.

de Kloet, E. R., Joels, M., & Holsboer, F. (2005). Stress and the brain: From adaptation to disease. *Nature Reviews. Neuroscience, 6*(6), 463–475. https://doi.org/10.1038/nrn1683.

Dong, E., Matsumoto, K., Uzunova, V., Sugaya, I., Takahata, H., Nomura, H., et al. (2001). Brain 5α-dihydroprogesterone and allopregnanolone synthesis in a mouse model of protracted social isolation. *Proceedings of the National Academy of Sciences of the United States of America, 98*(5), 2849–2854. https://doi.org/10.1073/pnas.051628598.

Dorovini-Zis, K., & Zis, A. P. (1987). Increased adrenal weight in victims of violent suicide. *American Journal of Psychiatry, 144*(9), 1214–1215. https://doi.org/10.1176/ajp.144.9.1214.

Drevets, W. C. (2003). Neuroimaging abnormalities in the amygdala in mood disorders. *Annals of the New York Academy of Sciences, 985*, 420–444.

Drevets, W. C., Price, J. L., & Furey, M. L. (2008). Brain structural and functional abnormalities in mood disorders: Implications for neurocircuitry models of depression. *Brain Structure & Function, 213*(1–2), 93–118. https://doi.org/10.1007/s00429-008-0189-x.

Drevets, W. C., Videen, T. O., Price, J. L., Preskorn, S. H., Carmichael, S. T., & Raichle, M. E. (1992). A functional anatomical study of unipolar depression. *The Journal of Neuroscience, 12*(9), 3628–3641.

Fanselow, M. S., & Dong, H.-W. (2010). Are the dorsal and ventral hippocampus functionally distinct structures? *Neuron, 65*(1), 7–19. https://doi.org/10.1016/j.neuron.2009.11.031.

Ferlisi, M., & Shorvon, S. (2014). Seizure precipitants (triggering factors) in patients with epilepsy. *Epilepsy & Behavior, 33*, 101–105. https://doi.org/10.1016/j.yebeh.2014.02.019.

Forsgren, L., & Nyström, L. (1990). An incident case-referent study of epileptic seizures in adults. *Epilepsy Research, 6*(1), 66–81. https://doi.org/10.1016/0920-1211(90)90010-s.

Gaitatzis, A., Carroll, K., Majeed, A., & Sander, L. (2004). The epidemiology of the comorbidity of epilepsy in the general population. *Epilepsia*, *45*(12), 1613–1622. https://doi.org/10.1111/j.0013-9580.2004.17504.x.
Gandy, M., Sharpe, L., Perry, K. N., Miller, L., Thayer, Z., Boserio, J., et al. (2015). Anxiety in epilepsy: A neglected disorder. *Journal of Psychosomatic Research*, *78*(2), 149–155. https://doi.org/10.1016/j.jpsychores.2014.12.002.
Garcia, C. S. (2012). Depression in temporal lobe epilepsy: A review of prevalence, clinical features, and management considerations. *Epilepsy Research and Treatment*, *2012*, 809843. https://doi.org/10.1155/2012/809843.
Gélisse, P., Genton, P., Coubes, P., Tang, N. P. L., & Crespel, A. (2015). Can emotional stress trigger the onset of epilepsy? *Epilepsy & Behavior*, *48*, 15–20. https://doi.org/10.1016/j.yebeh.2015.05.010.
Gloor, P., Olivier, A., Quesney, L. F., Andermann, F., & Horowitz, S. (1982). The role of the limbic system in experiential phenomena of temporal lobe epilepsy. *Annals of Neurology*, *12*(2), 129–144.
Goddard, G. V. (1983). The kindling model of epilepsy. *Trends in Neurosciences*, *6*, 275–279. https://doi.org/10.1016/0166-2236(83)90118-2.
Gold, P. W., Goodwin, F. K., & Chrousos, G. P. (1988). Clinical and biochemical manifestations of depression. *The New England Journal of Medicine*, *319*(7), 413–420. https://doi.org/10.1056/nejm198808183190706.
Goldstein, L. H., & Mellers, J. D. C. (2006). Ictal symptoms of anxiety, avoidance behaviour, and dissociation in patients with dissociative seizures. *Journal of Neurology, Neurosurgery & Psychiatry*, 77(5), 616. https://doi.org/10.1136/jnnp.2005.066878.
Groticke, I., Hoffmann, K., & Loscher, W. (2007). Behavioral alterations in the pilocarpine model of temporal lobe epilepsy in mice. *Experimental Neurology*, *207*(2), 329–349. https://doi.org/10.1016/j.expneurol.2007.06.021.
Gröticke, I., Hoffmann, K., & Löscher, W. (2008). Behavioral alterations in a mouse model of temporal lobe epilepsy induced by intrahippocampal injection of kainate. *Experimental Neurology*, *213*(1), 71–83. https://doi.org/10.1016/j.expneurol.2008.04.036.
Harris, E. C., & Barraclough, B. (1997). Suicide as an outcome for mental disorders. *British Journal of Psychiatry*, *170*(3), 205–228. https://doi.org/10.1192/bjp.170.3.205.
Herman, J. P., Cullinan, W. E., Morano, M. I., Akil, H., & Watson, S. J. (1995). Contribution of the ventral subiculum to inhibitory regulation of the hypothalamo-pituitary-adrenocortical axis. *Journal of Neuroendocrinology*, 7(6), 475–482. https://doi.org/10.1111/j.1365-2826.1995.tb00784.x.
Hesdorffer, D. C., Ishihara, L., Mynepalli, L., Webb, D. J., Weil, J., & Hauser, W. A. (2012). Epilepsy, suicidality, and psychiatric disorders: A bidirectional association. *Annals of Neurology*, 72(2), 184–191. https://doi.org/10.1002/ana.23601.
Himmerich, H., Zimmermann, P., Ising, M., Kloiber, S., Lucae, S., Künzel, H. E., et al. (2007). Changes in the hypothalamic-pituitary-adrenal axis and leptin levels during antidepressant treatment. *Neuropsychobiology*, *55*(1), 28–35. https://doi.org/10.1159/000103573.
Hitiris, N., Mohanraj, R., Norrie, J., Sills, G. J., & Brodie, M. J. (2007). Predictors of pharmacoresistant epilepsy. *Epilepsy Research*, *75*(2–3), 192–196. https://doi.org/10.1016/j.eplepsyres.2007.06.003.
Holsboer, F., & Barden, N. (1996). Antidepressants and hypothalamic-pituitary-adrenocortical regulation. *Endocrine Reviews*, *17*(2), 187–205. https://doi.org/10.1210/edrv-17-2-187.
Hooper, A., Paracha, R., & Maguire, J. (2018). Seizure-induced activation of the HPA axis increases seizure frequency and comorbid depression-like behaviors. *Epilepsy & Behavior*, *78*, 124–133. https://doi.org/10.1016/j.yebeh.2017.10.025.

Jackson, M. J., & Turkington, D. (2005). Depression and anxiety in epilepsy. *Journal of Neurology, Neurosurgery & Psychiatry*, *76*(Suppl. 1), i45. https://doi.org/10.1136/jnnp.2004.060467.

Jankord, R., & Herman, J. P. (2008). Limbic regulation of hypothalamo-pituitary-adrenocortical function during acute and chronic stress. *Annals of the New York Academy of Sciences*, *1148*(1), 64–73. https://doi.org/10.1196/annals.1410.012.

Jensen, J. B., Jessop, D. S., Harbuz, M. S., Mørk, A., Sánchez, C., & Mikkelsen, J. D. (1999). Acute and long-term treatments with the selective serotonin reuptake inhibitor citalopram modulate the HPA axis activity at different levels in male rats. *Journal of Neuroendocrinology*, *11*(6), 465–471. https://doi.org/10.1046/j.1365-2826.1999.00362.x.

Joëls, M., & de Kloet, E. R. (1992). Control of neuronal excitability by corticosteroid hormones. *Trends in Neurosciences*, *15*(1), 25–30. https://doi.org/10.1016/0166-2236(92)90345-9.

Jones, J. E., Jackson, D. C., Chambers, K. L., Dabbs, K., Hsu, D. A., Stafstrom, C. E., et al. (2015). Children with epilepsy and anxiety: Subcortical and cortical differences. *Epilepsia*, *56*(2), 283. https://doi.org/10.1111/epi.12832.

Jongsma, M. E., Bosker, F. J., Cremers, T. I. F. H., Westerink, B. H. C., & den Boer, J. A. (2005). The effect of chronic selective serotonin reuptake inhibitor treatment on serotonin1B receptor sensitivity and HPA axis activity. *Progress in Neuro-Psychopharmacology and Biological Psychiatry*, *29*(5), 738–744. https://doi.org/10.1016/j.pnpbp.2005.04.026.

Kanner, A. M. (2003). Depression in epilepsy: Prevalence, clinical semiology, pathogenic mechanisms, and treatment. *Biological Psychiatry*, *54*(3), 388–398. https://doi.org/10.1016/s0006-3223(03)00469-4.

Kanner, A. M. (2006). Depression and epilepsy: A new perspective on two closely related disorders. *Epilepsy Currents*, *6*(5), 141–146. https://doi.org/10.1111/j.1535-7511.2006.00125.x.

Kanner, A. M. (2009). Depression and epilepsy: Do glucocorticoids and glutamate explain their relationship? *Current Neurology and Neuroscience Reports*, *9*(4), 307–312. https://doi.org/10.1007/s11910-009-0046-1.

Kanner, A. M. (2011a). Anxiety disorders in epilepsy: The forgotten psychiatric comorbidity. *Epilepsy Currents*, *11*(3), 90–91. https://doi.org/10.5698/1535-7511-11.3.90.

Kanner, A. M. (2011b). Depression and epilepsy: A bidirectional relation? *Epilepsia*, *52*(s1), 21–27. https://doi.org/10.1111/j.1528-1167.2010.02907.x.

Kanner, A. M. (2016). Management of psychiatric and neurological comorbidities in epilepsy. *Nature Reviews. Neurology*, *12*(2), 106–116. https://doi.org/10.1038/nrneurol.2015.243.

Kanner, A. M., Barry, J. J., Gilliam, F., Hermann, B., & Meador, K. J. (2012). Depressive and anxiety disorders in epilepsy: Do they differ in their potential to worsen common antiepileptic drug-related adverse events? *Epilepsia*, *53*(6), 1104–1108. https://doi.org/10.1111/j.1528-1167.2012.03488.x.

Kanner, A. M., Byrne, R., Chicharro, A., Wuu, J., & Frey, M. (2009). A lifetime psychiatric history predicts a worse seizure outcome following temporal lobectomy. *Neurology*, *72*(9), 793–799. https://doi.org/10.1212/01.wnl.0000343850.85763.9c.

Kanner, A. M., Kozak, A. M., & Frey, M. (2000). The use of sertraline in patients with epilepsy: Is it safe? *Epilepsy & Behavior*, *1*(2), 100–105.

Koh, S., Magid, R., Chung, H., Stine, C. D., & Wilson, D. N. (2007). Depressive behavior and selective downregulation of serotonin receptor expression after early-life seizures: Reversal by environmental enrichment. *Epilepsy & Behavior*, *10*(1), 26–31. https://doi.org/10.1016/j.yebeh.2006.11.008.

Krystal, J. H., Sanacora, G., Blumberg, H., Anand, A., Charney, D. S., Marek, G., et al. (2002). Glutamate and GABA systems as targets for novel antidepressant and mood-stabilizing treatments. *Molecular Psychiatry*, 7(S1), 4001021. https://doi.org/10.1038/sj.mp.4001021.

Kwon, O.-Y., & Park, S.-P. (2014). Depression and anxiety in people with epilepsy. *Journal of Clinical Neurology*, *10*(3), 175–188. https://doi.org/10.3988/jcn.2014.10.3.175.

Lambert, M. V., & Robertson, M. M. (1999). Depression in epilepsy: Etiology, phenomenology, and treatment. *Epilepsia*, *40*(Suppl. 10), S21–S47.

Longone, P., Rupprecht, R., Manieri, G. A., Bernardi, G., Romeo, E., & Pasini, A. (2008). The complex roles of neurosteroids in depression and anxiety disorders. *Neurochemistry International*, *52*(4–5), 596–601. https://doi.org/10.1016/j.neuint.2007.10.001.

Lothman, E. W. (1994). Seizure circuits in the hippocampus and associated structures. *Hippocampus*, *4*(3), 286–290. https://doi.org/10.1002/hipo.450040311.

Maguire, J., & Salpekar, J. A. (2013). Stress, seizures, and hypothalamic-pituitary-adrenal axis targets for the treatment of epilepsy. *Epilepsy & Behavior*, *26*(3), 352–362. https://doi.org/10.1016/j.yebeh.2012.09.040.

Martin, E. I., Ressler, K. J., Binder, E., & Nemeroff, C. B. (2009). The neurobiology of anxiety disorders: Brain imaging, genetics, and psychoneuroendocrinology. *Psychiatric Clinics of North America*, *32*(3), 549–575. https://doi.org/10.1016/j.psc.2009.05.004.

Masia, S. L., & Devinsky, O. (2000). Epilepsy and behavior: A brief history. *Epilepsy & Behavior*, *1*(1), 27–36.

Matsumoto, K., Puia, G., Dong, E., & Pinna, G. (2009). GABAA receptor neurotransmission dysfunction in a mouse model of social isolation-induced stress: Possible insights into a non-serotonergic mechanism of action of SSRIs in mood and anxiety disorders. *Stress*, *10*(1), 3–12. https://doi.org/10.1080/10253890701200997.

Mazarati, A., & Sankar, R. (2016). Common mechanisms underlying epileptogenesis and the comorbidities of epilepsy. *Cold Spring Harbor Perspectives in Medicine*, *6*(7), a022798. https://doi.org/10.1101/cshperspect.a022798.

Mazarati, A. M., Shin, D., Kwon, Y. S., Bragin, A., Pineda, E., Tio, D., et al. (2009). Elevated plasma corticosterone level and depressive behavior in experimental temporal lobe epilepsy. *Neurobiology of Disease*, *34*(3), 457–461. https://doi.org/10.1016/j.nbd.2009.02.018.

Mazarati, A., Siddarth, P., Baldwin, R. A., Shin, D., Caplan, R., & Sankar, R. (2008). Depression after status epilepticus: Behavioural and biochemical deficits and effects of fluoxetine. *Brain*, *131*(Pt. 8), 2071–2083. https://doi.org/10.1093/brain/awn117.

Mitra, R., & Sapolsky, R. M. (2008). Acute corticosterone treatment is sufficient to induce anxiety and amygdaloid dendritic hypertrophy. *Proceedings of the National Academy of Sciences of the United States of America*, *105*(14), 5573–5578. https://doi.org/10.1073/pnas.0705615105.

Mitsueda-Ono, T., Ikeda, A., Inouchi, M., Takaya, S., Matsumoto, R., Hanakawa, T., et al. (2011). Amygdalar enlargement in patients with temporal lobe epilepsy. *Journal of Neurology, Neurosurgery & Psychiatry*, *82*(6), 652. https://doi.org/10.1136/jnnp.2010.206342.

Moore, J. L., Elliott, J. O., Lu, B., Klatte, E. T., & Charyton, C. (2009). Serious psychological distress among persons with epilepsy based on the 2005 California Health Interview Survey. *Epilepsia*, *50*(5), 1077–1084. https://doi.org/10.1111/j.1528-1167.2008.01996.x.

Moshe, S., Shilo, M., Chodick, G., Yagev, Y., Blatt, I., Korczyn, A. D., et al. (2008). Occurrence of seizures in association with work-related stress in young male army recruits. *Epilepsia*, *49*(8), 1451–1456. https://doi.org/10.1111/j.1528-1167.2008.01591.x.

Mula, M. (2013). Treatment of anxiety disorders in epilepsy: An evidence-based approach. *Epilepsia*, *54*(s1), 13–18. https://doi.org/10.1111/epi.12101.

Müller, C. J., Bankstahl, M., Gröticke, I., & Löscher, W. (2009). Pilocarpine vs. lithium-pilocarpine for induction of status epilepticus in mice: Development of spontaneous seizures, behavioral alterations and neuronal damage. *European Journal of Pharmacology*, *619*(1–3), 15–24. https://doi.org/10.1016/j.ejphar.2009.07.020.

Muller, C. J., Groticke, I., Bankstahl, M., & Loscher, W. (2009). Behavioral and cognitive alterations, spontaneous seizures, and neuropathology developing after a pilocarpine-induced status epilepticus in C57BL/6 mice. *Experimental Neurology*, *219*(1), 284–297. https://doi.org/10.1016/j.expneurol.2009.05.035.

Nakken, K. O., Solaas, M. H., Kjeldsen, M. J., Friis, M. L., Pellock, J. M., & Corey, L. A. (2005). Which seizure-precipitating factors do patients with epilepsy most frequently report? *Epilepsy & Behavior*, *6*(1), 85–89. https://doi.org/10.1016/j.yebeh.2004.11.003.

Nemeroff, C. B., & Owens, M. J. (2004). Pharmacologic differences among the SSRIs: Focus on monoamine transporters and the HPA axis. *CNS Spectrums*, *9*(S4), 23–31. https://doi.org/10.1017/s1092852900025475.

Neufeld, M. Y., Sadeh, M., Cohn, D. F., & Korczyn, A. D. (1994). Stress and epilepsy: The Gulf war experience. *Seizure*, *3*(2), 135–139. https://doi.org/10.1016/s1059-1311(05)80204-3.

Pariante, C. M. (2006). The glucocorticoid receptor: Part of the solution or part of the problem? *Journal of Psychopharmacology*, *20*(4_suppl), 79–84. https://doi.org/10.1177/1359786806066063.

Pariante, C. M., & Miller, A. H. (2001). Glucocorticoid receptors in major depression: Relevance to pathophysiology and treatment. *Biological Psychiatry*, *49*(5), 391–404. https://doi.org/10.1016/s0006-3223(00)01088-x.

Pepin, M.-C., Beaulieu, S., & Barden, N. (1989). Antidepressants regulate glucocorticoid receptor messenger RNA concentrations in primary neuronal cultures. *Molecular Brain Research*, *6*(1), 77–83. https://doi.org/10.1016/0169-328x(89)90031-4.

Pitkänen, A., Tuunanen, J., Kälviäinen, R., Partanen, K., & Salmenperä, T. (1998). Amygdala damage in experimental and human temporal lobe epilepsy. *Epilepsy Research*, *32*(1–2), 233–253. https://doi.org/10.1016/s0920-1211(98)00055-2.

Raadsheer, F. C., van Heerikhuize, J. J., Lucassen, P. J., Hoogendijk, W. J., Tilders, F. J., & Swaab, D. F. (1995). Corticotropin-releasing hormone mRNA levels in the paraventricular nucleus of patients with Alzheimer's disease and depression. *American Journal of Psychiatry*, *152*(9), 1372–1376. https://doi.org/10.1176/ajp.152.9.1372.

Reul, J. M. (1993). Chronic treatment of rats with the antidepressant amitriptyline attenuates the activity of the hypothalamic-pituitary-adrenocortical system. *Endocrinology*, *133*(1), 312–320. https://doi.org/10.1210/en.133.1.312.

Reul, J. M. H. M., Labeur, M. S., Grigoriadis, D. E., Souza, E. B. D., & Holsboer, F. (1994). Hypothalamic-pituitary-adrenocortical axis changes in the rat after long-term treatment with the reversible monoamine oxidase-A inhibitor moclobemide. *Neuroendocrinology*, *60*(5), 509–519. https://doi.org/10.1159/000126788.

Roberts, A. J., & Keith, L. D. (1994). Mineralocorticoid receptors mediate the enhancing effects of corticosterone on convulsion susceptibility in mice. *The Journal of Pharmacology and Experimental Therapeutics*, *270*(2), 505–511.

Rubin, R. T., Mandell, A. J., & Crandall, P. H. (1966). Corticosteroid responses to limbic stimulation in man: Localization of stimulus sites. *Science*, *153*(3737), 767–768. https://doi.org/10.1126/science.153.3737.767.

Rubin, R. T., Phillips, J. J., McCracken, J. T., & Sadow, T. F. (1996). Adrenal gland volume in major depression: Relationship to basal and stimulated pituitary-adrenal cortical axis function. *Biological Psychiatry*, *40*(2), 89–97. https://doi.org/10.1016/0006-3223(95)00358-4.

Rubin, R. T., Poland, R. E., Lesser, I. M., & Martin, D. J. (1987). Neuroendocrine aspects of primary endogenous depression—IV. Pituitary-thyroid axis activity in patients and matched control subjects. *Psychoneuroendocrinology*, *12*(5), 333–347. https://doi.org/10.1016/0306-4530(87)90062-x.

Salpekar, J. (2016). Mood disorders in epilepsy. *Focus*, *14*(4), 465–472.

Salpekar, J. (2017). Depression and epilepsy: We are no longer one direction. *Epilepsy Currents*, *17*(4), 223–224. https://doi.org/10.5698/1535-7597.17.4.223.

Salpekar, J. (2019). Seizures, nonepileptic events, trauma, anxiety, or all of the above. *Epilepsy Currents*, *19*(1), 29–30. https://doi.org/10.1177/1535759718822842.

Salpekar, J. A., Conry, J. A., Doss, W., Cushner-Weinstein, S., Pearl, P. L., Weinstein, S. L., et al. (2006). Clinical experience with anticonvulsant medication in pediatric epilepsy and comorbid bipolar spectrum disorder. *Epilepsy & Behavior*, *9*(2), 327–334. https://doi.org/10.1016/j.yebeh.2006.06.004.

Salpekar, J., & Dadson, N. (2012). Psychiatry and epilepsy: A century of evolving understanding. In R. E. Bautista (Ed.), *Epilepsy: A century of discovery* (pp. 233–255). Nova Science Publishers, Inc.

Salpekar, J. A., Mishra, G., & Hauptman, A. J. (2015). Key issues in addressing the comorbidity of depression and pediatric epilepsy. *Epilepsy & Behavior*, *46*, 12–18. https://doi.org/10.1016/j.yebeh.2015.02.036.

Seib, L. M., & Wellman, C. L. (2003). Daily injections alter spine density in rat medial prefrontal cortex. *Neuroscience Letters*, *337*(1), 29–32. https://doi.org/10.1016/s0304-3940(02)01287-9.

Serra, M., Pisu, M. G., Littera, M., Papi, G., Sanna, E., Tuveri, F., et al. (2000). Social isolation-induced decreases in both the abundance of neuroactive steroids and GABAA receptor function in rat brain. *Journal of Neurochemistry*, *75*(2), 732–740. https://doi.org/10.1046/j.1471-4159.2000.0750732.x.

Serra, M., Pisu, M. G., Mostallino, M. C., Sanna, E., & Biggio, G. (2008). Changes in neuroactive steroid content during social isolation stress modulate GABAA receptor plasticity and function. *Brain Research Reviews*, *57*(2), 520–530. https://doi.org/10.1016/j.brainresrev.2007.06.029.

Smith, D. F., Baker, G. A., Dewey, M., Jacoby, A., & Chadwick, D. W. (1991). Seizure frequency, patient-perceived seizure severity and the psychosocial consequences of intractable epilepsy. *Epilepsy Research*, *9*(3), 231–241. https://doi.org/10.1016/0920-1211(91)90057-m.

Smith-Swintosky, V. L., Pettigrew, L. C., Sapolsky, R. M., Phares, C., Craddock, S. D., Brooke, S. M., et al. (1995). Metyrapone, an inhibitor of glucocorticoid production, reduces brain injury induced by focal and global ischemia and seizures. *Journal of Cerebral Blood Flow & Metabolism*, *16*(4), 585–598. https://doi.org/10.1097/00004647-199607000-00008.

Stein, B. A., & Sapolsky, R. M. (1988). Chemical adrenalectomy reduces hippocampal damage induced by kainic acid. *Brain Research*, *473*(1), 175–180. https://doi.org/10.1016/0006-8993(88)90332-0.

Stevens, J. R. (1959). Emotional activation of the electroencephalogram in patients with convulsive disorders. *The Journal of Nervous and Mental Disease*, *128*(4), 339–351. https://doi.org/10.1097/00005053-195904000-00006.

Swaab, D. F., Bao, A.-M., & Lucassen, P. J. (2005). The stress system in the human brain in depression and neurodegeneration. *Ageing Research Reviews*, *4*(2), 141–194. https://doi.org/10.1016/j.arr.2005.03.003.

Swinkels, W. A. M., Engelsman, M., Trenité, D. G. A. K. N., Baal, M. G., Haan, G. J., & Oosting, J. (1998). Influence of an evacuation in February 1995 in the Netherlands on the seizure frequency in patients with epilepsy: A controlled study. *Epilepsia*, *39*(11), 1203–1207. https://doi.org/10.1111/j.1528-1157.1998.tb01312.x.

Tchekalarova, J., Pechlivanova, D., Atanasova, T., Markova, P., Lozanov, V., & Stoynev, A. (2011). Diurnal variations in depression-like behavior of Wistar and spontaneously hypertensive rats in the kainate model of temporal lobe epilepsy. *Epilepsy & Behavior, 20*(2), 277–285. https://doi.org/10.1016/j.yebeh.2010.12.021.

Temkin, N. R., & Davis, G. R. (1984). Stress as a risk factor for seizures among adults with epilepsy. *Epilepsia, 25*(4), 450–456. https://doi.org/10.1111/j.1528-1157.1984.tb03442.x.

Theodore, W. H., Hasler, G., Giovacchini, G., Kelley, K., Reeves-Tyer, P., Herscovitch, P., et al. (2007). Reduced hippocampal 5HT1A PET receptor binding and depression in temporal lobe epilepsy. *Epilepsia, 48*, 1526–1530.

Timofeev, I., Grenier, F., & Steriade, M. (1998). Spike-wave complexes and fast components of cortically generated seizures. IV. Paroxysmal fast runs in cortical and thalamic neurons. *Journal of Neurophysiology, 80*(3), 1495–1513.

van Elst, L. T., Woermann, F. G., Lemieux, L., & Trimble, M. R. (1999). Amygdala enlargement in dysthymia—A volumetric study of patients with temporal lobe epilepsy. *Biological Psychiatry, 46*(12), 1614–1623. https://doi.org/10.1016/s0006-3223(99)00212-7.

Vazquez, B., & Devinsky, O. (2003). Epilepsy and anxiety. *Epilepsy & Behavior, 4*, 20–25. https://doi.org/10.1016/j.yebeh.2003.10.005.

Vrinda, M., Sasidharan, A., Aparna, S., Srikumar, B. N., Kutty, B. M., & Rao, B. S. S. (2017). Enriched environment attenuates behavioral seizures and depression in chronic temporal lobe epilepsy. *Epilepsia, 58*(7), 1148–1158. https://doi.org/10.1111/epi.13767.

Vyas, A., Pillai, A. G., & Chattarji, S. (2004). Recovery after chronic stress fails to reverse amygdaloid neuronal hypertrophy and enhanced anxiety-like behavior. *Neuroscience, 128*(4), 667–673. https://doi.org/10.1016/j.neuroscience.2004.07.013.

Wassink, T. H., Piven, J., & Patil, S. R. (2001). Chromosomal abnormalities in a clinic sample of individuals with autistic disorder. *Psychiatric Genetics, 11*(2), 57–63.

Wellman, C. L. (2001). Dendritic reorganization in pyramidal neurons in medial prefrontal cortex after chronic corticosterone administration. *Journal of Neurobiology, 49*(3), 245–253. https://doi.org/10.1002/neu.1079.

Woolley, C. S., Gould, E., & McEwen, B. S. (1990). Exposure to excess glucocorticoids alters dendritic morphology of adult hippocampal pyramidal neurons. *Brain Research, 531*(1–2), 225–231. https://doi.org/10.1016/0006-8993(90)90778-a.

Wulsin, A. C., Franco-Villanueva, A., Romancheck, C., Morano, R. L., Smith, B. L., Packard, B. A., et al. (2018). Functional disruption of stress modulatory circuits in a model of temporal lobe epilepsy. *PLoS One, 13*(5). e0197955https://doi.org/10.1371/journal.pone.0197955.

Wulsin, A. C., Herman, J. P., & Danzer, S. C. (2016). RU486 mitigates hippocampal pathology following status epilepticus. *Frontiers in Neurology, 7*, 214. https://doi.org/10.3389/fneur.2016.00214.

Zeidler, Z., Brandt-Fontaine, M., Leintz, C., Krook-Magnuson, C., Netoff, T., & Krook-Magnuson, E. (2018). Targeting the mouse ventral hippocampus in the intrahippocampal kainic acid model of temporal lobe epilepsy. *eNeuro, 5*(4). https://doi.org/10.1523/ENEURO.0158-18.2018.

CHAPTER NINE

Stress and the vestibular system

Yougan Saman[a,b,*], Qadeer Arshad[a,b,c], Mayank Dutia[d], Peter Rea[a]
[a]ENT Department, Leicester Royal Infirmary, Leicester, United Kingdom
[b]inAmind Laboratory, Department of Psychology, Neuroscience and Behaviour, University of Leicester, Leicester, United Kingdom
[c]Academic Department of Neuro-otology, Charing Cross Hospital Campus, Imperial College London, London, United Kingdom
[d]Biomedical Sciences, Edinburgh Medical School, University of Edinburgh, Edinburgh, United Kingdom
*Corresponding author: e-mail address: yougans@gmail.com

Contents

Abstract

In this chapter we review the existing literature regarding the interactions between stress and the mechanisms that maintain balance. Evidence suggests that the interplay between neuro-endocrine and psychological factors may have a significant role in balance function. For example, in healthy individuals vestibular stimulation has been shown to trigger the stress response as indicated by increased blood cortisol levels, whereas in patients with vestibular pathology factors such as resilience and anxiety may be the key focus of interactions with stress. Critically, factors such as anxiety are known to influence clinical outcomes, despite our mechanistic understanding of these processes remaining in their infancy.

International Review of Neurobiology, Volume 152
ISSN 0074-7742
https://doi.org/10.1016/bs.irn.2020.03.013

1. Basic introduction to the anatomy and physiology of the vestibular system

The vestibular end organ is housed in the inner ear and consists of three semi-circular canals, and the otolith organs (i.e., utricle and saccule). It detects head acceleration and passes this information via the vestibular nerve to the brain where it is integrated with information from other sensory systems, including visual and proprioceptive cues in order to maintain balance and spatial orientation (Wuyts & Boudewyns, 2008).

Each semi-circular canal (SCC) contains the crista ampullaris which possesses sensory epithelium comprising hair cells or cilia that are adapted to detect angular head acceleration. The semi-circular canals can detect acceleration in different planes that are coded into an electrical signal and transmitted to the vestibular nuclei in the brainstem. Semi-circular canals on either side are paired, and at rest the information derived from both sides is balanced. During a head acceleration, the SCCs on one side are activated and on the contra-lateral side are inhibited (Baloh & Halmagyi, 1996).

Regarding the otolith organs, the maculae of the utricle and the saccule are the basic sensory units that detect linear acceleration and gravity. During head acceleration, hair cells in the crista ampullaris and macula are displaced resulting in ion channels between the cilia opening and closing, altering the flow of electrolytes such as calcium and potassium into the cells. This leads to the release of neurotransmitters and the generation of an electrical signal that is conducted via the vestibular nerve (Rask-Andersen & Bagger-Sjöbäck, 2008).

Functionally, the vestibular system sub-serves balance, locomotion, postural control, spatial orientation, spatial navigation, autonomic function and circadian rhythm. The vestibular system is also involved in generating reflexes, including the vestibulo–ocular, vestibulo–spinal and vestibulo-colic reflexes that are relayed via the vestibular nuclei in the brainstem. The vestibulo-ocular reflex controls eye movement and gaze stabilization while the vestibulo-colic and vestibulo-spinal reflexes are involved with neck and trunk stabilization and postural control (Luxon, 2003).

2. The physiology of vestibular compensation

Following an acute peripheral lesion to the vestibular system, there is an imbalance in the activity of the vestibular nuclei, resulting in the characteristic syndrome of static oculomotor (i.e., nystagmus) and postural (i.e., unsteadiness) signs (occurring in the absence of vestibular stimulation) accompanied with the sensation of vertigo. These signs and symptoms gradually dissipate over days to weeks as compensatory mechanisms come into effect. For example, the dynamic phase and gain abnormalities in the vestibulo–ocular and vestibulo-spinal reflexes which accompany a unilateral vestibular loss begin to improve over the coming weeks and months. This process (i.e., vestibular compensation) is a well described model of post-lesional central nervous system plasticity (Curthoys & Halmagyi, 1995, 1999; Straka, Vibert, Vidal, Moore, & Dutia, 2005; Vidal, de Waele, Vibert, & Muhlethaler, 1998).

Underlying vestibular compensation is a series of neuro-electrophysiological and neurochemical processes as well as longer-term changes in synaptic plasticity and neurogenesis at a local level. Further, these processes are associated with parallel changes at a network level involving other brain areas (Dutia, 2010).

For example, changes at the level of the medial vestibular nucleus (MVN) have been studied extensively in animal models (Straka et al., 2005). Unilateral vestibular deafferentation (UVD) results in a cessation of resting activity in the ipsilesional medial vestibular nucleus due to the absence of peripheral vestibular stimulation and commissural inhibition from contralesional neurons. Recovery involves an increase in the intrinsic excitability and a decrease in inhibitory receptor sensitivity in ipsilesional neurons. There is also an increase in cerebellar inhibition of contralesional neurons thereby decreasing commissural inhibition. This physiological recovery process is suggested to correspond with functional behavioral recovery (i.e., dissipation of nystagmus and unsteadiness). Further research has indicated that such behavioral recovery may even precede physiological recovery in the medial vestibular nucleus (Ris, Capron, de Waele, Vidal, & Godaux, 1997; Vidal et al., 1998), implying that plasticity mechanisms may additionally be influenced by higher-order brain mechanisms.

A similar compensatory process is likely to be the basis of vestibular recovery in patients with peripheral vestibular loss, most of whom go on to recover well and rarely complain of a significant clinical deficit. Yet, a significant minority of patients can go on to develop complaints of persistent symptoms and disability, despite objective physiological testing being normal, further supported by the poor correlation between objective recovery of balance function and symptom load (Bamiou, Davies, McKee, & Luxon, 1999).

Numerous factors influence the outcome of vestibular pathology, including not only central compensation (Helmchen et al., 2009, 2011;

Helmchen, Ye, Sprenger, & Munte, 2014; Kitahara et al., 2003; Ris et al., 1997) but also psychological factors, such as the degree of anxiety that individuals exhibit (Staab, Rohe, Eggers, & Shepard, 2014). Clinically, it has been reported that elevated anxiety levels, can predict the development of chronic neuro-psychiatric symptoms (i.e., Persistent Postural Perceptual Dizziness (PPPD)) (Cousins et al., 2017). PPPD is a newly defined functional vestibular disorder that unifies and reconciles three decades of research on previously described psychosomatic vestibular conditions including phobic postural vertigo, space-motion discomfort, visual-vertigo and chronic subjective dizziness. The definition of PPPD was promulgated by an international panel of experts convened by the Bárány Society and is included in the 11th edition of the International Classification of Diseases (Staab et al., 2017).

Clinically, patients with PPPD present with two key fluctuating or continuous symptoms; (a) a dizzy, not-truly-vertiginous sensation, with patients reporting that their head is swimming and/or (b) unsteadiness, such that patients report swaying, rocking or jelly legs. Symptoms can be exacerbated in visually complex environments, and during upright posture and head movements.

Typically, PPPD is triggered by an acute disruption to normal balance function; however, the development of PPPD is not attributable to the degree of otological dysfunction or failed ear recovery. Rather, psychological risk factors (anxiety related personality traits) and shifts in psychophysical functioning (space-motion perceptual style and visual dependence) are strong predictive factors for determining which patients will develop PPPD following vestibular dysfunction. This is in line with clinical histories that illustrate symptom severity is modulated by factors such as stressful life events, introspection, distraction, fatigue and alertness. A key challenge to researchers is to understand the physiological and psychological factors responsible for impaired vestibular compensation. Stress appears to be a key variable that exerts both physiological and psychological influences that can impact upon this process.

3. Stress and the vestibular system

Despite the proposition that activation of the stress response (i.e., via activation of hypothalamic pituitary adrenal HPA axis and the autonomic nervous system via the release of 'stress mediators' that include neuropeptides, steroid, and monoamine hormones) may be necessary for vestibular compensation, it needs to be considered that excessive or inappropriate stress can have a deleterious impact on compensation (Saman, Bamiou, Gleeson, & Dutia, 2012; Seemungal, Gresty, & Bronstein, 2001), suggesting a delicate balance between health and disease.

The classical stress response is a negative feedback system mediated by the hypothalamic–pituitary–adrenal (HPA) axis. It starts with activation of the paraventricular nucleus (PVN) in the hypothalamus and results in the synthesis of arginine vasopressin (AVP) and corticotrophin releasing factor (CRF). These compounds then enter the portal system vasculature and stimulate the release of adrenocorticotrophic hormone (ACTH) from the anterior pituitary, which then enters the systemic circulation and stimulates the synthesis and release of glucocorticoids from the adrenal cortex (Herman et al., 2003).

Accordingly, it follows that stress may affect central vestibular function either directly through the actions of glucocorticoids (cortisol and corticosterone) on ion channels and neurotransmission in the brain, or indirectly via the effects of stress-related neuro-active substances (Saman et al., 2012). Furthermore, artificial stimulation of the vestibular system via caloric stimulation has been shown to, (i) activate the HPA axis as illustrated by excitation of the paraventricular nucleus neurons (Azzena, Melis, Caria, Teatini, & Bozzo, 1993; Liu, Inokuchi, & Komiyama, 1997; Azzena et al., 1993; Liu et al., 1997) and (ii) increase plasma AVP levels (Horii, Koike, Uno, Uno, & Kubo, 2001).

4. Vestibular deafferentation activates the stress axis

Unilateral vestibular deafferentation (UVD) has been shown to activate the stress axis. Gliddon and colleagues assessed saliva samples for postoperative periodic cortisol levels in UVD guinea pigs and observed a significant increase in salivary cortisol 10h after surgery (Gliddon, Darlington, & Smith, 2003). Further, following UVD in both a rat and cat model, increased levels of Fos-like immunoreactivity in the PVN has been reported, implying activation of the stress (HPA) axis (Cameron & Dutia, 1999). Importantly, in these experiments a correlation was noted regarding behavioral improvements in postural/locomotor control and the expression of CRH and AVP neurons (Tighilet, Manrique, & Lacour, 2009).

5. Stress axis activation can facilitate or impair vestibular compensation

In one of the earliest studies that assessed the role of the HPA axis in vestibular compensation, hypophysectomy prior to UVL was found to delay compensation and further studies have demonstrated that administration of ACTH (4–10) post-UVL restored compensation (Flohr & Luneburg, 1982; Igarashi, Ishikawa, Ishii, & Schmidt, 1985). Moreover, administration of methylprednisolone perioperatively (Jerram, Darlington, & Smith, 1995) and ACTH (4–10) (Gilchrist, Smith, & Darlington, 1990) postoperatively after UVL were found to reduce spontaneous nystagmus frequency while dexamethasone had no effect on the resolution of spontaneous nystagmus but together with methylprednisolone had an effect on excitation of some MVN neurons (Alice et al., 1998). The synthetic ACTH (4–9) 51 fragment (Org 2766) was also shown to decrease the frequency of spontaneous nystagmus following UVL when administered peripherally (Gilchrist, Darlington, & Smith, 1994; Gilchrist et al., 1994) or directly into the MVN (Gilchrist, Darlington, & Smith, 1996).

Research suggests that there appears to be either an appropriate amount of stress or perhaps a specific type of stressor that is necessary for optimal vestibular compensation. Restraint stress applied to an animal undergoing vestibular compensation following UVD delays recovery in relation to the resolution of nystagmus. Further, administration of a glucocorticoid synthesis inhibitor, trilostane, together with restraint stress in a separate group restored compensation to previous levels (Yamamoto, Yamanaka, & Matsunaga, 2000). Taken together, the above reviewed findings indicate that stress via neuroendocrine pathways facilitates vestibular compensation however it has to be considered that excessive stress can impair compensation. Having illustrated functionally the role of stress via neuro-endocrine pathways upon vestibular function and compensation, we shall now turn to discuss the role of stress in vestibular compensation in humans.

6. Stress and vestibular function in man

The stress response evoked by vestibular stimulation has been studied in humans. In healthy males, artificial vestibular stimulation has revealed (i) a significant increase in salivary cortisol and an increase in perceived stress (assessed with questionnaires) which was found to induce corresponding changes in autonomic features (namely: Heart Rate, Systolic Blood Pressure and Mean Arterial Pressure) (Cozma et al., 2018), (ii) stimulation of ACTH, noradrenaline and adrenaline secretion (Kohl, 1985) and, (iii) significantly elevated serum cortisol levels and simultaneous decrease in glutamate, glycine and GABA (Dagilas et al., 2005).

More recent work has indicated that greater visual dependence (i.e., over-reliance on visual cues) during maintenance of postural control is associated with dysregulation of HPA axis activity as indicated by blunted cortisol reactivity to psychosocial stress and a blunted average cortisol awakening response (CAR). Healthy individuals with the largest cortisol response to psychosocial stress showed less deterioration in postural sway when exposed to visual motion (Smyth et al., 2019). In a separate study healthy participants with a smaller 4-day average CAR had greater increases in postural sway length when exposed to visual motion (Smyth et al., 2019). Both studies indicate dysregulation of the HPA axis as blunted cortisol reactivity to stress and smaller CARs have both been associated with poor psychosocial health. Therefore in healthy individuals it has been shown that vestibular stimulation can influence the stress response and that an individual's stress signature may have a role to play in multisensory integration that subserves balance function.

7. Stress and vestibular compensation in man

Steroids are often used to treat acute idiopathic vestibular dysfunction or vestibular neuritis. Two recent meta-analyses have assessed the literature in order to determine the role of steroids in the management of acute idiopathic vestibular dysfunction (Fishman, Burgess, & Waddell, 2011; Goudakos et al., 2010). Goudakos et al. (2010) found an improvement in clinical outcomes with steroid use using a fixed effect model whereas Fishman et al. (2011) found no-effect using a random-effects model. Both studies drew attention to the relatively small sample sizes of studies, and differences in the clinical, methodological, and statistical parameters between the studies.

More recent work, has suggested that early treatment with steroids within the first 24 h of onset results in better outcomes with respect to vestibular function (Sjogren, Magnusson, Tjernstrom, & Karlberg, 2019) and this would fit with facilitation of the early stage of vestibular compensation in animal models as has been discussed above.

8. Chronic stress and the poorly compensated patient

It has long been known that patients with chronic dizzy symptoms have difficulties in complex visual environments (Guerraz et al., 2001) and often have enhanced visual motion sensitivity. In these patients, there is an interrelationship between visual dependence, visual vertigo and anxiety (Staab, 2014; Zur, Dickstein, Dannenbaum, Carmeli, & Fung, 2014) as supported by a longitudinal studies demonstrating that poorer long-term outcome (i.e., development of chronic functional symptoms) is most aptly predicted by visual dependency (Cousins et al., 2014, 2017). Following an acute episode of vestibular neuritis, it was shown that worse recovery was associated with a combination of increased visual dependence, autonomic arousal, anxiety/depression, and fear of bodily sensations (Cousins et al., 2017).

Understanding the premorbid interaction between the cortisol response and visual dependence could potentially be predictive of patients that may go on to develop significant chronic dizziness following a vestibular insult and by extension those that go on to develop PPPD. However vestibular deafferentation in itself may reset baseline cortisol levels to a higher level as has been previously shown in a study comparing resting cortisol levels in bilateral vestibular loss patients (Martin et al., 2016).

Other findings have demonstrated increased cortisol levels in patients with chronic dizziness due to a defined diagnostic entity when compared with cortisol levels in a cohort with idiopathic unexplained dizziness, however, plasma AVP levels were increased in both groups (Horii et al., 2007). Elevated AVP levels have also been reported in patients during an acute attack of Menière's disease (Takeda, Kakigi, & Saito, 1995).

9. Balance, stress, anxiety and depression

It has been shown that following an earthquake, a significant traumatic life event, there was an increase in patients reporting dizziness and imbalance symptoms (Chen et al., 2007; Hasegawa et al., 2015; Honma, Endo, Osada, Kim, & Kuriyama, 2012) as well as significant objective disturbance of equilibrium (Honma et al., 2012). Further, there were robust correlations between the degree of disequilibrium and clinical questionnaires that measured anxiety (STAI) and depression (BDI). Namely they reported a significant relationship between the severity of anxiety and worsening balance.

Post-traumatic stress disorder in army veterans (Haber, Chandler, & Serrador, 2016) as well as psychological trauma related to various adverse life events (Radziej, Schmid, Dinkel, Zwergal, & Lahmann, 2015) have been shown to be associated with an increase in vestibular symptoms and may well help to explain in part the symptoms that have been described in earthquake victims. Understanding the relationship between the stress response and traumatic events that result in anxiety disorders and the relationship with balance dysfunction may also have significant implications for the ongoing management and rehabilitation of patients with chronic imbalance.

Eagger and colleagues in the early 90's reported on psychiatric morbidity in patients with a history of disequilibrium and objective evidence of a peripheral vestibular lesion. Patients were seen between 3 and 5 years following their initial referral, underwent a psychiatric assessment and completed the social stress and support interview schedule (Jenkins, Mann, & Belsey, 1981). Patients with vestibular symptoms and psychiatric morbidity were found to have significantly increased trait anxiety scores and reported a greater degree of social stress (Eagger, Luxon, Davies, Coelho, & Ron, 1992).

In a longitudinal study of 101 patients assessed at 7 month intervals, Yardley and coworkers found that autonomic symptoms or arousal and somatic anxiety were the best predictors of reported change in the severity of vertiginous symptoms. It was postulated that autonomic symptoms initially associated with vestibular dysfunction during a vertiginous attack cause anxiety and further physiological arousal, which amplifies the symptoms in a vicious circle (Yardley, Luxon, & Haacke, 1994). In line with this is, are the findings reported by Cousins et al., who observed that autonomic anxiety and visual dependence were significant predictors of a poorer clinical outcome after vestibular neuritis (Cousins et al., 2017).

The neuro-anatomical substrates underlying the relationship between balance, anxiety and stress have been described previously (Balaban, 2002; Balaban & Thayer, 2001). Anatomically, of particular relevance is the nucleus tractus solitarius that has been shown to have extensive connections with the vestibular nuclei both directly and via indirect projections through the parabrachial nucleus, which provides a major input into the limbic system including the extended central amygdaloid nucleus, the infralimbic cortex and the hypothalamus.

From clinical observations and empirical reports it is apparent that not all patients perceive or deal with stress in the same way. The ability to habituate or cope with differing stressors and repeated stress may be an important contributing factor that determines clinical outcome following balance dysfunction. Accordingly, long term outcome following vestibular disease or damage may be rather idiosyncratic, and likely to be influenced not only by the nature of the stress response in each subject but also by the coping mechanisms that they are able to adopt (Tschan et al., 2011).

10. Individuals factors, stress and balance

In the section above, we have just discussed how individual differences in resilience to stressors or traumatic stressful life events can impact upon the ability to recover from vestibular dysfunction. Human ability to be resilient is markedly idiosyncratic between individuals and in part differences in resilience are determined by our personality.

Personality traits, specifically those associated with neuroticism, have been shown to confer stress vulnerability, when conditions are felt to be threatening (Schneider, Rench, Lyons, & Riffle, 2012). Patients with poorly compensated dizzy symptoms appear to display high levels of neuroticism (Yan et al., 2017). That is higher levels of neuroticism, assessed using validated questionnaire measures, are significantly associated with the severity of symptoms in dizzy patients. Further, recent research has illustrated that neuroticism and introversion correlated with activity and connectivity in cortical and subcortical vestibular, visual and anxiety systems during vestibular stimulation. These personality-related changes in brain activity may represent neural correlates of threat sensitivity in posture and gaze control mechanisms in normal individuals. Neuroticism as well as introversion has also more recently been shown to confer increased risk for developing anxiety-related morbidity in patients that go on to develop chronic symptoms (e.g., Persistent Postural Perceptual Dizziness (PPPD)) following vestibular dysfunction (Indovina, Riccelli, Staab, Lacquaniti, & Passamonti, 2014).

Given the strong link between the development of PPPD and anxiety as well as visual dependence, it is also important to consider which factors give rise to individual differences in visual dependency and anxiety. The degree to which an individual relies on visual cues, as opposed to either vestibular or proprioceptive cues, is termed visual dependency (VD), which is normally distributed in the general population. Witkin and Asch first established this in 1948 using the 'Rod and Frame Test' by demonstrating that healthy individuals who were more visually dependent made greater errors in aligning a rod with the earth vertical axis when it was enclosed within a tilted frame (Witkin & Asch, 1948). Although increased visual dependence does not impair locomotor control, elevated visual dependence has been consistently demonstrated in patients with chronic visual vertigo, which typically develops as a sequela following peripheral or central vestibular dysfunction (Bronstein, 1995; Bronstein, Yardley, Moore, & Cleeves, 1996). We have recently demonstrated that an individual's degree of right-hemispheric vestibulo-cortical dominance determines their degree of visual dependency. That is, those individuals that exhibit greater right hemisphere dominance for vestibular processing are less visually reliant compared to individuals that are less right hemisphere dominant. In a separate study, we also showed that increased right hemispheric vestibulo-cortical dominance was associated with lower trait anxiety scores (Bednarczuk, Casanovas Ortega, Fluri, & Arshad, 2018). Taken together, the reviewed findings suggest that individual differences in the lateralization of vestibulo-cortical processing may play a role in predisposing individuals to the risk factors associated with the development of PPPD. These individual differences seemingly play an important role when considering the interaction between stress and balance. Further, the potential to utilize these differences should not be underestimated, and should be considered fully when developing customized rehabilitation strategies for patients.

11. Future directions for research

Previous research has shown that individuals with specific polymorphisms that affect the glucocorticoid receptor system recover more slowly following a stress-induced increase in levels of cortisol and have higher anxiety (Ising et al., 2008). It would be informative to assess whether in patients with chronic vestibular symptoms, a higher prevalence of polymorphism exists when compared to control groups. Another interesting area for future research is considering sex-related differences in the interaction between stress and balance. This is supported by previous research that has indicated that females are more susceptible to development of motion sickness (Dobie, McBride, Dobie, & May, 2001) and stress induced anxiety and depression (Bourke, Harrell, & Neigh, 2012). Finally, given that stress pathway dysregulation (Simeon, Guralnik, Knutelska, Hollander, & Schmeidler, 2001) may be related to states of depersonalization and derealization, notably which are also common symptoms in patients with peripheral and central vestibular dysfunction (Sang, Jauregui-Renaud, Green, Bronstein, & Gresty, 2006), it would be informative to assess whether stress exerts modulatory influences upon such symptoms in dizzy patients.

12. Concluding remarks

In this review we have extensively highlighted the complex and multifaceted interaction between stress and balance, from both a basic animal neuroendocrine model to the complex pathophysiological human model provided by patients with vestibular dysfunction. Specifically, as indicated in the future direction subsection above, research regarding the interaction of stress and balance in patients with vestibular dysfunction is in its infancy and much research is required in this area. This research has the potential to decipher the complex interrelationship between neurological, otological, psychiatric and endocrine function to better understand the impact of stress on the balance system.

References

Alice, C., Paul, A. E., Sansom, A. J., Maclennan, K., Darlington, C. L., & Smith, P. F. (1998). The effects of steroids on vestibular compensation and vestibular nucleus neuronal activity in the Guinea pig. *Journal of Vestibular Research: Equilibrium & Orientation*, *8*(3), 201–207.

Azzena, G. B., Melis, F., Caria, M. A., Teatini, G. P., & Bozzo, G. (1993). Vestibular projections to hypothalamic supraoptic and paraventricular nuclei. *Archives Italiennes de Biologie*, *131*(2–3), 127–136.

Balaban, C. D. (2002). Neural substrates linking balance control and anxiety. *Physiology & Behavior*, 77(4–5), 469–475.

Balaban, C. D., & Thayer, J. F. (2001). Neurological bases for balance-anxiety links. *Journal of Anxiety Disorders*, *15*(1–2), 53–79.

Baloh, R. W., & Halmagyi, G. M. (1996). *Disorders of the vestibular system*. New York; Oxford: Oxford University Press.

Bamiou, D. E., Davies, R. A., McKee, M., & Luxon, L. M. (1999). The effect of severity of unilateral vestibular dysfunction on symptoms, disabilities and handicap in vertiginous patients. *Clinical Otolaryngology and Allied Sciences*, *24*(1), 31–38.

Bednarczuk, N. F., Casanovas Ortega, M., Fluri, A. S., & Arshad, Q. (2018). Vestibulo-cortical hemispheric dominance: The link between anxiety and the vestibular system? *The European Journal of Neuroscience*, *47*(12), 1517–1524. https://doi.org/10.1111/ejn.13948.

Bourke, C. H., Harrell, C. S., & Neigh, G. N. (2012). Stress-induced sex differences: Adaptations mediated by the glucocorticoid receptor. *Hormones and Behavior*, *62*(3), 210–218. https://doi.org/10.1016/j.yhbeh.2012.02.024.

Bronstein, A. M. (1995). Visual vertigo syndrome: Clinical and posturography findings. *Journal of Neurology, Neurosurgery, and Psychiatry*, *59*(5), 472–476. https://doi.org/10.1136/jnnp.59.5.472.

Bronstein, A. M., Yardley, L., Moore, A. P., & Cleeves, L. (1996). Visually and posturally mediated tilt illusion in Parkinson's disease and in labyrinthine defective subjects. *Neurology*, *47*(3), 651–656. https://doi.org/10.1212/wnl.47.3.651.

Cameron, S. A., & Dutia, M. B. (1999). Lesion-induced plasticity in rat vestibular nucleus neurones dependent on glucocorticoid receptor activation. *The Journal of Physiology*, *518*(Pt 1), 151–158.

Chen, C. H., Tan, H. K., Liao, L. R., Chen, H. H., Chan, C. C., Cheng, J. J., et al. (2007). Long-term psychological outcome of 1999 Taiwan earthquake survivors: A survey of a high-risk sample with property damage. *Comprehensive Psychiatry*, *48*(3), 269–275. https://doi.org/10.1016/j.comppsych.2006.12.003.

Cousins, S., Cutfield, N. J., Kaski, D., Palla, A., Seemungal, B. M., Golding, J. F., et al. (2014). Visual dependency and dizziness after vestibular neuritis. *PLoS One*, *9*(9), e105426. https://doi.org/10.1371/journal.pone.0105426.

Cousins, S., Kaski, D., Cutfield, N., Arshad, Q., Ahmad, H., Gresty, M. A., et al. (2017). Predictors of clinical recovery from vestibular neuritis: A prospective study. *Annals of Clinical Translational Neurology*, *4*(5), 340–346. https://doi.org/10.1002/acn3.386.

Cozma, S., Ghiciuc, C. M., Damian, L., Pasquali, V., Saponaro, A., Lupusoru, E. C., et al. (2018). Distinct activation of the sympathetic adreno-medullar system and hypothalamus pituitary adrenal axis following the caloric vestibular test in healthy subjects. *PLoS One*, *13*(3), e0193963. https://doi.org/10.1371/journal.pone.0193963.

Curthoys, I. S., & Halmagyi, G. M. (1995). Vestibular compensation: A review of the oculomotor, neural, and clinical consequences of unilateral vestibular loss. *Journal of Vestibular Research: Equilibrium & Orientation*, *5*(2), 67–107.

Curthoys, I. S., & Halmagyi, G. M. (1999). Vestibular compensation. *Advances in Oto-Rhino-Laryngology*, *55*, 82–110.

Dagilas, A., Kimiskidis, V., Aggelopoulou, M., Kapaki, E., Fitili, C., Libitaki, G., et al. (2005). Changes in blood neurotransmitter and steroid levels during evoked vertigo. *Otology & Neurotology: Official Publication of the American Otological Society, American Neurotology Society [and] European Academy of Otology and Neurotology*, *26*(3), 476–480 (discussion 480).

Dobie, T., McBride, D., Dobie, T., Jr., & May, J. (2001). The effects of age and sex on susceptibility to motion sickness. *Aviation, Space, and Environmental Medicine*, *72*(1), 13–20.

Dutia, M. B. (2010). Mechanisms of vestibular compensation: Recent advances. *Current Opinion in Otolaryngology & Head and Neck Surgery*, *18*(5), 420–424. https://doi.org/10.1097/MOO.0b013e32833de71f.

Eagger, S., Luxon, L. M., Davies, R. A., Coelho, A., & Ron, M. A. (1992). Psychiatric morbidity in patients with peripheral vestibular disorder: A clinical and neuro-otological study. *Journal of Neurology, Neurosurgery, and Psychiatry*, *55*(5), 383–387.

Fishman, J. M., Burgess, C., & Waddell, A. (2011). Corticosteroids for the treatment of idiopathic acute vestibular dysfunction (vestibular neuritis). *Cochrane Database of Systematic Reviews*, *5*, CD008607. https://doi.org/10.1002/14651858.CD008607.pub2.

Flohr, H., & Luneburg, U. (1982). Effects of ACTH4-10 on vestibular compensation. *Brain Research*, *248*(1), 169–173.

Gilchrist, D. P., Darlington, C. L., & Smith, P. F. (1994). A dose-response analysis of the beneficial effects of the ACTH-(4-9) analogue, org 2766, on behavioural recovery following unilateral labyrinthectomy in Guinea-pig. *British Journal of Pharmacology*, *111*(1), 358–363.

Gilchrist, D. P., Darlington, C. L., & Smith, P. F. (1996). An in vitro investigation of the effects of the ACTH/MSH(4-9) analogue, org 2766, on Guinea pig medial vestibular nucleus neurons. *Peptides*, *17*(4), 681–688.

Gilchrist, D. P., Smith, P. F., & Darlington, C. L. (1990). ACTH(4-10) accelerates ocular motor recovery in the Guinea pig following vestibular deafferentation. *Neuroscience Letters*, *118*(1), 14–16.

Gliddon, C. M., Darlington, C. L., & Smith, P. F. (2003). Activation of the hypothalamic-pituitary-adrenal axis following vestibular deafferentation in pigmented Guinea pig. *Brain Research*, *964*(2), 306–310.
Goudakos, J. K., Markou, K. D., Franco-Vidal, V., Vital, V., Tsaligopoulos, M., & Darrouzet, V. (2010). Corticosteroids in the treatment of vestibular neuritis: A systematic review and meta-analysis. *Otology & Neurotology: Official Publication of the American Otological Society, American Neurotology Society [and] European Academy of Otology and Neurotology*, *31*(2), 183–189. https://doi.org/10.1097/MAO.0b013e3181ca843d.
Guerraz, M., Yardley, L., Bertholon, P., Pollak, L., Rudge, P., Gresty, M. A., et al. (2001). Visual vertigo: Symptom assessment, spatial orientation and postural control. *Brain: A Journal of Neurology*, *124*(Pt 8), 1646–1656.
Haber, Y. O., Chandler, H. K., & Serrador, J. M. (2016). Symptoms associated with vestibular impairment in veterans with posttraumatic stress disorder. *PLoS One*, *11*(12), e0168803. https://doi.org/10.1371/journal.pone.0168803.
Hasegawa, J., Hidaka, H., Kuriyama, S., Obara, T., Hashimoto, K., Tateda, Y., et al. (2015). Change in and long-term investigation of neuro-otologic disorders in disaster-stricken Fukushima prefecture: Retrospective cohort study before and after the great East Japan earthquake. *PLoS One*, *10*(4), e0122631. https://doi.org/10.1371/journal.pone.0122631.
Helmchen, C., Klinkenstein, J. C., Kruger, A., Gliemroth, J., Mohr, C., & Sander, T. (2011). Structural brain changes following peripheral vestibulo-cochlear lesion may indicate multisensory compensation. *Journal of Neurology, Neurosurgery, and Psychiatry*, *82*(3), 309–316. https://doi.org/10.1136/jnnp.2010.204925.
Helmchen, C., Klinkenstein, J., Machner, B., Rambold, H., Mohr, C., & Sander, T. (2009). Structural changes in the human brain following vestibular neuritis indicate central vestibular compensation. *Annals of the New York Academy of Sciences*, *1164*, 104–115. https://doi.org/10.1111/j.1749-6632.2008.03745.x.
Helmchen, C., Ye, Z., Sprenger, A., & Munte, T. F. (2014). Changes in resting-state fMRI in vestibular neuritis. *Brain Structure & Function*, *219*(6), 1889–1900. https://doi.org/10.1007/s00429-013-0608-5.
Herman, J. P., Figueiredo, H., Mueller, N. K., Ulrich-Lai, Y., Ostrander, M. M., Choi, D. C., et al. (2003). Central mechanisms of stress integration: Hierarchical circuitry controlling hypothalamo-pituitary-adrenocortical responsiveness. *Frontiers in Neuroendocrinology*, *24*(3), 151–180.
Honma, M., Endo, N., Osada, Y., Kim, Y., & Kuriyama, K. (2012). Disturbances in equilibrium function after major earthquake. *Scientific Reports*, *2*, 749. https://doi.org/10.1038/srep00749.
Horii, A., Koike, K., Uno, A., Uno, Y., & Kubo, T. (2001). Vestibular modulation of plasma vasopressin levels in rats. *Brain Research*, *914*(1–2), 179–184.
Horii, A., Uno, A., Kitahara, T., Mitani, K., Masumura, C., Kizawa, K., et al. (2007). Effects of fluvoxamine on anxiety, depression, and subjective handicaps of chronic dizziness patients with or without neuro-otologic diseases. *Journal of Vestibular Research: Equilibrium & Orientation*, *17*(1), 1–8.
Igarashi, M., Ishikawa, K., Ishii, M., & Schmidt, K. A. (1985). Effect of ACTH-(4-10) on equilibrium compensation after unilateral labyrinthectomy in the squirrel monkey. *European Journal of Pharmacology*, *119*(3), 239–242.
Indovina, I., Riccelli, R., Staab, J. P., Lacquaniti, F., & Passamonti, L. (2014). Personality traits modulate subcortical and cortical vestibular and anxiety responses to sound-evoked otolithic receptor stimulation. *Journal of Psychosomatic Research*, 77(5), 391–400. https://doi.org/10.1016/j.jpsychores.2014.09.005.
Ising, M., Depping, A. M., Siebertz, A., Lucae, S., Unschuld, P. G., Kloiber, S., et al. (2008). Polymorphisms in the FKBP5 gene region modulate recovery from psychosocial stress in

healthy controls. *The European Journal of Neuroscience, 28*(2), 389–398. https://doi.org/10.1111/j.1460-9568.2008.06332.x.

Jenkins, R., Mann, A. H., & Belsey, E. (1981). The background, design and use of a short interview to assess social stress and support in research and clinical settings. *Social Science & Medicine Part E, Medical Psychology, 15*(3), 195–203.

Jerram, A. H., Darlington, C. L., & Smith, P. F. (1995). Methylprednisolone reduces spontaneous nystagmus following unilateral labyrinthectomy in Guinea pig. *European Journal of Pharmacology, 275*(3), 291–293.

Kitahara, T., Kondoh, K., Morihana, T., Okumura, S., Horii, A., Takeda, N., et al. (2003). Steroid effects on vestibular compensation in human. *Neurological Research, 25*(3), 287–291.

Kohl, R. L. (1985). Endocrine correlates of susceptibility to motion sickness. *Aviation, Space, and Environmental Medicine, 56*(12), 1158–1165.

Liu, F., Inokuchi, A., & Komiyama, S. (1997). Neuronal responses to vestibular stimulation in the Guinea pig hypothalamic paraventricular nucleus. *European Archives of Oto-Rhino-Laryngology: Official Journal of the European Federation of Oto-Rhino-Laryngological Societies, 254*(2), 95–100.

Luxon, L. M. (2003). *Textbook of audiological medicine: Clinical aspects of hearing and balance.* London: Martin Dunitz.

Martin, T., Moussay, S., Bulla, I., Bulla, J., Toupet, M., Etard, O., et al. (2016). Exploration of circadian rhythms in patients with bilateral vestibular loss. *PLoS One, 11*(6), e0155067. https://doi.org/10.1371/journal.pone.0155067.

Radziej, K., Schmid, G., Dinkel, A., Zwergal, A., & Lahmann, C. (2015). Psychological traumatization and adverse life events in patients with organic and functional vestibular symptoms. *Journal of Psychosomatic Research, 79*(2), 123–129. https://doi.org/10.1016/j.jpsychores.2015.05.005.

Rask-Andersen, H., & Bagger-Sjöbäck, D. (2008). Anatomy and ultrastructure of the vestibular organ. In M. Gleeson & W. G. Scott-Brown (Eds.), *Vol. 3. Scott-Brown's otorhinolaryngology, head and neck surgery* (7th ed., pp. 3148–3157). London: Hodder Arnold.

Ris, L., Capron, B., de Waele, C., Vidal, P. P., & Godaux, E. (1997). Dissociations between behavioural recovery and restoration of vestibular activity in the unilabyrinthectomized Guinea-pig. *The Journal of Physiology, 500*(Pt 2), 509–522.

Saman, Y., Bamiou, D. E., Gleeson, M., & Dutia, M. B. (2012). Interactions between stress and vestibular compensation—A review. *Frontiers in Neurology, 3*, 116. https://doi.org/10.3389/fneur.2012.00116.

Sang, F. Y., Jauregui-Renaud, K., Green, D. A., Bronstein, A. M., & Gresty, M. A. (2006). Depersonalisation/derealisation symptoms in vestibular disease. *Journal of Neurology, Neurosurgery, and Psychiatry*, 77(6), 760–766. https://doi.org/10.1136/jnnp.2005.075473.

Schneider, T. R., Rench, T. A., Lyons, J. B., & Riffle, R. R. (2012). The influence of neuroticism, extraversion and openness on stress responses. *Stress and Health, 28*(2), 102–110. https://doi.org/10.1002/smi.1409.

Seemungal, B. M., Gresty, M. A., & Bronstein, A. M. (2001). The endocrine system, vertigo and balance. *Current Opinion in Neurology, 14*(1), 27–34.

Simeon, D., Guralnik, O., Knutelska, M., Hollander, E., & Schmeidler, J. (2001). Hypothalamic-pituitary-adrenal axis dysregulation in depersonalization disorder. *Neuropsychopharmacology, 25*(5), 793–795. https://doi.org/10.1016/S0893-133X(01)00288-3.

Sjogren, J., Magnusson, M., Tjernstrom, F., & Karlberg, M. (2019). Steroids for acute vestibular neuronitis-the earlier the treatment, the better the outcome? *Otology & Neurotology: Official Publication of the American Otological Society, American Neurotology Society [and] European Academy of Otology and Neurotology, 40*(3), 372–374. https://doi.org/10.1097/MAO.0000000000002106.

Smyth, N., Flynn, M., Milani, M., Hucklebridge, F., Thorn, L., Evans, P., et al. (2019). An attenuated cortisol awakening response in healthy participants visually dependent in postural control. *Psychoneuroendocrinology*, *107*, 18–19.

Smyth, N., Flynn, M., Rajcani, J., Hucklebridge, M. F., Thorn, L., Wood, C., et al. (2019). Attenuated cortisol reactivity to psychosocial stress is associated with greater visual dependency in postural control. *Psychoneuroendocrinology*, *104*, 185–190. https://doi.org/10.1016/j.psyneuen.2019.02.028.

Staab, J. P. (2014). The influence of anxiety on ocular motor control and gaze. *Current Opinion in Neurology*, *27*(1), 118–124. https://doi.org/10.1097/WCO.0000000000000055.

Staab, J. P., Eckhardt-Henn, A., Horii, A., Jacob, R., Strupp, M., Brandt, T., et al. (2017). Diagnostic criteria for persistent postural-perceptual dizziness (PPPD): Consensus document of the committee for the classification of vestibular disorders of the Barany society. *Journal of Vestibular Research: Equilibrium & Orientation*, *27*(4), 191–208. https://doi.org/10.3233/VES-170622.

Staab, J. P., Rohe, D. E., Eggers, S. D., & Shepard, N. T. (2014). Anxious, introverted personality traits in patients with chronic subjective dizziness. *Journal of Psychosomatic Research*, *76*(1), 80–83. https://doi.org/10.1016/j.jpsychores.2013.11.008.

Straka, H., Vibert, N., Vidal, P. P., Moore, L. E., & Dutia, M. B. (2005). Intrinsic membrane properties of vertebrate vestibular neurons: Function, development and plasticity. *Progress in Neurobiology*, *76*(6), 349–392. https://doi.org/10.1016/j.pneurobio.2005.10.002.

Takeda, T., Kakigi, A., & Saito, H. (1995). Antidiuretic hormone (ADH) and endolymphatic hydrops. *Acta Oto-Laryngologica Supplementum*, *519*, 219–222.

Tighilet, B., Manrique, C., & Lacour, M. (2009). Stress axis plasticity during vestibular compensation in the adult cat. *Neuroscience*, *160*(4), 716–730. https://doi.org/10.1016/j.neuroscience.2009.02.070.

Tschan, R., Best, C., Beutel, M. E., Knebel, A., Wiltink, J., Dieterich, M., et al. (2011). Patients' psychological well-being and resilient coping protect from secondary somatoform vertigo and dizziness (SVD) 1 year after vestibular disease. *Journal of Neurology*, *258*(1), 104–112. https://doi.org/10.1007/s00415-010-5697-y.

Vidal, P. P., de Waele, C., Vibert, N., & Muhlethaler, M. (1998). Vestibular compensation revisited. *Otolaryngology and Head and Neck Surgery*, *119*(1), 34–42.

Witkin, H. A., & Asch, S. E. (1948). Studies in space orientation; further experiments on perception of the upright with displaced visual fields. *Journal of Experimental Psychology*, *38*(6), 762–782. https://doi.org/10.1037/h0053671.

Wuyts, F., & Boudewyns, A. (2008). Physiology of equilibrium. In M. Gleeson & W. G. Scott-Brown (Eds.), *Vol. 3. Scott-Brown's otorhinolaryngology, head and neck surgery* (7th ed., pp. 3208–3244). London: Hodder Arnold.

Yamamoto, T., Yamanaka, T., & Matsunaga, T. (2000). The effect of stress application on vestibular compensation. *Acta Oto-Laryngologica*, *120*(4), 504–507.

Yan, Z., Cui, L., Yu, T., Liang, H., Wang, Y., & Chen, C. (2017). Analysis of the characteristics of persistent postural-perceptual dizziness: A clinical-based study in China. *International Journal of Audiology*, *56*(1), 33–37. https://doi.org/10.1080/14992027.2016.1211763.

Yardley, L., Luxon, L. M., & Haacke, N. P. (1994). A longitudinal study of symptoms, anxiety and subjective well-being in patients with vertigo. *Clinical Otolaryngology and Allied Sciences*, *19*(2), 109–116.

Zur, O., Dickstein, R., Dannenbaum, E., Carmeli, E., & Fung, J. (2014). The influence of visual vertigo and vestibulopathy on oculomotor responses. *Journal of Vestibular Research: Equilibrium & Orientation*, *24*(4), 305–311. https://doi.org/10.3233/VES-140519.

CHAPTER TEN

Social prescribing for stress related disorders and brain health

Alison Fixsen*, Marie Polley
University of Westminster, London, United Kingdom
*Corresponding author: e-mail address: A.Fixsen@westminster.ac.uk

Contents

Abstract

Social prescribing allows health professionals to refer at risk patients toward health and wellbeing interventions and activities in the local community. It is a key part of NHS (National Health Service) England health care policy, and schemes based on the social prescribing model have been developed in countries including Canada, New Zealand, the Netherlands and Singapore. In this chapter, we consider the role that social prescribing can play in reducing stress related problems and supporting and encouraging self-care and self-management of conditions for which conventional medicine may not be the only or the best option. Drawing on primary and secondary data sources, we examine the scope of social prescribing and professional and service users' perspectives concerning its strengths and limitations. Our findings suggest that link worker meetings within social prescribing schemes can motivate people to pursue activities with mental, physical and social benefits such as exercise, artistic pursuits and gardening. Problems within schemes included health provider engagement, recruiting those with low agency and communication between professionals and patients about social prescribing. Based on our findings, we propose a number of recommendations for enhancing social prescribing schemes. Professionals, including neurologists, we argue, can benefit from engaging in the concept and practice of social prescribing and referring patients and clients to social prescribing link workers where appropriate.

International Review of Neurobiology, Volume 152
ISSN 0074-7742
https://doi.org/10.1016/bs.irn.2019.11.005

Neurologists are also part of a larger team, as they work alongside allied health professionals such as occupational therapists and physiotherapists, some of whom are already performing aspects of the link worker role.

1. Introduction

Social prescribing offers an innovative solution to stress related disorders by creating referral pathways that direct people with non-clinical conditions to services and activities run within the community (Bertotti, Frostick, Hutt, Sohanpal, & Carnes, 2018). With its focus on the biopsychosocial model of illness, social prescribing acknowledges problems preceding or accompanying stress on health and wellbeing. Using a collaborative approach, it provides healthcare professionals with a referral option to address the psychosocial needs of patients more effectively. When the multidisciplinary model for the primary care networks lies firmly established, referrals to a link worker will be made by other professionals outside of the health system, e.g., social workers. By addressing problems such as social isolation and health anxieties potentially reduce the burden of ill-health on society. In this chapter, we consider the potential for social prescribing to reduce problems and disorders associated with stress, and to improve in brain health. We use examples from our own and other studies to illustrate the ways in which social prescribing equips professionals and lay people with a wide repertoire of health and wellbeing interventions, targeting those individuals who are experiencing social, emotional and practical stressors (Chatterjee, Camic, Lockyer, & Thomson, 2017). NHS (National Health Service) England is due to fund 1000 social prescribing workers by 2021 (Pulse, 2019) to work in primary care settings, in this case NHS funded GP(General Practitioner) practices—that is, primary care services for treating common medical conditions and referring patients onto secondary care services—in the United Kingdom. Social prescribing models are being adopted within a growing number of countries with different medical care models, such as Canada (Ontario Public Health, 2019), the Netherlands and Australia (Consumers Health Forum of Australia, 2019). Professionals, including neurologists, we argue, could benefit from actively engaging in the concept and practice of social prescribing by referring patients and clients suffering from stress disorders to social prescribing link workers, and thence on to engagement in community-based activities.

2. Stress and society

Like all organisms, humans must maintain a complex dynamic equilibrium or homeostasis, which is constantly challenged by internal or external adverse forces termed stressors. Since stress occurs when homeostasis is threatened or perceived to be so, optimal responsiveness of the stress system is essential for a sense of wellbeing, successful task performance, and appropriate social interactions (Chrousos, 2009). Studies of brain responses to emotional stimuli in healthy humans have reported heightened amygdala activation in response to disorder-relevant stimuli in people diagnosed with stress disorders and social or specific phobias (Shin & Liberzon, 2010). Pathological anxiety and chronic stress lead to structural degeneration and impaired functioning of the hippocampus and the prefrontal cortex, which may account for the increased risk of developing neuropsychiatric disorders, including depression and dementia (Mah, Szabuniewicz, & Fiocco, 2016). While the complexity of stress reactions and disorders is acknowledged by most health researchers, a tendency to separate biological and psycho-socio-cultural factors impacting on the severity and chronicity of these responses remains prevalent. Part of this is driven by the fact that the medical curriculum does not lead medical professionals to address social determinants of health; however, our recent study indicates the wish of medical students to include social prescribing as part of their medical curriculum (Santoni et al., 2019).

Even with the drive to embed a more public agenda within the role of primary and secondary care, the core business of physicians has remained that of providing individual care and treatment (South, Higgins, Woodall, & White, 2008). With mental and substance use disorders the leading cause of long-term disability worldwide, society-wide initiatives toward their prevention and treatment should be regarded as a public health priority (Vos et al., 2015; Whiteford et al., 2013). Conventional medicine has failed to halt the rising tide of depression and anxiety (World Federation For Mental Health (WFMH), 2012), which is unsurprising as an estimated 90% of the determinants of health derive from our lifetime social and physical environment, not from health care provision (Carod-Artal, 2017). Cumulative adversities that may cause or exacerbate stress reactions include; being economically impoverished, of a ethno-racial minority background and being socially isolated (Courtois & Ford, 2009, pp. 17–18). There is also growing policy concern about the extent of loneliness in advanced societies, and its prevalence among various social groups. In Kearns et al.'s study, feelings of loneliness were most strongly associated with poor mental health, and also long-term problems of stress, anxiety and depression and with low mental wellbeing (Kearns, Whitley, Tannahill, & Ellaway, 2015).

Patients typically bring a wide number of clinical and non-clinical issues to primary and even secondary healthcare consultations. These include long-term conditions such as diabetes, mild to moderate mental health problems such as anxiety and depression and psychosocial issues such as loneliness and isolation and welfare issues such as housing, employment and debt. For many people, and in particular those faced with disadvantages such as poor education low income and limited social network, medical providers may appear the only source of help and access to further services. Yet, too frequently the health impact of debt, poor housing, poor eating, irregular employment, rudimentary English language skills and so on is overlooked in the consulting or emergency room, as is the impact of loneliness and isolation on stress and mental wellbeing (Kawachi & Berkman, 2001). One study indicated that GPs rarely refer patients to local community groups or advice services, as they lack the knowledge about current local resources (Kilgore, Richter, Siler, & Sayre-Stanhope, 2008). In another study commissioned by Citizens Advice Board more than four in five (84%) of respondent GPs said they refer the patient with non-medical issues to an advice agency in the community. Just 6% of GPs said they referred patients to an expert advisor based in the surgery, while only 31% thought they were able to advise patients adequately themselves (Citizens Advice, 2015).

Due to the scope and time constraints of the consultation, where a GP is faced with a patient presenting with low mood and non-specific pains there is a risk of over-pathologizing what might be essentially a social isolation problem (Pescheny, Randhawa, & Pappas, 2018). Where a problem has its origins in socioeconomic deprivation or long-term psychosocial issues, both patient and physician may feel overwhelmed and reluctant to open up "a can of worms." Given that around 19% of patients attending GP surgeries do so for social rather medical reasons (Caper & Plunkett, 2015) a short-term fix approach can amount to a conspiracy of silence. In the Citizens Advice Bureau (CAB) study, 92% of GPs reported that their patients had raised issues about personal relationship problems, 77% had reported housing issues, 77% unemployment/work related issues and 75% had raised welfare benefits issues with them in the previous month.

For GPs working long hours and with large patient caseloads, dealing with patients' psychosocial needs adds additional stresses. High levels of burnout, anxiety, fatigue and other symptoms of distress among medical doctors have been revealed in prevalence studies (Medisauskaite & Kamau, 2019). In a recent systematic review, 31–54% of UK doctors had experienced burnout and 17–52% had experienced psychiatric morbidity (Imo, 2017). Health workers in secondary and tertiary care are similarly affected by stress and burnout (Busis et al., 2017; Zhou et al., 2017). One 2014 study found the prevalence of burnout to be higher in physicians than in other professions and particularly high in neurologists, amounting to a neurologic crisis (Sigsbee & Bernat, 2014). All this indicates the need for new approaches to health and wellbeing to address the psychosocial needs of patients and to relieve pressure on health services and health professionals ill-equipped to adequately deal with such problems. In the following section, we consider the potential of social prescribing to expand the options available within primary, secondary and tertiary care. We consider how such initiatives might relieve those working in medical and social services of pressure to unnecessarily prescribe or assume the role of wellbeing advisors in areas where a psychosocial intervention might be of greater long-term value.

3. Social prescribing

Social prescribing involves the referring of "at risk" patients or clients to non-clinical forms of intervention offered by local voluntary social and community enterprises (VCSEs) with the intention of enhancing the patient's health and wellbeing (Polley & Fleming, 2017). The typical pathway is that the public health and social care provider refers a client or patient to a social prescribing advisor or "link worker" who engages them in a conversation (typically for 40–60 min) in order to co-produce a personalized solution or "social prescription" (Polley & Fleming, 2017). The link worker then refers the client on to one of the local voluntary social and community enterprises (VCSEs) offering a suitable form of social and psychological support and/or to a suitable health care provider. The rationale for this is that a complex array of non-government directed institutions catering for divergent social and health requirements and interests is already in place in many urban and rural communities (Husk et al., 2016), and that social prescribing models can realize and build upon existing assets. Types of community referrals common in social prescribing schemes, at low or no cost to the patient, include healthy living (e.g., health checks, eating healthy); exercise and physical activity programs (e.g., gardening projects, walking and cycling groups); art workshops (e.g., painting, dance, music, poetry); self-help books; education; volunteering; and support with employment, finances and housing. The link worker then follows up the client and under many schemes their progress is also evaluated.

While its aims are diverse, initiatives such as those used in social prescribing can be viewed as practical responses to hard evidence concerning the accumulative, socially and economically determined nature of health (Carod-Artal, 2017; Marmot & Bell, 2012; Pescheny, Pappas, & Randhawa, 2018), the translation of unhealthy behaviors into often poorly managed diseases and the need for action to address health inequalities and the social gradient in health outcomes (Marmot & Bell, 2012). In 2008, the World Health Organization (WHO) called for initiatives aimed at closing the gap between health and lifespan inequalities within a generation (Commission on Social Determinants of Health, 2008) and conceptually social prescribing fits with that aim. Social prescribing could also contribute to the global "Triple Aim" (Berwick, Nolan, & Whittington, 2008) of applying an integrated approach to simultaneously improve care, enhance population health, and reduce costs per capita (Husk et al., 2016). On a social institution level, social prescribing help build relationships between primary care and community organizations. From governmental and cost saving perspectives, the widespread use of social prescribing has the potential to lessen expenditure related to multi-morbidities (Dayson, 2017; Naylor et al., 2012) and relieve pressure on main stream medical services (Torjesen, 2016). Models of social prescription can also modify the traditional boundaries within primary and secondary care, facilitating opportunities for contact with professionals who have more time and specific expertise (Pescheny, Pappas, et al., 2018), such as legal and welfare advice workers, counselors and art therapists.

From its small scale beginning as individual services delivered by local community organizations and volunteers working with single GPs, social prescribing has been widely promoted by successive UK Governments, culminating in the NHS Long-Term Plan announcement to recruit 1000 new SP link workers by 2020/21 (Cheater, 2019). Schemes based on the social prescribing model have been developed in countries including Canada (Ontario Public Health, 2019), New Zealand (CASPER, 2019), the Netherlands and Singapore. In the following section, we consider the role that social prescribing can play in reducing stress related problems and supporting and encouraging self-care and self-management of conditions for which conventional medicine may not be the only or the best option.

3.1 Social prescribing and mental wellbeing

Studies suggest that when stressors (negative events, chronic strains, and traumas) are considered together, their damaging impacts on physical and mental health are substantial and strong linked to accumulated health and social inequalities over the life course (Thoits, 2010). However, the impact of stressors on health and wellbeing appears to be reduced when people have higher levels of self-esteem and social support (Thoits, 2010). In 1 year, one in four people in the United Kingdom have some kind of mental health problem, with high cost to both the individuals, communities and health services (Kimberlee, 2015). Among the ways in which social prescribing seeks to address the wider determinants of mental health is by supporting better access to psychological treatments and to positive health interventions. Possible outcomes claimed by advocates of social prescribing include reduction of mental illness related to loneliness and isolation (Friedli, Jackson, Abernethy, & Stansfield, 2008; South et al., 2008) and reduction of age related illnesses including dementia (Baker & Irving, 2016). A sense that one is living a worthwhile and meaningful life is a key component of individual wellbeing and known to be related to a range of social, economic, and health factors (Steptoe & Fancourt, 2018). Meetings with the link worker can contribute to autonomy, positive affect and self-efficacy and address circumstances such as housing, employment, mobility, which impact on people's psychological and emotional lives.

Subjective wellbeing and health are also closely linked to age (Steptoe, Deaton, & Stone, 2015). With aging, psychological wellbeing tends to decline due to events such as worsening health, partner's death or disease, living alone, low income levels and lessening social interactions (NICE, 2019). Around 18% of the UK population are aged 65 years and over and 2.4% 85 years and over (Office for National Statistics, 2018). According to the Office of National Statistics the number of people aged 85 years and over is set to double in the 25 years (Office for National Statistics, 2019) making the need to explore sustainable ways of promoting wellbeing in later life a strategic priority. The full scope of social prescribing for different population groups remains to be documented; however, allied studies suggest that many of the activities supported and promoted through stakeholders involved in social prescribing are beneficial to older people. One longitudinal study investigated whether frequency of arts engagement over 10 years contributes to different types of wellbeing in older adults. The study found that long-term, frequent engagement with certain arts activities is associated with higher levels of happiness, life satisfaction, self-realization, and control/autonomy in this cohort (Tymoszuk, Perkins, Spiro, Williamon, & Fancourt, 2019).

Much of the data linking social prescribing with improved mental health is prospective or anecdotal; however, research on this topic is rapidly expanding. The Glasgow Deep End link worker randomized controlled trial (RCT) study found that, when compared to controls, anxiety symptoms, depressive symptoms and self-reported exercise levels were all significantly better among those who saw a link worker more than twice (Mercer et al., 2017). Small scale studies of schemes suggest that these initiatives can not only benefit individual recipients' mental health and wellbeing, but lead to a more appropriate use of health care professionals' time and reduce unnecessary medical prescribing (Brandling & House, 2009; Chatterjee et al., 2017; Loftus, McCauley, & McCarron, 2017; South et al., 2008). Data from another study of social prescribing interventions (exercise gyms, walking projects, arts etc.) showed statistically significant improvement in the following areas; a Public Health Questionnaire, Generalized Anxiety Disorder, the Friendship Scale scores for social isolation, the Office National Statistics' Wellbeing measures and the International Physical Activity Questionnaire item for moderate exercise. A social return on investment of £2.90 for every £1 invested was also detected (Kimberlee, 2015).

Arts for health interventions are an accepted option for medical management of mental wellbeing in health care (Crone et al., 2018). In one qualitative study of an Arts on Prescription scheme, participants who had or were using mental health services reported experiencing the program as a therapeutic environment, and felt a sense of pride in their creative work (Stickley & Hui, 2012). Another prospective longitudinal follow-up observational study of an arts on referral program in UK General Practice measured changes in mental wellbeing and factors associated with those outcomes in over 1200 patients over a 7-year period (2009–16). Those defined as multiple health care service users, although fewer than for the group as a whole, had a significant increase in wellbeing. The authors conclude that art interventions can be effective in the promotion of wellbeing for those that complete them, including those referred with multi-morbidity, with significant changes in wellbeing evident across the intervention periods (Crone et al., 2018). Qualitative data from the study on participant reactions suggested that the program provided a range of personal and social benefits, with participants more able to self-manage aspects of their health-related conditions, and make progress toward a better physical and/or mental health (Redmond, Sumner, Crone, & Hughes, 2018).

One social prescribing service in Rotherham, England was specifically developed to help people with mental health conditions overcome the barriers obstructing discharge from secondary mental health care services. Initially scheduled to run from April 2015 to March 2016, the service was extended to March 2017. By the end of March 2016, 156 mental health service users had been referred to the pilot, and 136 (87%) had taken-up a service on an individual or group basis. An evaluation of the service undertaken by a team from Sheffield University found a range of positive health and wellbeing impacts on mental health service users following their engagement with social prescribing. Ninety-three percent of service users made progress against at least one of the one following outcomes: positive feelings, improved lifestyle, improved self-care, managing symptoms, engagement in work and leisure activities, living conditions, relationships and going out. Qualitative findings indicated that social prescribing had provided service users with "something to look forward to that prevented isolation, increased confidence and improved happiness in a way that developed and utilized skills in new and interesting areas" (Dayson & Bennett, 2018, p. 32). More than half of service users eligible for a discharge review have been discharged from secondary mental health services. It was estimated in terms of the wellbeing benefits experienced by service users, a social return on investment of £2.19 for every £1 invested in the pilot (Dayson & Bennett, 2018).

3.2 "Green medicine"

Green space has been associated with a wide range of health benefits, including stress reduction in deprived areas (Ward Thompson et al., 2012). Increasing people's exposure to, and use of, green spaces has been linked to long-term reductions in reported health problems including heart disease, cancer and musculoskeletal conditions, reduced levels of obesity, increased physical activity, social contacts, and higher self-rated mental health (Alcock, White, Wheeler, Fleming, & Depledge, 2014; Nieuwenhuijsen, Khreis, Triguero-Mas, Gascon, & Dadvand, 2017). In one field experiment, gardening was found to promote neuroendocrine and affective restoration from stress more rapidly than other calming activities such as reading (Van Den Berg & Custers, 2011).

Horticultural therapy (HT) has been used as a subgroup of occupational therapy (OT) for some years (Detweiler et al., 2015), but more recently gardening on prescription schemes have been established across the United Kingdom. In older age, gardening becomes not just a means of physical activity, but a source of social viability, independence and an amelioration of loneliness. There is emerging evidence that gardening may also be important in falls prevention (helping to maintain good gait and balance) and also in dementia prevention and cognitive decline (Buck, 2016). A report commissioned by the National Gardens Scheme looked at the impact of gardens and gardening on health and wellbeing and considered what the NHS and the wider health and social care system could do to maximize these effects (Buck, 2016). It concluded that the evidence gathered for the report was overwhelming in illustrating the diverse and holistic ways in which gardens and gardening can positively influence our health during every phase of life (Buck, 2016).

In the United Kingdom there are a growing number of gardening on prescription schemes offered to people experiencing mild to moderate mental health problems such as depression, anxiety, post-traumatic stress disorder (PTSD) and other such conditions. A 2019 episode of the BBC program "Gardeners World" looked at the role that plants and gardens can play in promoting good mental health and wellbeing where traditional approaches were proving less than ineffective (BBC, 2019). For 30 years, Blackthorn Trust Garden in Kent has been helping people with mental health issues and now offers rehabilitation and support into employment for patients suffering with chronic pain which has proven unresponsive to hospital treatment (Blackthorn Trust, 2019). Suzanne, a therapy coordinator at the garden, explained on the program how people referred to the scheme are often very depressed and have lost their confidence; "Most people are in a very dark place when they arrive." Alongside receiving counseling and therapy, people are encouraged to spend time with others in the garden. "Working in the garden enlivens the senses again," Suzanne explained.

L, who had been coming to the garden for 2 years after a car accident spoke of how she was never really interested flowers or plants before her accident; "It's wonderful, beautiful, the peace it brings to me—I get lost in it" she said. When asked what she would be going if she was not at Blackthorn, L. replied that she would probably be at home lying on her back and on lots of medication—not really doing anything. "Coming to the garden has brought me back to life," she said, "there's only one way to say it—gardening is good for you!" P., another member of the garden, explained how when he first came, he was quite broken with depression, while now he is working as a self-employed gardener. Another person explained how as someone disabled and in constant pain, daily living had become:

> *...Just about managing my pain and the anxieties and problems that come with it. And I came to Blackthorn and it's giving me a bit of space from me, so I have got something I can look forward to. There's something about being sort of close to plants that really makes you feel sort of so grounded and connected.*
>
> ***(BBC, 2019)***

Another community garden was set up by volunteers to help people with Parkinson's disease (Parkinsons.me Community Garden, 2019). As one member explained in the BBC program:

> *You have been given this diagnosis of this progressive degenerative disease... but the acceptance of the diagnosis is huge.*
>
> ***(BBC, 2019)***

When asked what he felt the garden had given to the community another member of the garden thought it was both therapeutic and interactive:

> *[The garden] brings a space for people to relax, the opportunity to talk about Parkinson's ... interacting with people with similar disabilities.*
>
> ***(BBC, 2019)***

4. Social prescribing: Service users' perspectives

Our work as researchers at the University of Westminster has entailed gathering quantitative data on social prescribing schemes and also eliciting multiple stakeholder perspectives of prescribing processes and outcomes. In one such study, a single arm mixed methods study of a Shropshire social prescribing scheme (Polley, Seers, & Fixsen, 2019), we conducted semi structured interviews with 12 service providers and 12 service users about their experiences of social prescribing. Topic guides rather than set questions were employed so that participants could feel free to relate their stories and unanticipated issues could emerge. This section focuses on the service users' experiences on visiting a link worker and subsequent actions.

Service users in this study who had visited a link worker identified a number of aspects specifically relating to the meeting which had value. These included the person-centered approach to the meeting, the opportunity to talk and the importance of follow-up calls and meetings for increasing motivation. Service users as a body were appreciative of the time allocated for this meeting and for the opportunity to explore their personal health needs in a relaxed way, e.g.,

The times were very convenient and the ease in which dates could be altered was really helpful. The room and atmosphere in the appointments were very relaxed which made 'opening up' so much easier.

(Social prescribing service user, Shropshire)

Several users acknowledged how rare it was to have an in-depth conversation about their health and wellbeing needs in a GP practice setting. As one user remarked; "How often do you get offered an hour's chat about a particular problem in a doctor's medical center? You don't, and I have to say that was quite an incentive." Another user described how the link worker has listened carefully and then "came up with some good answers and suggestions." Participants appreciated the fact that their meeting with the link worker had been "more a sort of discussion" about what they wanted and what was going to work for them. User responses also suggest that the empathic nature of the consultation process played its part, "The advisor (link worker) has been a sensitive and helpful advisor, who's given me enough latitude to open up about what are quite private matters." Some issues were, a service user thought, "difficult to talk about," however, in their meeting with the link worker they had "felt able to share things not shared before." A further example of this sharing effect was exemplified when a patient who was being supported to make behavior changes to support diabetes was asked about their experience of seeing the link worker. Not only did they report that they were reducing their alcohol intake as a result, but the patient reported:

I had a lot of trauma toward the end of last year. I lost my brother who I was very close and just in conversation with the link worker, they have pointed me in quite a lot of different directions in that respect as well.... Yeah, they have helped me, yeah.

(Social prescribing service user, Shropshire)

Participants in the Shropshire study had been referred by the link worker to a range of services. Several had opted to pursue an activity or intervention of a social as well as physical nature, such as a weekly walking group, a leisure activity or a swimming class:

[I met a] really nice lady who...said, 'Come along to the to the pool and see what we do and if you think it's for you...all you have to do is get your GP to sign a form'...So, I went and saw what it was it like and thought, yes, that's, I could cope with that.

(Social prescribing service user, Shropshire)

I opted for Help to Slim and Help to Swimming...I was certainly on Help to Slim... And then I was introduced to a lady.... who organized for me to have swimming lessons with a buddy.

(Social prescribing service user, Shropshire)

Social prescribing may not suit or be needed by everyone. We also interviewed service users who had experienced problems accessing services or had not felt that they needed to see a link worker in the first place. Social prescribing is not conceived of as a long-term intervention and one user complained that their program had finished before they could gain much benefit from it, e.g., "I really enjoyed it, felt so much better for it, but it wasn't enough." There was the travel and expense to consider. Even at a reduced cost, gym membership had been too much for one participant to afford on a regular basis. Other barriers mentioned by service users included poor physical and mental health, and work or family commitments. Later, in our recommendations, we will consider some of these concerns.

5. Social prescribing: Service providers' experiences

In the course of our studies we spoke with and interviewed a range of UK-based professionals working in health and auxiliary services about their experiences of social prescribing. GPs we interviewed agreed that social prescribing steers the patient away from being over medicalized and reminds doctors that there were other ways of working. One of the GPs, L., interviewed for the Shropshire study (Polley et al., 2019), described the types of patients he might refer to a social prescribing link worker:

> *I'll tend to refer people with low level depression, anxiety, loneliness, isolation, people with chronic disease management problems to try and encourage them to take ownership in certain cases. But also, you do have a number of patients that are, frequent attenders with perhaps no confirmed diagnosis but they've had lots of tests, seen lots of specialists and just not getting anywhere.*
>
> ***(GP, Shropshire)***

Another GP, D., spoke of the frustration of having someone come to you with five or six issues; "You want to be able to help them all but in 10 min you don't physically have the time, you have to tackle what is the most medically relevant and urgent." While GP L. was aware that not all her colleagues were as enthusiastic as herself about social prescribing, she believed it could have a "huge impact on workload to GPs" and practice in general, because social prescribing was like saying, "You don't need to come here... We've absolutely got to own this and validate it and say this is a really viable option for people." L. went on to tell the story of how one elderly patient had benefited from social prescribing:

> *Actually, he was about 80, but he was quite lonely by his own admission. And he was still quite able, and just talked about not getting out enough. So, I referred him*

to social prescribing and through that he did the Elevate exercise program…And then he [came to] tell me how much he was enjoying it, and what a difference it was making to him, and he was meeting people. And he seemed a lot happier, and a lot less lonely and isolated as a result…And so we avoided a pill, and yeah, and he gained an awful lot from it.

(GP, Shropshire)

GP D., who was newly qualified, thought that a lot of social prescribing initiatives were presently focused on the older population, yet social prescribing held the potential to fill the "huge gap" in mental health provision for younger people; "There are so many young people suffering with depression, anxiety and stress" who stay, "in their bedrooms, never come out." Young men were a particularly hard group for GPs to reach and one solution lay in developing an NHS online app through which people could self-refer to a social prescribing advisor. This fits with a recent report by the Young Women's trust (2019) suggesting an epidemic of loneliness among the young, with one on four young people reporting feeling lonely.

Professionals in what can be termed "bridging organizations" (Brown, 1991) play an important role in identifying issues of concern, build trusting relationships and connecting people who might not otherwise be connected to appropriate services (Stewart & Tyler, 2019). In addition to interviewing GPs, we have spoken with a range of non-clinical partners working in bridging organizations about their experiences of social prescribing. In the Shropshire study, for example, we interviewed staff from the Department of Health and Pensions and from Adult Social Care who had referred a number of people to a link worker in the last 12 months. Many of the Department of Health and Pensions customers were on universal credit and had a form of disability or a long-term health problem and the young and isolated who stay, "in their bedrooms, never come out." Social prescribing fitted well with the universal credit aim of motivating people toward better self-care, but it remained a big task:

North West Shropshire is one of the most deprived areas of the county, there are a lot of people who…have got compromised health or disabilities because of their level of income opportunities, health, there's a lot of people that don't work, smoke heavily, have got poor diets, overuse of alcohol, so [this is] definitely one of the right areas to target.

(DWP staff, Shropshire)

What might be presented as physical symptoms to a health provider may well be associated with psychosocial issues which are better supported through physical or creative activity than any medical invention. In another study, one link worker who worked for a cancer charity was well aware of this in older clients who had undergone chemotherapy:

What I often see [is] where a GP refers them to me because they know there's something else going on, but they [the patient] present with very physical aches and pains. And, yeah, you just go on the journey with them to try and identify what's going on and also give them other options to deal with aches and pains rather than just medication [when] the answer is a bit of stretching, a bit of exercise.

(Link worker, London)

A similar viewpoint was expressed by a paramedic who frequently attended calls from people who were primarily suffering from loneliness or anxiety, "It was mainly anxiety that seemed to be prompting the calls, but again, it was the element of isolation [and] loneliness." As a result of following a model of social prescribing, his service had seen a big reduction in call outs; "As much as between 60% and 70% reduction in calls over the space of about 4 or 5 months."

Service providers interviewed also identified a number of barriers and challenges to social prescribing. These included engaging all GPs, recruiting the young and socially isolated and informing patients about social prescribing when capacity was limited. GPs and practice managers all knew of doctors who had expressed doubts about the skills and training of the link workers and generally resisted directives coming from authorities. Another issue was that of service duplication, with some community care professionals concerned that their roles might be usurped by link workers. Not all patients were enthusiastic either; while some people really latched onto social prescribing, others "don't want to move out of the medical model because they're still not [ready]" (GP N.). On the other hand, professionals we spoke to agreed that joined-up solutions were urgently required to health and social problems and that social prescribing presented one way to achieve this goal.

5.1 Link workers

The link workers in the United Kingdom we interviewed expressed satisfaction in their work and commitment to their clients' wellbeing. They were trained in behavioral, motivational and listening skills and had a broad knowledge of available, low or no cost, quality-assured health and wellbeing interventions in their local area. An important function of the link worker role was explaining the social prescribing service in simple terms to clients:

I always explain that it's not so much the medical side of things, because you see your doctor and nurse for that. This is a bit more about how you feel in yourself... what's going on and anything to do with your lifestyle that we can support.... trying to link you up with things that are already going on in the community.

(Link Worker, Shropshire)

In practice, the link workers were frequently dealing with multiple issues and it was often a case of deciding which to tackle first; "Do we try and boost your confidence…or go for something that helps with depression and anxiety?" (Link Worker). Following up with clients was seen as a vital part of the service. Some clients had a lot of things going on in their lives (financial problems, domestic problems, etc.) and a person might need several meetings with a link worker to share their whole narrative. In cases that were "too challenging" or "too severe" for the SP service, the link worker would find clients other support such as emergency intervention from mental health services. As trained advisers, the link workers understood that you were not going to help people change their behavior by telling them what to do or trying to fix it for them, but by "trying to tease from the person what they would like to do, what they feel confident to do."

6. Conclusions

In this chapter, we have considered the potential of social prescribing to support and assist people suffering from stress and distress, where medical intervention is at best a partial solution. Our findings confirm that social prescribing schemes can and do motivate some people to take up activities which can have measurable benefits on people's mental and physical health such as exercise, artistic pursuits and gardening; however, much work is still needed to ascertain the scope of social prescribing. Determining the overall "value" of social prescribing vs other interventions has proved difficult, as many of its returns are of the social and emotional, rather than the economic calculable, variety (Kimberlee, 2015; Wildman et al., 2019). More studies are needed to discover which aspects of social prescribing work best and in which localities, so that long-term strategies to support and expand the voluntary and community services at the delivery end of social prescribing can be developed.

Health care professionals provide, for some individuals, their most trusted relationships and connections to a wider community. Specialist health providers—including neurologists whose time-limited focus is predominantly on medical diagnosis and treatment—may overlook the importance of non-medical factors, such as finding a sense of purpose and better social connectedness for illness prevention and recovery. A further issue is reluctance on the part of health care practitioners who know little about the operational level and delivery of social prescribing to refer patients to social prescribing local link workers. Our real-life studies of social prescribing schemes indicate that lack of "buy in" from bridging organizations and auxiliary services remains an obstacle to the success of projects in terms of joined-up care. Rightly or not, patients place great trust in health professionals and without their approval or support, a referral to a community activity is less likely to be taken-up by the patient (Kimberlee, 2015; Pescheny, Pappas, et al., 2018; Pescheny, Randhawa, et al., 2018).

Social prescribing, as with all interventions, is unlikely to suit everyone. Identifying which categories of individuals could most benefit from being referred into schemes remains a challenge for policy makers, local teams and practitioners, but should become easier, in the United Kingdom, under the new NHS England long-term plan (NHS, 2019). Research on social prescribing within primary care (GP practices) is relatively developed and the advent of "read codes" on the electronic medical record, will allow nationwide data analysis to be henceforth possible. There is, however, a lack of research into how social prescribing can be implemented successfully into secondary care (e.g., hospital settings, out-patient care), where there are consultants and allied health care professionals already working as a team. Research also needs to review the longitudinal effects of social prescribing to understand how initial short-term effects are best sustained. An on-going challenge is the limited resourcing and funding of the voluntary and community sector in times of relative austerity. While assisting in community ventures can bring its own rewards, is it important that professionals and volunteers supporting local schemes are given the credit and support they deserve. Based on our findings, we propose the following recommendations for enhancing social prescribing schemes:

- Ensure that information on social prescribing is made widely available to health professionals and the public through stakeholder meetings and through material in service venues.
- Actively seek out and prioritize support for the isolated and vulnerable, and for those with complex social and health needs of all generations.
- Referral and audit processes should be regularly reviewed to ensure that people are clear about who to refer to a link worker and that numbers and outcomes of referrals and recommendations are recorded.
- Expand the routes of referral into social prescribing, including in secondary and tertiary care.
- Link workers should receive where needed specialized training in areas such as mental health issues and long-term neurological conditions.
- Provide scope for home visits from a link worker where travel is difficult or not possible or provide weekend or early evening appointments.

- Consider introducing an online self-referral system/app for social prescribing.
- Use of electronic mapping tools to enable better signposting to local community-based services.

References

Alcock, I., White, M. P., Wheeler, B. W., Fleming, L. E., & Depledge, M. H. (2014). Longitudinal effects on mental health of moving to greener and less green urban areas. *Environmental Science and Technology*, *48*(2), 1247–1255.

Baker, K., & Irving, A. (2016). Co-producing approaches to the management of dementia through social prescribing. *Social Policy and Administration*, *50*(3), 379–397.

BBC. (2019). *Gardeners' world episode*. (p. 29). Retrieved October 3, 2019, from https://www.bbc.co.uk/iplayer/episode/m0008rws/gardeners-world-2019-episode-29.

Bertotti, M., Frostick, C., Hutt, P., Sohanpal, R., & Carnes, D. (2018). A realist evaluation of social prescribing: An exploration into the context and mechanisms underpinning a pathway linking primary care with the voluntary sector. *Primary Health Care Research and Development*, *19*(3), 232–245.

Berwick, D. M., Nolan, T. W., & Whittington, J. (2008). The triple aim: Care, health, and cost. *Health Affairs (Project Hope)*, *27*(3), 759–769.

Blackthorn Trust. (2019). Retrieved October 4, 2019, from https://www.blackthorn.org.uk.

Brandling, J., & House, W. (2009). Social prescribing in general practice: Adding meaning to medicine. *British Journal of General Practice*, *59*(563), 454–456.

Brown, L. D. (1991). Bridging organizations and sustainable development. *Human Relations*, *44*(8), 807–831.

Buck, D. (2016). *Gardens and health: Implications for policy and practice*. Retrieved from https://www.kingsfund.org.uk/sites/default/files/field/field_publication_file/Gardens_and_health.pdf.

Busis, N. A., Shanafelt, T. D., Keran, C. M., Levin, K. H., Schwarz, H. B., Molano, J. R., et al. (2017). Burnout, career satisfaction, and wellbeing among US neurologists in 2016. *Neurology*, *88*(8), 797–808.

Caper, K., & Plunkett, J. (2015). *A very general practice: How much time do GPs spend on issues other than health? Citizens Advice policy briefings public services in a constrained spending environment*. Retrieved from https://www.citizensadvice.org.uk/Global/CitizensAdvice/Publicservices publications/CitizensAdvice_AVeryGeneralPractice_May2015.pdf.

Carod-Artal, F. J. (2017). Social determinants of mental health. In S. Bahrer-Kohler & F. J. Carod-Artal (Eds.), *Global mental health: Prevention and promotion*. Cham Switzerland: Springer Nature.

CASPER. (2019). *Social prescribing for depression*. Retrieved November 6, 2019, from http://sundancedx.com/social-prescribing/.

Chatterjee, H. J., Camic, P. M., Lockyer, B., & Thomson, L. J. M. (2017). Non-clinical community interventions: A systematised review of social prescribing schemes. *Arts and Health*, *10*(2), 97–123.

Cheater, S. (2019). The NHS long term plan. *International Journal of Health Promotion and Education*, *57*(3), 174–175. https://doi.org/10.1080/14635240.2019.1595526.

Chrousos, G. P. (2009). Stress and disorders of the stress system. *Nature Reviews Endocrinology*, *5*(7), 374–381.

Citizens Advice. (2015). *A very general practice: How much time do GPs spend on issues other than health?* Retrieved from https://www.citizensadvice.org.uk/Global/CitizensAdvice/Public%20services%20publications/CitizensAdvice_AVeryGeneralPractice_May2015.pdf.

Commission on Social Determinants of Health. (2008). *Closing the gap in a generation: Health equity through action on the social determinants of health.* World Health Organization. Retrieved from https://www.who.int/social_determinants/final_report/csdh_finalreport_2008.pdf.

Consumers Health Forum of Australia. (2019). Retreived from https://chf.org.au/blog/social-prescribing-new-way-think-about-healthcare.

Courtois, C. A., & Ford, J. D. (2009). *Treating complex traumatic stress disorders: Scientific foundations and therapeutic models.* New York: The Guildlford Press.

Crone, D. M., Sumner, R. C., Baker, C. M., Loughren, E. A., Hughes, S., & James, D. V. B. (2018). "Artlift" arts-on-referral intervention in UK primary care: Updated findings from an ongoing observational study. *European Journal of Public Health, 28*(3), 404–409.

Dayson, C. (2017). Evaluating social innovations and their contribution to social value: The benefits of a "blended value" approach. *Policy and Politics, 45*(3), 395–411.

Dayson, C., & Bennett, E. (2018). *Evaluation of the Rotherham mental health social prescribing pilot.* Retrieved from https://www4.shu.ac.uk/research/cresr/sites/shu.ac.uk/files/eval-rotherham-mental-health-social-prescribing.pdf%0Ahttp://www4.shu.ac.uk/research/cresr/sites/shu.ac.uk/files/eval-rotherham-health-social-prescribing-2015-2017.pdf.

Detweiler, M. B., Self, J. A., Lane, S., Spencer, L., Lutgens, B., Kim, D. Y., et al. (2015). Horticultural therapy: A pilot study on modulating cortisol levels and indices of substance craving, posttraumatic stress disorder, depression, and quality of life in veterans. *Alternative Therapies in Health and Medicine, 21*(4), 36–41.

Friedli, L., Jackson, C., Abernethy, H., & Stansfield, J. (2008). *Social prescribing for mental health—A guide to commissioning and delivery. Vol. 9.* Care Services Improvement Partnership. Retrieved from https://www.centreforwelfarereform.org/uploads/attachment/339/social-prescribing-for-mental-health.pdf.

Husk, K., Blockley, K., Lovell, R., Bethel, A., Bloomfield, D., Warber, S., et al. (2016). What approaches to social prescribing work, for whom, and in what circumstances? A protocol for a realist review. *Systematic Reviews, 5*(1), 93. Retrieved from https://doi.org/10.1186/s13643-016-0269-6.

Imo, U. O. (2017). Burnout and psychiatric morbidity among doctors in the UK: A systematic literature review of prevalence and associated factors. *BJPsych Bulletin, 41*(4), 197–204.

Kawachi, I., & Berkman, L. F. (2001). Social ties and mental health. *Journal of Urban Health, 78*(3), 458–467.

Kearns, A., Whitley, E., Tannahill, C., & Ellaway, A. (2015). Loneliness, social relations and health and wellbeing in deprived communities. *Psychology Health and Medicine, 20*(3), 332–344.

Kilgore, C., Richter, R. R., Siler, W. L., & Sayre-Stanhope, D. (2008). Psychosocial issues in primary care physician assistant practice: A descriptive study. *The Journal of Physician Assistant Education, 19*(4), 4–19.

Kimberlee, R. (2015). What is social prescribing? *Advances in Social Sciences Research Journal, 2*(1), 102–110. https://doi.org/10.14738/assrj.21.808.

Loftus, A. M., McCauley, F., & McCarron, M. O. (2017). Impact of social prescribing on general practice workload and polypharmacy. *Public Health, 148*, 96–101.

Mah, L., Szabuniewicz, C., & Fiocco, A. J. (2016). Can anxiety damage the brain? *Current Opinion in Psychiatry, 29*(1), 56–63.

Marmot, M., & Bell, R. (2012). *Fair society, healthy lives. Public Health.* Retrieved from: http://www.instituteofhealthequity.org/resources-reports/fair-society-healthy-lives-the-marmot-review.

Medisauskaite, A., & Kamau, C. (2019). Reducing burnout and anxiety among doctors: Randomized controlled trial. *Psychiatry Research, 274*, 383–390.

Mercer, S. W., Fitzpatrick, B., Grant, L., Chng, N. R., O'Donnell, C. A., Mackenzie, M., et al. (2017). The Glasgow 'deep end' links worker study protocol: A quasi-experimental evaluation of a social prescribing intervention for patients with complex needs in areas of high socioeconomic deprivation. *Journal of Comorbidity*, 7(1), 1–10.

Naylor, C., Parsonage, M., McDaid, D., Knapp, M., Fossey, M., & Galea, A. (2012). *Long term conditions and mental health: The cost of co-morbidities* (pp. 1–32). The king's fund and centre for mental health. https://doi.org/ISBN9781857176339.

NHS. (2019). *NHS long term plan.* Retrieved from https://www.longtermplan.nhs.uk.

NICE. (2019). *Older people.* Retrieved January 1, 2019, from https://www.nice.org.uk/guidance/population-groups/older-people.

Nieuwenhuijsen, M. J., Khreis, H., Triguero-Mas, M., Gascon, M., & Dadvand, P. (2017). Fifty shades of green. *Epidemiology*, *28*(1), 63–71.

Office for National Statistics. (2018). *Population estimates for the UK, England and Wales, Scotland and Northern Ireland: Mid-2018.* Retrieved from https://www.ons.gov.uk/peoplepopulationandcommunity/populationandmigration/populationestimates/bulletins/annualmidyearpopulationestimates/mid2018.

Office for National Statistics. (2019). Retrieved from https://www.ons.gov.uk.

Ontario Public Health. (2019). Connected communities: Healthier together. In *Annual report 2017 of the chief medical officer of health of Ontario.* Retrieved from http://www.health.gov.on.ca/en/common/ministry/publications/reports/cmoh_19/default.aspx.

Parkinsons.me Community Garden. (2019). Retrieved October 4, 2019, from https://www.parkinsons.me/community-garden-west-lockinge.

Pescheny, J. V., Pappas, Y., & Randhawa, G. (2018). Facilitators and barriers of implementing and delivering social prescribing services: A systematic review. *BMC Health Services Research*, *18*(1), 86.

Pescheny, J., Randhawa, G., & Pappas, Y. (2018). Patient uptake and adherence to social prescribing: A qualitative study. *BJGP Open, 2.* https://doi.org/10.3399/bjgpopen18x101598. Retrieved from.

Polley, M., & Fleming, J. (2017). *Making sense of social prescribing.* Retrieved from https://doi.org/10.1038/scientificamerican0906-84sp.

Polley, M., Seers, H., & Fixsen, A. (2019). *Evaluation report of the social prescribing demonstrator site in Shropshire—Final report.* Retrieved from https://westminsterresearch.westminster.ac.uk/item/qx18z/evaluation-report-of-the-social-prescribing-demonstrator-site-in-shropshire-final-report.

Pulse. (2019). *NHS England will fund 1,000 social prescribing workers to support practices.* Retrieved February 7, 2019, from http://www.pulsetoday.co.uk/news/practice-news-/nhs-england-will-fund-1000-social-prescribing-workers-to-support-practices/20038156.article.

Redmond, M., Sumner, R. C., Crone, D. M., & Hughes, S. (2018). 'Light in dark places': Exploring qualitative data from a longitudinal study using creative arts as a form of social prescribing. *Arts and Health*, *11*(3), 232–245.

Santoni, C., Chiva Giurca, B., Li, T. M., Mulligan, H., Chilaka, J., Lazzereschi, L., et al. (2019). Evaluating student perceptions and awareness of social prescribing. *Education for Primary Care*, 1–7.

Shin, L. M., & Liberzon, I. (2010). The neurocircuitry of fear, stress, and anxiety disorders. *Neuropsychopharmacology*, *35*(1), 169–191.

Sigsbee, B., & Bernat, J. L. (2014). Physician burnout: A neurologic crisis. *Neurology*, *83*(24), 2302–2306.

South, J., Higgins, T. J., Woodall, J., & White, S. M. (2008). Can social prescribing provide the missing link? *Primary Health Care Research & Development*, *9*(4), 310–318.

Steptoe, A., Deaton, A., & Stone, A. A. (2015). Subjective wellbeing, health, and ageing. *The Lancet*, *385*(9968), 640–648.

Steptoe, A., & Fancourt, D. (2018). Leading a meaningful life at older ages and its relationship with social engagement, prosperity, health, biology, and time use. *Proceedings of the National Academy of Sciences of the United States of America, 116*(4), 1207–1212. Direct Submission.

Stewart, J., & Tyler, M. E. (2019). Bridging organizations and strategic bridging functions in environmental governance and management. *International Journal of Water Resources Development, 35*(1), 71–79.

Stickley, T., & Hui, A. (2012). Social prescribing through arts on prescription in a UK city: Participants' perspectives. *Public Health, 126*(7), 574–579.

Thoits, P. A. (2010). Stress and health. *Journal of Health and Social Behavior, 51*, S41–S53.

Torjesen, I. (2016). Social prescribing could help alleviate pressure on GPs. *BMJ [British Medical Journal], 352*, i1436. https://doi.org/10.1136/bmj.i1436.

Tymoszuk, U., Perkins, R., Spiro, N., Williamon, A., & Fancourt, D. (2019). Longitudinal associations between short-term, repeated, and sustained arts engagement and well-being outcomes in older adults. *Journals of Gerontology: Series B, Psychological Sciences and Social Sciences*, 1–11.

Van Den Berg, A. E., & Custers, M. H. G. (2011). Gardening promotes neuroendocrine and affective restoration from stress. *Journal of Health Psychology, 16*(1), 3–11.

Vos, T., Barber, R. M., Bell, B., Bertozzi-Villa, A., Biryukov, S., Bolliger, I., et al. (2015). Global, regional, and national incidence, prevalence, and years lived with disability for 301 acute and chronic diseases and injuries in 188 countries, 1990-2013: A systematic analysis for the Global Burden of Disease Study 2013. *The Lancet, 386*(9995), 743–800.

Ward Thompson, C., Roe, J., Aspinall, P., Mitchell, R., Clow, A., & Miller, D. (2012). More green space is linked to less stress in deprived communities: Evidence from salivary cortisol patterns. *Landscape and Urban Planning, 105*(3), 221–229.

Whiteford, H. A., Degenhardt, L., Rehm, J., Baxter, A. J., Ferrari, A. J., Erskine, H. E., et al. (2013). Global burden of disease attributable to mental and substance use disorders: Findings from the Global Burden of Disease Study 2010. *The Lancet, 382*(9904), 1575–1586.

Wildman, J. M., Moffatt, S., Steer, M., Laing, K., Penn, L., & O'Brien, N. (2019). Service-users' perspectives of link worker social prescribing: A qualitative follow-up study. *BMC Public Health, 19*(1), 98.

World Federation For Mental Health (WFMH). (2012). Depression: A global crisis. World Mental Health Day. In *382*, (pp. 1575–1586). [9904].

Young Women's Trust. (2019). Retrieved from https://www.youngwomenstrust.org/what_we_do/media_centre/press_releases/904_lifetime_of_loneliness_one_in_four_young_people_feels_lonely.

Zhou, X., Pu, J., Zhong, X., Zhu, D., Yin, D., Yang, L., et al. (2017). Burnout, psychological morbidity, job stress, and job satisfaction in Chinese neurologists. *Neurology, 88*(18), 1727–1735.

Further reading

Public Health England. (2019). *Effectiveness of social prescribing: An evidence synthesis. London.* Retrieved from http://www.gov.uk/phe.

Kolibri 360 GmbH
Ferdinand-Jühlke-Straße 7
99095 Erfurt, Deutschland
produktsicherheit@kolibri360.de